The Frontiers of Knowledge

The Frontiers of Knowledge

What We Now Know about Science,

History and the Mind

A. C. GRAYLING

VIKING

an imprint of

PENGUIN BOOKS

VIKING

UK | USA | Canada | Ireland | Australia
India | New Zealand | South Africa

Viking is part of the Penguin Random House group of companies
whose addresses can be found at global.penguinrandomhouse.com.

First published 2021
001

Copyright © A. C. Grayling, 2021

The moral right of the author has been asserted

Set in 12/14.75 pt Bembo Book MT Std
Typeset by Jouve (UK), Milton Keynes
Printed and bound in Great Britain by Clays Ltd, Elcograf S.p.A.

The authorized representative in the EEA is Penguin Random House Ireland,
Morrison Chambers, 32 Nassau Street, Dublin D02 YH68

A CIP catalogue record for this book is available from the British Library

HARDBACK ISBN: 978–0–241–30456–3
TRADE PAPERBACK ISBN: 978–0–241–30458–7

www.greenpenguin.co.uk

MIX
Paper from
responsible sources
FSC® C018179

Penguin Random House is committed to a
sustainable future for our business, our readers
and our planet. This book is made from Forest
Stewardship Council® certified paper.

To the founding students, faculty and staff of NCH
and to all who carry its legacy forward:
Animi Cultura Gaudere

Dear Mike
On the occasion of your
40th Birthday
With all our love
Dad & Mum xx

Contents

Figures

Preface

In very recent times humanity has learned a vast amount about the universe, the past, and itself. Since the nineteenth century it has unearthed thousands of years of history forgotten or wholly unknown beforehand: the history of the great pre-classical civilizations, and before that the story of human evolution. Since the beginning of the twentieth century it has made hitherto inconceivable discoveries about the physical universe at the smallest and largest scales we can so far reach, from quantum theory to cosmology and the origins of space and time. And in just the past few decades it has been able to look inside the brain to begin finer-grained mapping of its structures and to observe them actually at work.

These advances have been enormous, exhilarating, and consequential. We occupy a different and much richer universe than our forebears living as recently as the nineteenth century. Yet a remarkable fact attends these developments: whereas it was once believed that every advance in knowledge diminishes our ignorance, these recent giant strides have shown us just how little we know. Enquiry thus generates a paradox: increasing knowledge increases our ignorance. So – what do we know? And what do we now know that we do not know? And what have we learned about the nature of enquiry itself – the barriers and difficulties that have to be overcome or taken into account as our increasing knowledge increases our ignorance?

This book seeks to answer these questions in three crucial areas at the frontiers of knowledge, these being science, history, and psychology – more particularly: fundamental physics and cosmology, the discovery of the pre-classical past and human evolution, and the new neurosciences of brain and mind.

In writing often and variously about the growth of ideas and the history of philosophy, I have been deeply intrigued by questions about humanity's labours on the frontiers of knowledge, and about the nature, methods and problems of enquiry. These are questions that lie at the heart of philosophy – understood in its broadest sense as reflection on what we know, how we know it, and why it matters – because they lie at the heart of human endeavour itself. My aim in these pages is to illustrate and explore three of the most important frontiers of this endeavour, to describe where they lay and how they were advanced to where they lie now; and to discuss what their current position teaches us about what we have yet to learn.

A. C. Grayling
New College of the Humanities
London, 2021

Acknowledgements

I have learned much over many years from friends and colleagues, too many for all of them to be listed, but some merit particular mention for conversations and their writings over the years: Tejinder Virdee and Lawrence Krauss on physics and the universe, Adam Zeman, Daniel Dennett, and Patricia Churchland on the brain, mind and consciousness, Richard Dawkins on evolution, Steven Pinker on Enlightenment and the progress of ideas, Simon Blackburn, Peter Singer, and Alex Orenstein on philosophical dimensions. My thanks go also to Dr Ron Witton for comment on the History Wars, to Caroline Williams and Dr John Gribbin for advice on technical points, to Daniel Crewe my perceptive and knowledgeable editor at Viking, to Mollie Charge for her bibliographical researches, and to my colleagues at the New College of the Humanities for the variety and insight of our Ottoline Club faculty seminars.

Introduction

What do we know about the world, the past, and ourselves? Very recently, just in the course of the last century and a half, there have been spectacular advances in our enquiries into these topics. Using the most general labels for them, we call them *science*, *history*, and *psychology* respectively, but the labels do little justice to what has been achieved in them, nor what the achievements mean, nor where they are or might be leading us. They are the result of rapidly evolving technologies of enquiry that have vastly extended humanity's observational reach, both backwards in time and across previously inaccessible scales of distance from the remotest galaxies to the intricacies of the human brain, and yet further down to the inner architecture of the atom. Each step in these advances has in its turn raised new questions, questions that beforehand were not possible to ask; and one of the chief results has been to expose a paradox: the *paradox of knowledge*, which is that *the more we know, the more we realize the extent of our ignorance*, not least in these three crucial areas of enquiry about the world, the past and the mind.

The paradox of knowledge has been made familiar by these recent advances of knowledge, but until they became so rapid and far-reaching the belief was that knowledge was *reducing* the domain of ignorance. Humankind believed that knowledge was accumulating in such a way that the prospect of reaching the frontier of knowledge itself, enabling humankind to know everything there is to know, seemed to be implicit in the unfolding success of enquiry. The dramatic reversal in this perspective is no longer a surprise, but the implications of this fact, including those that raise questions about the nature of enquiry itself, have yet to be fully grasped.

Every stage in the growth of knowledge over the course of human history has had its frontiers, and for the pioneers venturing across them the frontiers defined the *terra incognita* that lay on their other side. Quite often the direction of travel they seemed to indicate turned out to be wrong. One of the greatest questions therefore about today's frontiers is whether the directions of travel they indicate are the right ones. Of course the proper response is: who can say until you try? But it might be that there are some clues both in the history of past frontiers and in the approach to present frontiers that could help.

As it happens, in an importantly different sense of the word 'knowledge', our ancestors have known a great deal not just for thousands but for millions of years. It appears that the earliest stone tools date to 3.3 million years ago, halfway back to the point in evolutionary history when the ancestral trees of chimpanzee and human forebears diverge. The knowledge they had is *knowledge how* – that is, practical knowledge, from making tools to building shelters, mastering fire, creating cave art, domesticating animals and plants, shaping and moving huge stones, digging irrigation canals, fashioning textiles and pottery, casting bronze from copper and tin, smelting iron, and so onwards to the advanced technologies of today.

'Knowledge how' almost certainly came to be accompanied, at some perhaps quite early point, by efforts to achieve *knowledge what* – that is, theoretical knowledge, *explanations* of *why* the *how* works. The frameworks of explanation devised by our ancestors almost certainly involved imputing agency to natural forces. To explain thunder, wind, rain, and the movements of the heavenly bodies, our ancestors are likely to have inferred from their own powers of agency – the feeling of 'I caused that', as when one throws a stone into water and makes a splash – that anything that moves, emits noise, changes in any way, must have an agent, a mover, behind or within it. Moreover, the appearance of intentional behaviour in animals also doubtlessly made our ancestors think that animals have mental lives similar to their own; what looks like timidity in the deer and ferocity in the lion were assumed to mirror their own feelings: if a deer ran away, it was because of fear; if a lion attacked, it was because it was enraged. The animistic sources of religious beliefs remain apparent in some of the earliest-known efforts to form theoretical explanations of the world. The Presocratic philosopher Thales, for example, hypothesized that 'everything is full of souls' – by 'soul' he meant an animating principle – to explain such phenomena as the magnet's ability to attract iron.[1]

History tells us that these kinds of 'knowledge what' explanations consisted principally in what we now call 'religious' beliefs. These in turn contributed further kinds of supposed 'knowledge how' by suggesting forms of interaction with aspects of nature, or the agencies that control nature, hoping to influence or propitiate them through ritual, prayer, and sacrifice. It is an interesting speculation that, as liturgical (religious, ritualistic) means of influencing nature came to be displaced by more practical and mundane expertise, so the interest in effecting control transferred itself from nature to society; perhaps, as suggested by the concept of 'taboo', when controlling certain kinds of behaviour was no longer regarded as necessary for influencing nature or nature's gods, the social control – in the form of conceptions of 'morality' – endured. Whether or not this is the case, the main point remains that until very recently in human history 'knowledge how' has been far in advance of 'knowledge what', and the effort to provide the latter has until very recently rested chiefly on imagination, fancy, fear, and wishful thinking.

As suggested by reference to Thales above, the story of humankind's efforts to 'know what' in addition to 'how', but without relying on imagination and traditional beliefs, first comes fully into view with the philosophers of Greek classical antiquity from the sixth century BCE onwards. Thales, who flourished around 585 BCE in Ionia on the east coast of the Aegean, is often cited as 'the first philosopher', because he is the first person known to have asked and answered a question about the nature and source of reality without recourse to myth. In desiring a more intellectually plausible account than was offered by mythographers and poets, he sought to identify the cosmos's *arche* ('principle'), defined by Aristotle as 'that of which all existing things are composed . . . the element and principle of the things that are', by working it out from what he saw around him. His choice of candidate for the *arche* was: *water*. His thinking can be reconstructed as follows. Water is everywhere, and it is essential. It is in the sea, it falls from the sky, it runs in your veins, plants contain it, all living things die without it. Water can even be said to produce earth itself; look at the vast quantities of soil disgorged by the Nile in its annual floods. And as the clincher: water is the only substance Thales knew that can occupy all three material states of solid (when frozen), liquid (the basic state), and gas (when boiling away as steam). So, it is ubiquitous, essential, productive, and metamorphic; it is the only thing he knew to be so; it must therefore be the substance from which all other things come and on which they depend: the *arche* of the universe.[2]

This is ingenious thinking, in the context of the time. But the really important point about it is that it relies on *observation and reason alone*, not on myths, legends, or imagination. This is why Thales is nominated as the first philosopher. No doubt plenty of other people thought in similar ways before him, but we have no record of them. As a result we identify him as the first figure in a new phase of history, for whereas *technology* – 'knowledge how' – had been in development for millions of years, *science* – 'knowledge what' – began at that moment.

Note, however, that observation and reason require a context of enquiry, and an accumulation of results corrected by tests. They are not enough just by themselves. Observation and reason gave our ancestors the view that the Sun moves across the sky, for indeed it appears to move – that is the observation we make – and everything on the Earth around us remains still. Therefore the reasonable conclusion is that the Sun is the mobile object, not the Earth. We think the same about the Moon and for the same reason, and in that case we are right. It took repeated and deeper applications of observation and reason to arrive at the counterintuitive result that it is the Earth that moves relative to the Sun, not the other way round.

This is one indication of the general point that the history of 'knowledge what' started slowly, increasing falteringly until a body of context and test had built up – and also because it was too often opposed by powerful traditionalist interests, principally religious, that felt threatened by it. It began its swiftest climb only at the beginning of modern history, in the sixteenth and seventeenth centuries.[3] Since the nineteenth century, however, the growth of 'knowledge what' has been meteoric.

But note that this knowledge is still growing, is still in an incomplete state; perhaps much of it is in a very early state, and perhaps some of it will be adjusted, corrected or discarded as evidence accumulates and the methods and technologies of enquiry continue to improve, which they do all the time. The questions about our world and ourselves that the explosive growth in knowledge *so far* make us ask can therefore only suggest tentative answers – though the human hunger for answers will seek them regardless.

In asking 'What do we know?', we are naturally led to ask, 'How do we know it?' and 'Are there limits to what can be known?' These questions involve others: what do we mean by 'knowledge' in contrast to 'belief' and 'opinion'; and, if there is a strict definition of 'knowledge' that makes a sharp contrast with 'mere' belief and opinion, are we not forced to ask, 'Isn't it belief, rather than knowledge *as such*, that we have? For if we define "knowledge" very strictly as *what is true and accepted on indubitable grounds*, is knowledge even possible at all – for what is there outside mathematics that is indubitable?' Some preliminaries help to sort through this set of important questions, and a quick way to do so is as follows.

One of the central areas of philosophy is *epistemology*, or 'theory of knowledge'.[4] A pointed way of carrying out this task is to show how we can respond to sceptical challenges to our claims to knowledge. In philosophy's technical debates about this matter, very simple knowledge claims such as 'I know that there is a laptop in front of me now', and very abstruse possibilities of being wrong when we make such claims, such as 'I could be dreaming or hallucinating; how can I rule out that possibility?', are the staples of epistemology. This makes us ask, 'Do we know *anything*? Can we *know* anything?' If the simplest and most direct claims to know something cannot be defended against sceptical challenges, even of the most *outré* kinds, then – obviously – we have a problem.

And, as it happens, we do indeed have a problem. It might be that what we learn from sceptical challenges, however bizarre they seem (like Descartes's 'I might be deluded by an evil demon' example, which he used in a purely heuristic way – that is, as a merely enabling device – to explore whether we can know anything *with certainty*), is that in the strictest sense we do not in fact know anything, at least outside mathematics and logic, which are the only enquiries where certainty is attainable.[5] This means that instead of *knowledge* in the strict sense, we have to accept that the best we can achieve is *highly credible and well-supported belief*; and correlatively that any of our beliefs, however powerfully supported by our best evidence, could turn out to be wrong.

This is exactly the view that science is based upon. Science recognizes itself as *defeasible*, that is, as subject to adjustment or revision in the light of new evidence if that new evidence calls current theory into question. Science is arguably humanity's greatest intellectual achievement; the scientific method is the paradigm of responsible, careful, scrupulous investigation into its various subject-matters, and it is acutely self-critical and controlled by the empirical data of experiment – which is to say, by the way the world is and not by how we wish it to be. In line with this deep sense of epistemological responsibility, scientists do not claim to *know*, but they do ensure that their theories are secured to the utmost degree by rigorous testing and evaluation. It is standard practice in high-energy physics experiments, for example, not to publish a result unless it achieves the degree of confidence known as 5-sigma, that is, that the chance that the results achieved over the course of all the experimental runs amount merely to a statistical fluctuation is only 1 in 3.5 million. The journal *Physical Review Letters* regards 5-sigma results as 'discoveries'.

This kind of intellectual responsibility characterizes all serious enquiry, in history and the social sciences as well as the natural sciences. Techniques and methodologies may vary according to the subject-matter being studied, but the *ethics* of enquiry apply to them all, not least in dealing with the problems that all forms of enquiry encounter, and that I will describe shortly.

Note that *scientism* – the view that science can and will ultimately explain everything – is not the same thing as science. Particle physics does not pretend to explain political systems; inorganic chemistry does not pretend to explain the qualities of Romantic poetry. Science is subject-specific – its researches are individually focused on the fundamental structure of matter, the evolution of biological species, the nature of distant galaxies, the development of vaccines against viral infections, and so on. It is an acutely self-aware enterprise, governed always by the scrutiny to which scientists subject their own and others' work long before venturing to publish it.[6] Its example is general. History, along with other social sciences and the humanities, offers more commentary on society and the human condition, but the same considerations about intellectual integrity apply.

These considerations oblige us to confront the problems – sceptical ones, methodological ones, cautionary ones – that beset enquiry, and that are brought more clearly to mind by the recent dramatic advances in knowledge precisely because of the extent of ignorance they reveal. I identify a dozen such, and raise them in their relevant places in the discussions to follow. I name as follows:

The Pinhole Problem. Our starting point in all our enquiries is the very limited and highly circumscribed data available to us locally in space and time, and, from our finite point of view, allowing us a view of the universe and the past as if through a pinhole positioned at just our restricted scale. Do our methods successfully carry us through and beyond the pinhole?

The Metaphor Problem. What metaphors and analogies are invoked to make sense of what these enquiries are telling us, and might they mislead?

The Map Problem. What is the relation between theories and the realities they address, given the analogous differences between a map and the country of which it is a map?

The Criteria Problem. What are the justifications and, where necessary, correctives for the application of criteria such as 'simplicity', 'optimality', even 'beauty' and 'elegance', in the formulation of research programmes and approval of results? Do appeals to these 'extra-theoretical criteria' help or distort enquiry?

The Truth Problem. Given that empirical enquiry gives us defeasible probabilities, what are the standards (such as the sigma scale in science) that can be regarded as satisfactory short of certainty? Does this imply that we have to treat the concept of truth pragmatically, as a (possibly unattainable) goal of enquiry upon which, in the ideal, enquiry strategically converges? Where does this leave the concept of 'truth' itself?

The Ptolemy Problem. Ptolemy's geocentric model of the universe 'worked' in a number of ways, permitting the successful navigation of the oceans and prediction of eclipses, thus showing that a theory can be efficacious in some respects while still being incorrect. How do we avoid being misled by pragmatic adequacy?

The Hammer Problem. Summed up pithily as 'If your only tool is a hammer, everything looks like a nail', this reminds us that we tend only to see what our methods and equipment are capable of revealing to us.

The Lamplight Problem. One searches for one's lost keys under the street lamp at night, because it is the only place where one can see. We enquire into what is accessible to enquiry, for the obvious reason that we cannot access what is inaccessible.

The Meddler Problem. Investigating and observing can affect what is being investigated or observed. When one studies animals in the wild, is one studying them as they would be if unobserved, or is one studying behaviour influenced by their being observed? This, accordingly, is known as the 'Observer Effect'. Can the disruption caused by slicing and staining a specimen for microscopic examination be reliably excluded? Can smashing subatomic particles reliably reveal how they formed in the first place?

The Reading-in Problem. A problem mainly for history and the psychological sciences, in which interpretations of data are often made according to assumptions local in time and experience to the investigators. Can we guard against this as a source of distortion?

The Parmenides Problem. The danger implicit in *reductionism*: reducing everything to a single ultimate causal or explanatory principle, which on the face of it looks like the worst kind of elementary mistake, but which, remarkably, is a characteristic of hard science.

And, finally, *the Closure Problem.* The desire to reach a conclusion, to have a completed explanation or story, to tidy up and sign off. It is a natural human impulse to have satisfying narrative explanations, 'this because that' where 'that' does the job of terminating the explanatory chain, closing down the need for a further 'that'.

Putative explanations of the 'god of the gaps' kind provide classic examples. But so does what is implicit in the Parmenides Problem.

The three areas of enquiry canvassed in the pages to follow are affected by various of these problems with different degrees of force. I discuss the most salient of them in each area, and in the Conclusion.

These problems make some thinkers say that there are things we can never know. They say, for example, that questions about the nature of consciousness will never be answered because trying to answer them is like an eye trying to see itself. This is a counsel of despair that no enquirer should accept. If the question 'Are there limits to knowledge?' is meaningful, it is at best a defeatist one in implying that there might be such limits. But it is not a meaningful question, because it is not an answerable one – it could only, *per impossibile*, be answered when we reach the contradictory position of transcending the limits of knowledge and being able to look back at them to see where they lie. So the agnoiological ('cannot know') position is untenable as a general theory of enquiry and its aims. Instead, a commitment to the unlimited possibilities of knowledge is key; it is what motivates us in the continuing search for greater understanding of the universe and ourselves. But we learn from a consideration of the Pinhole Problem and the others how we should enquire, what we must avoid or take into account; we learn what we must do to try to advance knowledge and diminish ignorance, given the challenges that face the endeavour.

This book is not about epistemology in the narrow philosophical sense of answering sceptical challenges to our most basic knowledge-claims in an effort to see what can be known on the *strictest* definition of 'knowledge' as *what is true and accepted on indubitable grounds*. Instead it is about exploring and understanding, in a broader philosophical sense, the *highly credible and well-supported beliefs* that we *informally* call 'knowledge'. Indeed from now on I shall use the term 'knowledge' in this latter sense, which in any case is its mainstream sense: it is the sense of 'knowledge' in which what encyclopaedias contain is 'knowledge'. And it is about knowledge and ignorance, in this sense, regarding the science, history, and psychology we have so recently learned.

I ask the following questions about these fields of enquiry. What do we know in these areas of enquiry, and what did we once think we knew? How do we know it now, and are there any questions that arise, or reservations we might have, about the claims, methods, and assumptions at work in this knowledge and its acquisition? One of the constructive tasks of philosophy is the conceptual housekeeping it can offer by the kinds of questions it asks, a task described by John Locke in his *Essay Concerning Human Understanding* – written in support of the burgeoning genius of seventeenth-century science – as that of an 'underlabourer' helping to clear the path along which enquiry progresses. In the general sense of philosophy as reflection and the quest for understanding, this metaphor is apt as a description of the task here.

The three areas of enquiry canvassed in the following pages – the world, the past, and the mind – are: (Part I) particle physics and cosmology; (Part II) history, archaeology, and palaeoanthropology; and (Part III) the investigations of mind and brain in neuroscience and cognitive neuroscience. It is not, of course, possible to be comprehensive; I focus on central aspects of each field.

These are not the only new and recent areas of knowledge that have appeared and grown with dizzying rapidity in recent times, but they are arguably the ones that make the greatest difference to our self-understanding. In another time and place one would add areas of science that are destined to have a major effect on the future of humanity in their different ways. One is gene therapy, 'genetic engineering' (as benignly envisaged to protect, for example, against inherited disease), and applications of stem cell research in medical science. These developments are imminent but not yet fully arrived, and their impact highly speculative; one can hope that they will bring benefits in relation to many of the diseases that plague humanity now that lifespans are so much longer – cardiovascular disease and the cancers chief among them – and to ageing itself. But very little thought has yet been given to the social, psychological, and economic impacts of lives even longer and much healthier than now.

The other set of developments destined to affect the future of humanity relates to artificial intelligence and its applications. Perhaps saying that these developments will affect the future is already out of date: AI is here and already at work in many ways, most of them beneficial. How far the developments will go and what their combined effect will be are questions currently open for debate.[7]

'Recent' is the significant word in connection with the three areas of enquiry I consider here. Only think: the first observation of a subatomic particle occurred in 1897, when J. J. Thomson discovered the electron. The atomic nucleus was first described in 1909 by Hans Geiger and Ernest Marsden working in Ernest Rutherford's laboratory. Einstein's Special Theory of Relativity was published in 1905, his General Theory in 1915. Quantum theory developed in the first decades of the twentieth century, receiving a form of official endorsement by physicists at the Solvay Conference of 1927; the photon had received its name just the year before. The 'Standard Model' of the atom had become widely accepted by the 1970s, and confirmation that the Higgs field exists completed the model in July 2012.

It was not until the work of Edwin Hubble in the 1920s that the Milky Way Galaxy in which our solar system is located was recognized as just one of a vast number of galaxies, and not, as previously thought, itself the entire universe. That was in 1924; in 1929 Hubble observed that the universe is expanding. That led to the formulation of the 'Big Bang Theory'; in 1992 NASA's Cosmic Background Explorer (COBE) confirmed the existence of the background radiation left over from the Big Bang, now calculated to have occurred 13.72 billion years ago.[8]

Intimations and hypotheses that led to these discoveries existed beforehand, of course: ancient Greek philosophers had suggested that matter must be made of smallest parts (which is what 'atom' means – indivisible, uncuttable); seventeenth-century thinkers such as Pierre Gassendi and Robert Boyle speculated about *corpuscles* ('little bodies') as the constituents of matter and gases; and on an even more secure observational basis John Dalton and Robert Brown suggested the same in the nineteenth century. The philosopher Immanuel Kant proposed in the eighteenth century that the universe is expanding; as an originator of the 'Kant–Laplace Nebular Hypothesis' he has credentials in the field. And none of the work of Thomson, Rutherford,

Einstein, and their successors in twentieth-century science could have been possible without such predecessors as Galileo, Newton, Faraday, Maxwell, and others. But it remains that by far the greater part of physics and cosmology as we now have them is of very recent date; the advances have all been made in the course of the last hundred years.

Yet the most amazing thing about this growth of knowledge is that it has revealed to us that we have access only to about 5 per cent of physical reality. It is less than a century since humanity arrived at a disciplined, evidenced view of the history of the universe itself from the Big Bang to the present – an immense achievement – but already the puzzles are prompting more exotic possibilities: that the universe is only one among many universes, or one phase in an unimaginable set of universal histories, or just the best explanation from a limited virtual-reality construct from our pinhole perspective on reality – these promptings courtesy of the hypothesized existence of *dark matter* and *dark energy*, and the highly speculative nature of suggestions about how relativity theory and quantum theory might be unified.

A different set of problems besets our knowledge of the deeper historical past. There has always been fairly extensive knowledge of classical antiquity and what has followed it up to the present day, because classical antiquity itself has survived to us, both in physical remains and in some of its literature, in a continuous line from its own time until now.[9] But all that was known in addition was suppositious knowledge of an earlier past, in the Homeric poems and in the histories and legends of the Hebrew bible (Christianity's 'Old Testament'). The latter purported to stretch history back to the creation of the universe about six thousand years prior to the period when the Old Testament histories were formulated. In those writings reference to the pharaohs of Egypt, Ur of the Chaldees, the empire of Babylon, and other places and features hinting at a past remoter than the classical period, together with their associated legends and myths, kept alive a sense of deeper historical time than was positively known about. Renaissance collectors of antiquities and curiosities stimulated interest by their activities in what lay beyond the pale of familiar history, but it was chiefly from the late eighteenth and – mainly – nineteenth centuries that more systematic efforts to dig into that deeper historical past began; quite literally so, in the form of archaeology. And only then did that deeper past start to come into view.

When Napoleon invaded Egypt in 1798, he took with him two hundred scholars to study that country's topography, botany, zoology, mineralogy, society, economy, and history. The temples and monuments of Luxor, Dendera, Philae, and the Valley of the Kings were measured and drawn. Within a decade the findings of the scholars began to be published in the first volumes of what became, by 1828, the encyclopaedic twenty-three-volume *Description d'Égypte*, unleashing an international mania of interest in all things Egyptian and by extension Levantine. Translation of the hieroglyphic inscriptions on the Rosetta Stone was painstakingly begun by a number of scholars, the breakthrough coming in the early 1820s when Jean-François Champollion successfully identified some of the language's phonetics by means of the names figuring within cartouches both in the Rosetta Stone inscriptions and in other sources such as the Philae Obelisk.

A rapidly increasing interest in digging up the past, in the literal archaeological sense, seized a number of nineteenth-century amateurs. A significant part of the motivation for some was to find confirmation of Old Testament history; for others it was the search for curios and collectibles; for thieves alerted by the amateurs' interest it was profit. The first major site discovered in Mesopotamia, Nineveh, was a trigger for the two latter kinds of activity. Paul-Émile Botta, France's Consul General at Mosul, made some excavations of a mound on the east bank of the Tigris and uncovered significant-looking structures. They later turned out to be the palace of Sargon II. That was in 1842; five years later a young British diplomat called Austen Layard set to work on the mound with a view to collecting as many objects of artistic or historical interest as he could find, 'at the least possible outlay of time and money', as he put it. But it was an Homeric impulse that led to the century's best-publicized excavation: Heinrich Schliemann's dig for the city of Troy, beginning in 1870. This famous endeavour did far more harm than good, because of the destructive methods Schliemann employed, cutting a huge crude slice into the many layers of archaeology at the Hissarlik site, and because of the over-ambitious claims he made about his findings there and at Mycenae later in the 1870s. His insensitive archaeological methods were alas par for the course with his predecessors and most of his contemporaries; they did much harm to fragile sites, annihilating evidence that time itself had not been able to efface.

In the succeeding decades a vastly more careful and systematic approach to archaeology began, among other things bringing the early civilizations of the Near East into sharper and more copious view. As the twentieth century progressed, so did archaeological methods and the contributions to them of science. Radiocarbon dating began in the 1940s, followed by advances in geochemistry and geophysics, with various forms of remote sensing including radar and lidar, 3-D laser scanning, aerial archaeology, Raman spectrometry, portable X-ray fluorescence, medical analyses of teeth and bones and examination of ancient DNA, forensic examination of the treasure-house of information in ancient middens and toilets, and more – all greatly enhancing the investigative capacities of archaeology. These developments have not been uncontroversial: debate over 'processual' and 'post-processual' methodologies and tensions between scientific and humanistic approaches in archaeology continue, even as archaeology progressively strips away more layers of past time and adds more layers of understanding.

Major mysteries remain. What caused the collapse of Bronze Age civilization in the period around 1200 BCE, plunging what had been the highly advanced civilizations of the eastern Mediterranean and Near East into a 'Dark Age' of several centuries? Egyptian records blamed successive invasions by an unknown group described as the 'Sea Peoples', but historians largely agree that the causal factors were much more complicated – among them climate change, famine, and breakdown of the complex trade routes running from as far east as the Indus Valley to as far west as Britain. That Dark Age drew a blind across the past until archaeology removed it; it is remarkable to think that the impressive architecture and exquisite art of Mesopotamia, the Levant, the Aegean, and Egypt were almost completely unknown until so recently.

But these discoveries relate only to the last six thousand years or so, although also offering some routes into the twelve thousand years since the inception of the Neolithic Period, when systematic agriculture and urbanization began. Before that the history of *Homo sapiens* and its relatives and predecessors tails evermore thinly and ambiguously into a complex and vastly remote past. Science has been a major boost here too, but, although assiduous searching has brought increasing amounts of evidence into palaeoanthropology and anthropogeny, what it tells us about human origins grows evermore tantalizingly inconclusive; every new discovery of teeth, bones, and tools seems to complicate rather than clarify the picture of our deep ancestral past. An example is the remarkable discovery in South Africa, less than a decade before these words were written, of *Homo naledi*, whose puzzling mixture of characteristics – its primitive head, upper body, hips, and curved fingers – are reminiscent of australopithecines, which lived on either side of 3 million years ago, but its advanced hands and feet are similar to Neanderthals and modern humans. Meticulous dating of the remains produced the astonishing result that *naledi* is recent, living about three hundred thousand years ago, which makes it contemporary with early-modern *Homo* and itself a member of the *Homo* clade.

It is no surprise that the largest and smallest scales of the universe, and the buried past both of civilization and our species, should present the investigative challenges they do. What is striking is how vividly they present that evermore familiar and challenging paradox of knowledge: *the more we know, the greater the extent of our ignorance becomes apparent.* But what of the third area of enquiry, brain science and psychology? Knowledge of ourselves, our minds, consciousness, human nature – is this not something we are intimately close to, and obsessively interested in, as our literature, entertainment, gossip, meditation, anxieties, hopes, loves, dreams, and fears unremittingly tell us? And yet even here the paradox is repeated, of an explosion of knowledge creating yet deeper mystery. For all the devotion given by philosophy, art, literature and our other self-reflective endeavours to the question of who and what we are, we still do not fully understand – even, perhaps, yet half understand – human nature and psychology, still less the complex material reality that underlies them, namely, the brain.

It is a matter of mere decades since it became possible to view brain activity non-invasively and in real time to try to correlate areas of the brain with functional and psychological capacities by means of *functional* magnetic resonance imaging, 'fMRI'. Prior to the advent of fMRI as a neuropsychological tool, most reliance had to be placed on 'lesion studies', matching injury or disease in parts of the brain to loss or disruption of such various functions as speech, movement, vision, hearing, memory, and emotional control. Brain research has an important practical application to the task of finding ways of repairing damaged brains, preventing or reversing dementias, and curing epilepsy. For obvious reasons these tasks go hand in hand with understanding the brain localization of mental capacities. But brain studies by themselves might not say everything we wish to know about human nature and psychology. Evolutionary psychology, and its more inclusive forerunner sociobiology, offer perspectives in these respects – controversially, as indeed is the case with neuropsychology too; for both sciences are nascent, their methods and equipment still in development, the opposing weight of traditional views and beliefs still great.

The intractable nature of human psychological material poses a formidable challenge to understanding it. But that is not the only barrier facing enquiry: there is anxiety too, fears that a Pandora's box might here be opened – in the worst case, as conceived by science-fiction writers who, either alarmingly or helpfully, have a propensity to identify thought-provoking scenarios, including the following: chips implanted in the brain to control behaviour and thought, the complete invasion of privacy this could be imagined to involve, the usurpation of humanity by artificial intelligence incorporated into what, by comparison, is the Model T Ford of the evolved primate brain, which the Ferrari of intelligence technologies might overwhelm; and more.

What we know matters a great deal, obviously. It might seem to matter less that we understand why the Bronze Age Collapse occurred three thousand years ago than to understand the structure and function of the brain, because this latter guides us in treating its diseases and injuries. The former may seem a matter of mere curiosity, though in fact it could teach valuable lessons, to anyone wishing to learn them, about the factors that lead to economic and social problems, even to civilizational catastrophe, as has happened more than once in recorded time. As this shows, all knowledge is useful and much of it is vital.

But it also matters that we understand *how we know*. When we see how scientific and historical knowledge is acquired, what problems are overcome in acquiring it, and what questions are raised about the assumptions and methods involved, we not only learn how to evaluate what we know, but we also learn a great deal about responsible thinking and the demands of intellectual honesty. These things matter in every sphere of human activity, and they are at a premium. The arts of persuasion, of redirecting attention, of magnifying or masking facts, of influencing and manipulating opinion, are commonplace everywhere from politics to advertising – and they all rest on the truth that, as Bertrand Russell once remarked, 'Most people would rather die than think, and most people do.' For, alas, persuasion and manipulation are made to seem to matter far more than the effort to be truthful. Therefore to know about what we know and how we know it is a grand corrective to the virtual realities and semi-realities that are continually being flashed before our minds by partisan causes and agencies wishing to sell us something – a product, an idea, a policy, a lie.

The following discussions are arranged as follows. I begin with a survey of what *was* the frontier of knowledge in the process – not always linear or smooth – that led to the bodies of knowledge that are current, at this time of writing, in physics, ancient history, and studies of the brain and mind. I survey the main discoveries that have so recently been made in these regions of enquiry, and consider some of the questions, problems, and promises associated with each. As this is a book for the interested reader, no prior knowledge of these fields is assumed. Those with expertise in one of the fields might wish to go directly to the sections of the relevant part where questions and problems are discussed. Clarity and accuracy have been the goals throughout, but as each of these fields is an actively developing and vigorously contested domain, I do not expect that any of the views mentioned will command universal assent. But debate is a good thing; it is the motor that drives the wheels of progress.

PART I

Science

In plain sober truth, without overstatement, science is humanity's greatest intellectual achievement. 'Science' is a capacious term but clear enough: most people get some elementary physics, chemistry, and biology at school, but less familiar and far deeper are the parts and combinations of these broad domains and their subject-matters, which range from enquiry into the fundamental elements of physical reality, through the complexities of life, to the furthest reaches of the cosmos. In the very recent past these enquiries have burgeoned exponentially. The applications of many of these discoveries in technology and medicine have been, in the literal sense of both terms, revolutionary and transformative.

Yet it is still the case that the great majority of people on the planet know little about what the sciences have so far revealed, and they adhere to a picture of the world similar in many respects to the one that had been dominant before the scientific revolution of the sixteenth and seventeenth centuries CE. That world-view – of a deity-created universe centred upon mankind physically and morally – was functionally dominant then, but, although it is still the majority belief, it has become functionally marginal, for in almost all practical respects the world runs on science and technology.

Given the success of science, it is remarkable that, even as it progresses with such giant strides, it should at the same time more vividly exemplify the paradox of knowledge – that every gain in knowledge multiplies the sense of our ignorance. This is most true in fundamental physics and cosmology, less so in the biological and medical sciences, the latter in particular showing how in applied science the horizon of competence in controlling aspects of the world has advanced beyond the imagination of previous generations. I therefore focus on physics and cosmology in the following pages but add two ingredients: a sketch of the technological prehistory of science, and what might be described as a superposition of the structures of thought that shape scientific enquiry, to illustrate that assumptions concerning 'what the world must be like', and what shape a satisfactory explanation of it should take, are surprisingly persistent features of our sense of reality and might explain some of the perplexities generated by science's very successes.

1. Technology before Science

Although the distinction between 'knowledge how' and 'knowledge what' (or 'why') is not a mutually exclusive one, it is significant. Human history is the history of an ingenious, exploratory, fix-it species, which for most of the time it has existed has invented technologies that not only served its survival needs but, in a feedback loop, positively drove its evolution. Theodor Adorno remarked that humanity has grown cleverer over time, as evidenced by the development of the spear into the guided missile, but no wiser, as evidenced by the guided missile itself. This is an educative thought, but it is well to remember that most technology has been invented for the peaceful purposes of surviving and flourishing, even though now the budgets for technologies of war are as large or larger than those for most other activities of interest to humanity.

Almost all technology is *what works* in satisfaction of a need. Understanding why it works can matter in some cases, but this is not always, and even perhaps not often, necessary. This can sometimes be true of science itself: allegedly the advice of the celebrated physicist Richard Feynman was to ignore the difficult question of how to interpret quantum physics, and instead just to 'shut up and calculate'. Such robust views aside, science is chiefly the effort to understand, to know the reason why, to discern the principles of things. Technology is about getting things done, whatever the explanation. The philosophical nature of the French is summarized in a witticism attributing the contrary view to them: they ask, 'It works in practice, but does it work in theory?' This is said of economists too. For technology, however, theory is not the point; practicality is all.

The history of technology is long and impressive. The history of science is short and even more impressive. Since the history of technology is rarely told, and usually only of interest to people caricatured as having a row of pens in their top pockets, the following survey provides a background to the general question of knowledge both in its *how* and *what* senses.

It was once thought that tool-use is a distinctively human trait. But many other species make and use tools, even if, generally speaking, they are rudimentary ones. So this no longer serves as a demarcating feature of humanity. But the kind, quality and variety of tools associated with the human lineage, and their development into the sophisticated technologies of today, are most certainly a differentiating feature. Even more so is the fact that human tool-use became increasingly committed and, eventually, almost obligatory in the evolution of *Homo sapiens*, such that without tools our species would not have flourished as it did, and would now find it difficult to survive. One might therefore say that one of the distinctive things about humans is that they are technological creatures, not in the adventitious manner of the termite-fishing chimp with its stripped stick, or the shellfish-cracking sea otter with its pebble, but *systematically* technological.

Some hominin, or perhaps hominid – taking the latter term to denote the more inclusive class of primates of which hominins are the human and human-related subset – might have begun shaping stone tools as early as 3.3 million years ago, as somewhat controversially suggested by the discovery, at the Lomekwi 3 archaeological site in Kenya's West Turkana region, of stones that appear to have been worked. According to their discoverers, some rocks found there display evidence of *knapping*, the process by which flakes are chipped or broken from a core. The Lomekwian tools were large, the rocks used as anvils especially so. Since they predate by half a million years the first *Homo* fossils (the time of the Lomekwi tools is that of the australopithecines) and by seven hundred thousand years the first tools to be securely associated with hominins – viz. those of the Oldowan stone industry – the Lomekwi finds are intriguing.

At Lomekwi the assemblage of about one hundred and fifty artefacts suggests employment of a technique observable in chimpanzees today: the block-on-block hammerstone method of fracturing stones. But evidence of *double* knappings suggests intentional working of the stones, though if they are indeed tools their purpose is not known, because they were not found in association with animal bones showing cut or pounding marks. Although chimpanzees strip leaves from a twig and chew its end into a brush for fishing termites, they have never been observed to use stones for butchering meat or cracking open bones and skulls to access the soft tissue within.

Oldowan tools are unequivocally hominin tools. They are named for the place of their first discovery, Olduvai Gorge in Tanzania, but earlier examples were found at Gona in Ethiopia, and the oldest at Ledi-Geraru in Ethiopia's Afar region, dating from more than 2.6 mya ('mya' is the standard shorthand for 'million years ago'). The characteristic Oldowan tool is a 'chopper', a stone with one side sharpened into an edge by the percussive removal of flakes, making it serviceable for slicing and scraping as well as chopping. Micro-evidence from Oldowan cutting edges shows that they were used both on plants and on meat and bone. The Oldowan tool-kit included pounders also, for softening plant fibres and cracking open bones to expose the marrow. These types of tools remained in use for a million years, examples of them being found throughout eastern and southern Africa, the Near East, Europe, and South Asia, associated with *Homo habilis* ('Handy Man') and early *Homo erectus*.[1]

These tools are evidence of a change in the dietary and social habits of the hominins who made them. The knapping techniques were skilled, producing sharp edges whose marks are evident on animal bones found alongside the tools at Oldowan sites. Both stones for tool-making and food were transported to these sites from their source localities, demonstrating that hominins congregated there to share the work and the benefits.

Increased sophistication in tools is represented by the Acheulean stone industry beginning 1.76 mya. The time gaps cited here are extremely large; the period between the Lomekwian and Oldowan stone industries seems to imply low levels of usage and scarcely any development, but development between the Oldowan and Acheulean industries was no more rapid. This is not surprising, for, although tools reduce the energy-investment in tasks, this is only the case once the tools exist; there is a significant cost to getting them in the first place. Suitable raw materials have to be found, which then have to be worked into configurations suitable for the purpose envisaged, and the skills required for identifying appropriate materials, creating tools out of them, and then employing them effectively, have to be developed. It has been calculated that hundreds of hours of experience are required for the manufacture and expert use of Acheulean-level artefacts. Laziness is the easier option, encouraging adherence to a familiar technology if it does the job more or less satisfactorily. Certainly the elegant, symmetrical, and diverse Acheulean tools required much higher levels of skill and planning than the Oldowan tools, and that speaks volumes about the development of their makers' minds.

Acheulean tools are fashioned from stones selected for their desirable fracture properties: chalcedony, jasper, and flint, and in some places quartzite. Suitable stones were transported considerable distances from where they were found to where the tool-makers camped, and the instruments made from them were worked into bifaced handaxes and cleavers. The tools display increasing elaboration over the 1.3 million years during which the industry flourished. Early handaxes were made by hitting them against a stone serving as an anvil; later on wooden hammers were used to produce smaller, more slender axes with sharper and straighter edges. There is evidence of hafting in the Acheulean industry; the wood of hafts does not survive, but traces of adhesive material such as bitumen and conifer resin are found on some axes and hammers, and, together with the impact marks they display, this suggests that they were wielded as tools with handles.

About 300 kya ('kya' is 'thousand years ago') tool-makers evolved the Levallois technique, characterized by careful preparation of cores. This involved taking a lump of stone and fashioning it into a tortoise-like shape, flat on the bottom and humped above; that is the core. A skilful blow on a selected striking point produces a flake that can be worked further, using bone, an antler tip or soft stone, to achieve the desired result. Use of this technique by Neanderthals characterizes the Mousterian stone industry, named for Le Moustier in France where examples were first found, though the technique was evident throughout much of Africa at the same period.

The development of Levallois techniques is contemporary with the appearance of 'anatomically modern humans' in Africa at about the same date of 300 kya. Around 100 kya art appears in the human record – the discoveries in Blombos Cave in South Africa provide some of the earliest evidence – and then, both in and outside Africa from about 60 to 50 kya, increasingly rapid technological changes began, leading to the Aurignacian tool industry from 40 kya onwards, typified by blades, burins, needles, and scrapers made from bone and antler as well as stone. Given that the Aurignacian is also character-ized by cave art, sculpture (for outstanding examples, the Venus of Hohle Fels and the Lion-Man Figurine of Hohlenstein-Stadel), dec-orative items such as necklaces, and musical instruments (such as the bone flute also found at Hohle Fels), the tools made were not restricted to subsistence activities. These developments signify another large stride in the human story.

Before 12 kya microlithic tools – small sharp flakes fixed into a haft to serve as a saw or scythe – and polished tools made their appear-ance. Polishing stones by careful grinding improves their strength and effectiveness both as tools and weapons, making them less prone to fracture. It also doubtless enhanced their aesthetic qualities in the opinion of those who used them, as suggested by the fact that pol-ished stone axes were included among other grave goods with their deceased owners.

Just how advanced human technology had become by the Neo-lithic Period – commencing 12 kya – can be inferred from Oetzi the Iceman, whose glacier-preserved body was discovered in the Alps in 1991. Although he lived much later, towards the end of the fourth millennium BCE, Oetzi's tools and equipment were little different from what would have been available at the beginning of the Neo-lithic, except in one respect: he had a copper-headed axe. Both his arrowheads and his dagger were made from knapped flint, he wore clothes made of different kinds of leather, and he had a bearskin cap with a leather chinstrap. His cloak was made of woven grass, and he had waterproof shoes with bearskin soles and uppers of stitched deer hide. His tool-kit included an awl or burin for punching holes in leather, scrapers, and flint flakes, and an instrument possibly for sharpening arrows. Some of the arrows in his quiver were fletched (had feathers fixed into the rear of the shaft for accuracy and stability in flight) and some not, suggesting that he made and repaired his equipment as he went along. He had been killed in some sort of fracas: an arrow was lodged in his left shoulder, and he probably died of blood loss because the wound is close to the site of an artery. His copper axe-head – Oetzi's period in history is known as the Chalcolithic, or Copper, Age, roughly 6.5 to 3.5 kya, the period immediately before the Bronze Age – was fixed to a yew handle with leather thongs. That is also how he secured his flint arrowheads to their shafts.

It is not so fanciful to image that one of Oetzi's ancestors at the beginning of the Neolithic Period might have been clothed and equipped much as he was; stone tools were still used in butchery in the Near East until around 1200 BCE, and throughout the Bronze Age flint daggers mimicked bronze daggers and vice versa, showing what it is anyway plausible to expect: that stone technologies and the development of metallurgy overlapped for a long time.

As interesting as tools themselves is what they say about their makers. The activities associated with more than 3 million years of tool-making are evidence of planning based on experience. Think of what that means: remembering, pondering, coming to a realization, experimenting repeatedly, making improvements – these are the deliberate and purposive applications of intelligence, and, even though the greater part of those three million years saw very slow development, the contrast between the worked stones of the Oldowan industry 2.6 mya and the use of stones to crack open nuts by some species of primates today is speaking.

Better tools made for more and better-quality food. For hominin evolution that meant keeping step with the increasing energy needs of bigger and more active brains. Indeed the relationship is reciprocal; it is a feedback loop involving a suite of adaptations – the intelligence to visualize a tool and then create it, with development of the associated manual dexterity, matched with the resulting increased quantity and quality of nutrition to fuel the whole process, input and output mutually fostering each other. Rising intelligence in the human lineage is therefore intimately connected with tool technology and the social and dietary advances it made possible.

A key development in much of this story is control of fire, which provided warmth, light, and safety from predators, and greatly increased the availability – because of digestibility and safety – of foods. Take the point about safety: the earliest of our ancestors who supplemented their diet of roots and fruits with animal food were almost certainly scavengers, making use of leftover carcasses once predators had taken their fill. There is much evidence of bone marrow being consumed, a highly nutritious food, and – more to the present point – safe to eat, because less likely to have rotted. Some meat might have been sun-dried or even preserved with salt before cooking became possible, and doubtless remained an option afterwards. But cooking the meat of carcasses makes it safe as well as more palatable; this is what all human meat-eaters do today with most meats, given that meat is the already-decaying flesh of dead animals.[2] Fire also aided our ancestors in tool-making, for example, by hardening the points of wooden spears, and by rendering certain types of stone easier to flake.

No doubt hominins profited from adventitious occurrences of wildfires whenever they could, but control of fire – being able to start one at will, to contain it in a space, and to transport it from place to place – is what really counted. Profiting from the aftermath of a forest fire would have revealed to our ancestors the benefits it conferred – roasted carcasses, more accessible and digestible tubers, for example – and today's chimpanzees are observed to make use of burned landscapes likewise. A gradual process of taking advantage of naturally occurring fires, preserving a fire for a period of time, and eventually discovering how to start one – all this signifies observation and mastery of an energy-source that could be dangerous if mishandled, but was powerful when governed. There is evidence dating from as early as 1.7 mya that *Homo erectus* made systematic use of fire; certainly anatomically modern *Homo* had inherited control of fire from before 200 kya, as evidenced by finds in South Africa at the Cave of Hearths in Limpopo Province and the Klasies River Mouth Caves in Eastern Cape Province.

All the fires detected at earlier and later sites could have been taken from wildfires by holding a branch in the flames, say. It is hard to find a secure date for the first systematic control of fire. But, even before that happened, the inclusion of fire among the resources of human ancestors made a big difference to them and therefore their descendants.

Empowered by the skills and capacities, not least social ones, entailed by tool-making and control of fire, humans had reached Australia by 40 kya and the Americas by 15 to 12 kya. Every clime had been colonized, demonstrating not just the colonizers' ingenuity but also their adaptability. The knowledge and skill required for a hunter-gatherer life had to be prodigious for them to establish themselves in environments as different as the Arctic and the Australian outback, and to flourish there. One of today's human beings unexpectedly teleported back to 40 kya would not last long, unless he or she had military-level survival training – and even then would be a tyro in comparison to the human ancestors at home in those times and places.

From the beginning of the Neolithic Period around 12 kya the story of technology enters a new phase. Agriculture and animal domestication, urbanization, engineering, metallurgy, the wheel, and writing are among the most salient of the developments in the period between 12 kya and 1200 BCE. It is a lengthy period in its own right, but next to nothing in comparison to the hundreds of thousands of years that had passed in the course of the developments described above.

One suggestion is that the choice made by some people to settle and farm in one place, instead of continuing to hunt and forage nomadically, was forced on them by circumstance, and in certain respects represented a retrograde step. For one thing, skeletal remains of farmers from the Early Neolithic Period show them to be less healthy than their hunter-gatherer contemporaries. For another, population increases in urban settlements led to further division of labour and more hierarchical social organization, representing a loss both of equality and liberty, and promoted communicable diseases. The circumstance in question may have been that populations were

already growing, and competition over hunting grounds and foraging had increased the frequency of conflict between groups. Depletion
of resources would have been a likely reason for this, leading some
to rely on the cultivation of edible grains and the domestication of
animals instead.

But when agriculture and settled living began, the disadvantages
came to be balanced, perhaps indeed outweighed, by advances that,
as the millennia passed, gathered pace and brought a new dispensation into existence for those in what are called the first civilizations.

Settled and systematic farming began in the 'Fertile Crescent', a
geographical curve stretching from the southern plains of Mesopotamia, up the Tigris and Euphrates rivers, and round through Syria and
the eastern Mediterranean coast to Palestine. Some make the crescent
stretch from the junction of the Tigris and Euphrates to Egypt's
Nile Valley. The last Ice Age, or the Last Glacial Maximum, which
occurred about 20 kya, was followed by a warming of the global
climate and a correlative retreat of ice sheets, until a sudden return
of cold weather lasting 1,300 years, a period known as the Younger
Dryas (12,900 to 11,600 years ago), plunged the world – or at least the
Northern Hemisphere – into glacial conditions again, perhaps precipitated by a south-to-north reversal of major fresh water flows into
the Atlantic from North America. Its ending in the middle of the
twelfth millennium allowed global warming to resume, marking the
end of the Pleistocene Epoch, which had commenced 2.8 mya, and
the beginning of the age we occupy now, the Holocene. It was in the
warmer conditions following the Younger Dryas that more groups
of people began to settle and farm in the Fertile Crescent.

'More groups': the phrase acknowledges the existence of evidence
for some degree of settlement, at least for part of the year, in the
Levant as early as 20 to 22 kya; hunter-gatherer lifestyles were not
necessarily nomadic, if sources of food to be hunted and gathered were
sufficient and renewable. Moreover, people had doubtless harvested,
milled into flour, and cooked the grains of wild grasses for a long
time before they began to cultivate them deliberately. Cultivation
rather than opportunistic harvesting involved reserving some of
the grain to plant later on, bringing water to it, and weeding out

competing plants. This activity was the immediate precursor of agriculture as such. Developing an ability to choose strains of grain that grew taller, with heavier ears and therefore a higher yield, was a natural consequence of observation and need combined. Associated activities quickly followed, chief among them land clearance, making pots, or at least larger pots, to carry and store grain and protect it from rodents, fashioning equipment for harvesting and grinding, building ovens, and accumulating reserves of fuel.

It is a ready inference that the first settlers started to keep the more docile young of the animals they hunted, thus domesticating them. It was easier to access their meat, hides, and wool, and later their milk also, when they were quietly at hand rather than having to be pursued and caught – this was long before horses were domesticated. It is also readily inferable that it was women who took charge of planting, harvesting, milling, and baking, while men continued to hunt and herd, for skeletons of women found at Abu Hureyra in Syria, dated to 9700 BCE, show the effects on toes, knees, and spines of much laborious kneeling to weed crops and grind the harvested grain. Moreover, agricultural and related tasks were easier to combine with pregnancy and childcare than was hunting or, later, herding, which typically involved seasonal movement of livestock between winter and summer pastures. The transition from gathering and hunting to tending crops and herding evidently followed gender lines, which with equal probability were long established.

Farming and animal husbandry in the Fertile Crescent predated its appearance elsewhere, starting soon after 12,000 BCE and spreading to Egypt, and to nearby parts of Europe with similar climates such as Greece, by about 6000. Between 8000 and 6000 the Indus Valley peoples had begun cultivating wheat and barley also, and domesticating goats, sheep, and cattle. From around 6500 people living along China's Yellow River – so called because of the rich loess silt washed down in its floods and colouring it accordingly – had begun to grow millet first, then sorghum, soybeans, and hemp. Further south in China and across South-East Asia the cultivation of rice, taro, and banana began soon afterwards, and the farmers there domesticated water buffalo as draught animals, and chickens and pigs for food.

Agriculture came later to Central and South America. From about 5000 BCE people in Mexico grew teosinte (an early form of maize), chilli peppers, tomatoes, beans, and squash as a supplement to their main sources of food, which were hunted and gathered. They did not rely on systematic farming until about 1500. The only kinds of animals available for domestication were dogs and guinea pigs. In South America potatoes, beans, and quinoa were grown, and alpacas and llamas domesticated. The latter were too small for riding or use as draught animals, so they served chiefly as pack animals and as sources of wool and meat.

All these developments in all these regions depended upon observation and experience, and the resultant knowledge and skills were of an advanced order. They involved social organization, planning and structure. Division of labour meant that those who produced food had to ensure a surplus to support those who performed other tasks, so a system of exchange was required. The potter needed flour and the miller needed pots, and they could come to terms; both desired meat, and the herdsman who supplied it in turn needed flour and a pot to carry it away in − and so on; the situation speaks for itself as one necessitating both records and tokens of exchange, and eventually money.

In Mesopotamia the seeds of wheat and barley were the seeds of civilization, providing conditions that would carry humanity into new arrangements. One of the first consequences was the appearance of villages and towns. Among the earliest was Jericho, dating perhaps from 10,000 BCE. Signs of even earlier occupations of the area, by Natufian hunter-gatherers, suggest that the site's constant supply of spring water made it a natural place of resort, and then a natural place for a settled community to grow. By the middle of the eighth millennium Jericho had a population of two thousand; over the whole Fertile Crescent similar, although typically smaller settlements, had likewise emerged.

Settled life made possible much technological novelty. A roaming life limits the size, weight, and number of artefacts that can be carried, while the tools required for farming are different from those used for foraging and hunting. Both invention and experimentation flourish in more permanent conditions. Pottery was one of the first innovations, appearing in the Near East in the seventh millennium BCE (it had been independently invented by the Jomon of Japan long before). Making and operating mills for grinding grain, and looms for weaving textiles from hemp, flax, cotton, and wool, likewise necessitated settled living.

As the word 'Neolithic' attests, the first several thousand years of farming were still in the Stone Age, albeit the New Stone Age. The most emphatic statements of this period's culture are the stone monuments characteristic of it, some of them enormous, the earliest known being Göbekli Tepe in Turkey, dating from around 9000 BCE, the best known being Stonehenge in England. The latter, as it now appears, dates from about 2500 BCE, but it occupies a site where one or more earlier henges stood. There are massive standing stone monuments, *megaliths*, all over Europe, thirty-five thousand of them, from the remotest parts of Scotland and Scandinavia to the shores and islands of the Mediterranean Sea and Anatolia.

The sarsen stones of Stonehenge on England's Salisbury Plain weigh 25 tons, and were transported from quarries in the Marlborough Downs 15 miles away. The monument's smaller bluestone megaliths weigh between two and five tons, were quarried in Pembrokeshire in Wales, 140 miles away, and transported to the site without, it seems, benefit of wheels and pulleys. This is remarkable. Human muscle power, hauling and pulling and perhaps rolling the megaliths on roadways of logs, might have been the only resource available for moving them. Such a major investment of time and manpower testifies to Stonehenge's significance for those who built it. The purpose of megalithic monuments is unknown, though speculation abounds, with religious and astronomical (and usually their combination) interpretations being the favourites.[3]

Perhaps megalithic monuments, or the building of them, contributed in some way not just to the social bonds of the communities that created them but to their livelihood also. That is unknown. But in the farming communities of Mesopotamia there was another and different engineering challenge, this time connected directly to food production: watering the crops. Mesopotamia is a low-rainfall region, its water coming principally from the great rivers Tigris and Euphrates and their tributaries. The Tigris flows more swiftly than the Euphrates, which lies to the west of it, but because both pass through low-lying land they are apt to change course over time, and some of the great cities of the Bronze Age that once stood on their banks are now at a distance from them.

The need for water might make it seem obvious that farms should be situated close to rivers, but the risk of flooding is great unless the floods are regular, predictable, and occur after the harvest season – which is how things are with the Nile. Early Neolithic farmers had no choice but to risk planting near rivers, and because serious floods did not occur every year they were able to manage. But as soon as they devised methods for controlling the waters, great benefits followed. Irrigation canals, flood barriers, and drainage systems meant that crops could be protected or – thus bringing more land into production – grown further away from rivers themselves, while securing a consistent supply of water. Up to forty grains of barley could be grown from one grain in these conditions, as compared with five grains per grain in an area reliant only on adventitious rainfall.

The digging of canals and ditches and the building of dykes, together with the constant maintenance of all three, take the labour of thousands. Planning, the logistics of supplying food, housing, tools, and payment to work-gangs, and an organization to supervise the maintenance of the resulting irrigation system, are required to supplement the engineering feat of preventing floods, raising water above the river levels for higher ground, and achieving a good balance in the water level of the farmland itself. The Mesopotamians achieved all this with spectacular success.

But it took several millennia. Farming on the river banks was established by the sixth millennium BCE, the first irrigation systems were dug in the fifth millennium, by the fourth millennium marshes were being drained and reservoirs built, and the network of irrigation canals and dykes had grown to the point where serious floods were more a matter of legend than familiarity. The first flood story is told in *The Epic of Gilgamesh*, written down in the early centuries of the second millennium BCE, and is repeated with cultural variations in the Old Testament story of Noah and in several Greek myths, principally those of Ogyges and Deucalion.

By the time that *The Epic of Gilgamesh* was written down, Mesopotamian farmers were ploughing their fields with oxen and sowing their seeds with seed-drills to ensure the right depth and spacing for a good crop. On the slopes of hills at a distance from the rivers, farmers lifted water from the irrigation canals with a *shaduf*, a mechanism consisting of a bucket with a counterweight at the opposite end of a pole on a fulcrum, which enabled them to scoop up water and swing it round for emptying into channels between the planted rows. In addition to wheat and barley the farmers grew peas, beans, lentils, onions, and dates, and raised sheep, goats, pigs, cattle, and donkeys. They harvested reeds from the river shallows and marshes to build huts and make boats and baskets, and they fished the abundantly stocked rivers and canals. In all, Mesopotamia had become a land of plenty, so it is no surprise to find Hammurabi, King of Babylon (reigned 1792–1750 BCE), naming one of the canals 'Hammurabi spells Abundance'. The canals merited the praise thus bestowed: 75 feet wide and many miles long, they were a testament to engineering skills that by then were also raising great palaces and temples in the cities that were mushrooming from this bounteous plenty.

The Nile, the Indus, and the Yellow River in China also fed the rise of farming and urbanization along their banks. Whereas the Nile was predictable and easy to control, the Yellow River was aptly described as 'China's Sorrow', because of the mighty floods that so often burst its high silt-raised levees and drowned hundreds of miles of land on either side, people and livestock with them, and buried the fields in deep layers of mud. But the intense fertility of that mud – the soft loess – along the Yellow River's course kept bringing the people back, as attested by the archaeological remains of thousands of settlements dating to the legendary Xia Dynasty in the third millennium BCE. The supposed founder of this dynasty, King Yu, is credited with having built the Yellow River's first flood defences, but it took further millennia before the sorrows caused by the river became less frequent.

The Harappan civilization of the Indus Valley rose in the fourth millennium BCE and flourished in the third millennium, before beginning to collapse in the first half of the second millennium for reasons still unclear. Theories about its collapse include invasion by Aryan peoples, a great flood, the drying of the Saraswati River, earthquakes, climate change, drought. Most likely a cumulative combination of several of these factors is to blame. But the disappearance of Harappan writing (still undeciphered) and the sophisticated system of the culture's standardized weights and measures – necessary both for the reliable functioning of markets and trade, and for taxation – was complete by 1300 BCE, a harbinger of the Dark Age that was also to befall the Near East and eastern Mediterranean lands from 1200 BCE onwards.[4] Yet, on the agricultural bounty made possible by the waters of the Indus River, Harappan civilization attained remarkable heights: many houses had indoor bathrooms and their own wells, the cities were provided with underground drainage systems and paved streets, trade between the cities was lively and spread from today's north-western Afghanistan to north India. Harappa and Mohenjo-daro are the best-known archaeological sites, but thousands of settlements of various sizes have been identified, signalling the vigour of the civilization in its heyday. Harappan technology of building, drainage, communications, and much besides, was markedly advanced; so was the culture's elegant art.

But the earliest civilization, the first civilization, to rise from river-watered soil was Sumer in Mesopotamia. By the end of the fourth millennium BCE Sumerians had established a constellation of city-states along the Tigris and Euphrates, including Ur, Kish, Eridu, Lagash, and Nippur, with Uruk as the greatest of them. It was almost certainly the largest in the world, with as many as 80,000 people living within the six-mile circumference of its walls. To the physical technologies described above the Sumerians added an even greater: writing – an art necessitated by the increased complexity of life, requiring records of goods traded and debts owed and a means of communication between people separated by days or weeks of journeying.

The preservation of cuneiform texts owes itself to a circumstance that in other respects put Mesopotamia at a disadvantage relative to

Egypt. The Nile is bordered by limestone hills, which provided the durable stone for its pyramids, palaces, temples, and monumental statuary. The Mesopotamians had only mud and clay. From this material they constructed their buildings, and on this material they wrote their texts. Over time mud buildings decayed, and the Sumerians and their successors simply erected new mud buildings on top of them. The result is the *tells* – the mounds or artificial hills – dotted around Iraq and its neighbourhood today, indicating the existence of layers of archaeology dating back thousands of years, each layer a slice of the history of a city or settlement. But the artefacts – and not least the clay tablets on which texts were inscribed – were, if not needed, abandoned to the mud layer on which they had been tossed, as the next layer was built on top of it; and thus were they preserved.

Cuneiform – from Latin *cuneus*, 'wedge'; the stem *cun-* appears in various words relating to wedge-shaped items – is a writing system consisting of marks impressed into wet clay by a reed stylus. The Sumerians invented it between 3500 and 3000 BCE, in the first instance as a means of recording inventories or trade orders. Some take the view that cuneiform preceded Egyptian hieroglyphics and was an influence on their development. But, whereas hieroglyphics preserved pictographic elements, cuneiform soon transformed pictograms into abstract representations, combining syllabic, abjadic, and logophonetic elements.[5] Of the estimated 1 to 2 million cuneiform tablets that have been unearthed so far in excavations, about one hundred thousand have been read by scholars. The biggest collection is in the British Museum; other substantial collections, but each less than half as many, exist in Berlin, Paris, and Baghdad.

The Sumerians' cuneiform writing system was adopted by the successor Akkadian Empire and the various versions of Akkadian spoken in it, including Old Babylonian and Assyrian. Later the Hittites, whose empire occupied the Anatolian Plateau to the north of Mesopotamia, adopted it also, probably as a result of the influence of traders from Assyria. In its later history cuneiform served not only for business but also for personal and diplomatic correspondence, medical treatises, mathematical works, astronomical records, and literature, *The Epic of Gilgamesh* being the world's first major example of the last.

It scarcely needs saying that, of all the technologies developed in the period from the fourth to the second millennium BCE, writing is the most significant. But it is not the only significant one: the pen might be mightier than the sword, but the sword – or, rather, the metals from which swords, daggers, shields, breastplates, greaves, spearheads, and arrowheads were made (and also of course ploughs, axes, awls, knives, and spades) – had a great part to play too. It might be said that the development of metallurgy is second in significance only to writing, both as a product of human intelligence and in its civilizational consequences.

Native copper and gold – by 'native' meaning uncombined and occurring naturally – were the first metals worked into axes, knives, arrowheads, and ornaments. Native metals are uncommon; most metals exist in combination with other substances, either in ores or mixed with various minerals. The desirability of copper and gold, which are both readily workable and attractive, made them objects of particular attention, and people were constantly in hopes of coming across them, and were careful to learn where they were likely to be found. But then it was noticed that copper emerges from certain stones, such as azurite and malachite, when these are heated. A metal industry began, and by the fifth millennium BCE copper-bearing ores were being traded from locations where they were more abundant – the Arabian Peninsula, Iran, and Anatolia, and the island of Cyprus, whose name *kupros* is either the source or a derivation of the metal's name – to places including Egypt, the Indus Valley, and China.

The Copper Age, also called the Chalcolithic Age, 6.5 to 3.5 kya, is marked by the development of techniques for mining, crushing, and smelting copper-containing ore. For most of this period furnaces could not be raised to temperatures above copper's melting point, which is 1,200 degrees Celsius, and therefore the work of shaping it demanded great physical effort, with constant reheating and hammering. When blast furnaces were developed, so called because their fires were fanned by forced blasts of air from bellows, molten copper was produced for pouring into moulds. This technology arrived in the late second millennium, already well into the Bronze Age.

A copper industry required specialists: mining engineers and miners, smelters and smiths, designers and craftsmen. At first the products of the industry were reserved to the use of elites, because of their expense. Everything from axes to swords to drainpipes to kitchenware came from the coppersmith's forge. For a considerable period during the Copper Age less well-off people continued to use wood and stone implements.

Because they were valuable, copper items could be returned to the furnace and then the smith's anvil for reshaping, but a greater degree of durability was desired. It was found that mixing a little soft metal with copper would achieve durability, the soft metals being either arsenic or tin. Arsenic was used to begin with, but the dangers of working with it made tin the preferred alternative. A mixture of 90 per cent copper and 10 per cent tin is *bronze*. Bronze is hard, durable, workable, and attractive; tools and weapons made from it were far more effective than those made from stone, wood, or pure copper, and art and artefacts made from it admitted of great intricacy in patterning and design. Bronze therefore became the material of choice for almost all these purposes.

Tin was not abundant in the Near East, and had to be imported from distant places – the north-east of what is now Afghanistan was one source, Cornwall in England another. The design of bronzeware was influenced by the long-standing sophistication of decorated pottery, and everywhere that bronze manufacture flourished its artistic as well as utilitarian value grew. In China bronze objects of extraordinary intricacy in design, and sometimes of enormous size, constitute some of the high points of the Bronze Age (3500–1200 BCE).

The Bronze Age Collapse – the rapid simultaneous fall of the Near Eastern and eastern Mediterranean civilizations around 1200 BCE (Egypt survived, but in very diminished fashion) – was followed by a Dark Age of several centuries. After 900–800 BCE a revival of literacy, organization, and culture began; this is the period of archaic Greece, immediately preceding the classical period. Rome did not yet exist, and the Old Testament stories of Saul, David, and Solomon are placed in the middle of the Dark Age (1000–900 BCE), with the Babylonian captivity and Second Temple era occupying the sixth century BCE. Accompanying the emergence from this period was the replacement of bronze by iron – the substance giving its resonant and intimidating name to the Iron Age.

In fact iron had been smelted and fashioned as early as 1500 BCE by the Hittites, but the process they used was rather ineffectual because of the insufficient temperatures of their furnaces. What came out of their furnaces was *bloom,* a mess of iron and slag that had to be heated and hammered again and again to expel the slag. Tools and weapons made out of the resulting wrought iron were not as hard as bronze, were brittle, and rusted easily. Sharpened edges quickly became blunt. The only reason its use continued was that iron was plentiful and easily available; it did not need deep mines because it lay close to the surface, and nor did it require the expensive importation of tin.

If iron could be made more serviceable, it would have much to recommend it. And indeed, doubtless because tin for bronze-making was hard to get after the disruption of trading networks caused by the Bronze Age Collapse, the technology of ironwork advanced dramatically. Iron turned out to be better than bronze if processed effectively: it made sharper and stronger instruments far more cheaply than bronze. Furnaces improved, and the high temperatures required were reached. It was found that iron would be less brittle if cooled, reheated, and cooled again, repeatedly, by immersion in water. Heating iron in contact with charcoal, for the carbon effect, turned it into steel. A steel sword took a fine edge, and could slice easily through bronze sword blades and breastplates. Armies with iron weapons were therefore more formidable because of them. Iron axes felled trees more efficiently; iron ploughshares turned over the soil more effectively. Because iron products were cheaper than bronze, they were available to everyone: farmers, soldiers, carpenters, builders, cooks. When the Iron Age shook off the shadows that had fallen in the Bronze Age Collapse, a broader and more robust set of technological possibilities existed.

★

Two other technologies, perhaps more important than all the others so far mentioned, and especially in their combination, are the wheel, and the domestication of the horse.

It is a piety that the wheel is one of the greatest inventions of human ingenuity. A salutary corrective is provided by Richard Bulliet.[6] 'In 1850, the steam engine was ranked as the world's greatest invention. By 1950, the wheel, a much older invention, had surpassed it. The advent of the electric motor and internal-combustion engine partly explains the decline of the steam engine; but the spread of automobiles, trucks, and buses – not to mention grocery carts, bicycles, and roll-aboard luggage – played a greater role. For in 1850 the wheeled vehicles that rumbled over the cobblestones of city streets and jounced along the rutted dirt roads of the countryside seemed neither new nor particularly ingenious.'[7]

Although this reminds us how invidious fashions can be – no doubt the marvels that clockwork and steam once seemed will take computers along with them into the dusty cupboard of history when technology has moved on again – it remains that the wheel had a great impact on history from Neolithic times onwards. History might, however, have moved on even if the wheel had never been invented; in the Americas the wheel did not exist – except, amazingly, in children's toys – until the conquistadores brought it. Loads were carried by people or llamas, or dragged on a *travois*, a pair of sticks angled away from each other to form a sled. In Africa, and in Egypt until the middle of the second millennium BCE, the wheel was known but not used. This might seem remarkable, given that the Egyptians were in contact with the Mesopotamian civilizations where wheels had been in widespread use since before 3000 BCE.

Theories to the effect that wheels were not invented or used in Central and South America because there were no large animals to pull vehicles mounted on them overlook the fact that humans can push or pull carts, wheels relieving their handlers of the weight and friction of the load being moved. So Central and South Americans could have made use of wheels if they had so chosen. But they did not so choose: loads were carried by people or pack animals. The terrains of South America, high mountains and dense forests, did not lend themselves to wheeled transport. In Egypt wheels were also not deemed necessary, because most loads of significant weight could be transported by river, and could be got to and from the river by the kind of concerted manpower that built the pyramids.

These points undermine the generally accepted view that the wheel is one of humanity's most significant technological advances.[8] Bulliet also shows that the assumption that the wheel was invented in Mesopotamia – prompted by the fact that so much else was invented there – is incorrect. The wheel appeared in Sumer around 3000 BCE, but had already been in use in the copper mines of the Carpathian Mountains for hundreds of years, back to the first half of the fourth millennium BCE.[9] On the steppe north of the Black Sea the wheel and wagon came into existence together as mobile homes for the nomads of the region, their importance signified by wagons being interred with their owners in graves dating from the millennium after 3000 BCE. Both climatic and linguistic evidence, the latter drawn from the Proto-Indo-European roots of words for *wheel*, *axle*, *cart*, and related items, suggests the steppes north of the Caucasus as the place where the wheel first came into its own. The nomads' rolling towns existed into modern times, until in the nineteenth century CE the Noghays – last inheritors of the tradition, if one excludes Roma, who had a different origin – were forced by the Tsarist government to settle or flee into the Ottoman Empire.

Because of the later evolution of the cart into the chariot, and depictions of war chariots dashing into battle behind galloping horses, domestication of the horse is often associated with the evolution of wheel-associated technologies. In fact the first draught animals were oxen or onagers (a bad-tempered relative of the donkey), and the heroic image of the dashing chariot has to be supplemented by the earlier reality, portrayed on a stele (an upright slab or column of inscribed stone) from the third millennium, of a king being drawn to the battlefield in a slowly lumbering wagon behind donkeys. As Bulliet has shown, in Mesopotamia wagons drew royalty or images of gods in processions; for most of their history – well into the Renaissance – riding in wagons or carts was reserved to women and lower-class men, because it was regarded as *infra dignitatem* for a man of status to be seen in one. For a long time after horses began to be ridden, it was manly to ride horses, and women did so in very few cultures. It was an act of great significance for Sir Lancelot of the Round Table to agree to ride, demeaningly, as a passenger in a cart in exchange for information about his beloved Guinevere.[10]

Horses originated in the Americas, migrated west across the Bering land-bridge before melting glaciers raised sea levels, populated the vast expanse of the Eurasian steppes – and shortly thereafter went extinct in their continent of origin. They were reintroduced to the Americas by the conquistadores several millennia later. The original steppe horses were small and tough, and even after domestication were rarely used as draught animals, lacking the strength of oxen and even donkeys until bred to compete in that department – as was, millennia later, the magnificent Shire horse. The practice of keeping horses does not seem to have occurred before 6000 BCE at the earliest. The first evidence of horse-keeping is associated with the Botai people of the steppes north-west of the Caspian Sea – today's Kazakhstan – who in the first instance used them for milk and meat. From about 3500 BCE, as the evidence of bit-wear on the teeth of horse skulls suggests, horses began to be put under harness and ridden.

Images in artworks surviving from Mesopotamia and Egypt in the period from 2000 BCE onwards show horses pulling war chariots, and from about 1600 BCE being ridden. One theory is that most horse breeds were not strong enough to carry the weight of a man unless he sat well back over the horse's haunches, from which position it is not so easy to control the animal or to keep a secure seat. Once horses were strong enough to be ridden in the now familiar way, they became formidable partners to humanity, giving speed and mobility even greater than the chariot, which is why – eventually, a considerable time later – the chariot gave way to cavalry. A memorable example of the latter's superiority over the former is Alexander the Great's cavalry defeat of the Persian King Darius's chariot forces at the Battle of Gaugamela in 331 BCE.[11]

Horses were, however, being ridden long before the events just mentioned. Note that taming horses and domesticating them are two different things; some scholars insist, very plausibly, that 'domestication' requires deliberate physiological changes brought about by selective breeding to enhance some characteristic chosen by humans, for – as it might be – more milk, more wool, more meat, greater docility for draught work. Horses were certainly tamed before they were domesticated in this sense. It would not have been possible for the Botai to corral and tame (and eventually to domesticate) horses unless they could ride, so we must imagine that they caught and tamed horses, and mounted them to pursue, catch, corral and tame more horses.

As will be discussed in Part II below, the movement of Indo-European language speakers westward into Europe and south-eastward towards India from their place of origin on the steppes – these being people of the Yamnaya culture – replacing, according to controversy-promoting genetic studies, the hunter-gatherer populations then living there, was almost certainly facilitated by the mobility provided by wheels and domesticated animals, technologies that distinguished the migrants from those they displaced. Accordingly, the wheel and control of animals have to be regarded as significant technologies indeed, in the places where the difference they made was so large.

Engineers and builders of the Bronze Age achieved remarkable things, among the most notable being the pyramids of Dynastic Egypt. The temples and palaces of Mesopotamia (for example, Uruk and Babylon) and later the Levant (for example, Ugarit) are evidence of high skill. For grace, symmetry, and proportion, there is scarcely anything superior to the temples of the classical Greeks. But the great builders of antiquity were the Romans. They introduced the arch and its extension into barrel vaulting, further developed into intersecting vaults that could support a large dome. In Rome itself two outstanding examples of engineering are the Pantheon, built between 110 and 125 CE, whose dome, spanning a diameter of 43.3 metres, remains a miracle of ancient engineering; and the Colosseum, built between 70 and 80 CE on the site of a filled-in lake that had once been part of the pleasure grounds of the Emperor Nero's Domus Aurea. The Colosseum is 189 metres long by 156 metres across the widest part of the ellipse it forms, covering 24,000 square metres (six acres) and seating 50,000 spectators. The underground tunnels and spaces contained machinery for scene changes, including lifts for bringing wild-animal cages to the arena floor and (in the early part of the Colosseum's existence) a means of flooding the arena for sea-battle displays.

The structural principles evolved by Roman engineers served them excellently in one of their greatest achievements: the aqueducts that brought fresh water into their cities, and the sewage systems that took waste water away. The dizzying height of some of the majestic rows of arches taking water across valleys still provokes admiration; think of the double-arched aqueduct of Segovia. A significant factor in their success as builders was their cement, *pozzolana*, a mixture of volcanic earth and lime that was immune to damage by fire and water, explaining why some of their great structures stand to this day. The recipe for Roman cement, the knowledge of how to mount a large dome over a space – and much else – were among the technologies lost following the fall of the western Roman Empire in the fifth century CE. Recovering the knowledge of how to raise a dome had to wait for Florence's Brunelleschi a thousand years later, in the mid fifteenth century. The cement recipe was not rediscovered until the nineteenth century.

Roman roads were another triumph. They were meticulously engineered: flat stones were laid on a bed of sand, then two layers of gravel set in clay or concrete were laid over them, followed by a final cambered surface consisting of cobbles in concrete. These roads took the wear of heavy traffic from wagons and armies so well that some of the Romans' 50,000 miles of highway are still in use today. The roads were as straight as possible, crossed rivers on spanned bridges, had postal relay stations every 10 to 12 miles and inns every 30 to 40 miles in which the same official language and the same currency were in use; they united the empire from the borders of Scotland to the deserts of Egypt and Arabia.

Boats were obvious means of transport on the rivers of Egypt and Mesopotamia in the Neolithic Period and Bronze Age. In Egypt boats of reed could deploy their simple square sails to travel southwards up the Nile, and then could furl the sail and simply float back downriver with the current. On the Euphrates the first boats were made of animal skins. Sailing on the sea was a perilous, shoreline-hugging affair for a long time; the respect for the sea displayed by Odysseus in Homer's eponymous epic is doubtless an accurate reflection of persisting attitudes. But there was an established maritime trade in the Bronze Age, knitting the lands of the eastern Mediterranean closely together, the commodity at its heart being the all-important tin which has been likened to the ancient world's version of oil.

In the centuries after the Bronze Age Collapse the masters of the sea were the Phoenicians. Their first homeland was the eastern Mediterranean shore, until the politics of the region prompted them to move their base of operations to the safer and more convenient location of what until then had merely been their trading post on the north African coast. This was Carthage, which grew from its utilitarian beginnings to be the centre of an immensely wealthy commercial empire. The magnificent city was itself a wonder of its time. Carthaginians were not alone in their naval skills: the island-dwelling Greeks were also at home on the water. Athens' Themistocles built a strong navy in the face of Persia's threat in the fifth century BCE, and with it won a famous victory over Xerxes's ships at Salamis in September 480.

Later, in the conflict between the nascent power of Rome and the established power of Carthage, which eventually took the form of a series of wars (the Punic Wars, beginning in the early third century BCE and ending with the literal obliteration of Carthage in 146 BCE), Rome became a ship-building nation too. It is a remarkable thought that the sea battles of the Greeks, Persians, Romans, and Carthaginians, fought by triremes and similar oared ships, set a pattern still in place two thousand years later: the Battle of Lepanto in 1571 CE took place between two navies of rowing vessels. The author of *Don Quixote*, Miguel de Cervantes, lost the use of his left hand in that battle.

The Phoenicians may have been masters of the Mediterranean Sea in their day, but their day was late, and local. Long before them, beginning in the period between 3000 and 1500 BCE, Austronesian peoples were sailing vast distances across the Pacific, a region encompassing half the planet. According to maritime historians it was the

invention of the outrigger and the catamaran that made these feats
possible – but without observation of the stars, currents, prevailing
winds, and weather systems, not even the most stable of boats would
have permitted these extraordinary migrations. The Western world
saw nothing comparable, until Prince Henry, 'the Navigator', of Portugal, in the fifteenth century CE, set himself the task of finding sea
routes south into the Atlantic and along the west African coast. The
volta do mar ('return of the sea') technique of using the prevailing trade
winds blowing westwards at the equator and eastwards in mid-north
Atlantic eventually encouraged ideas about venturing further west;
the first to do so from Iberia was Christopher Columbus, though the
Vikings had got to North America five hundred years earlier.

Henry the Navigator is credited with launching 'the Age of Discovery', leading, via the explorations of Bartolomeu Dias and Vasco
da Gama round the Horn of Africa – in da Gama's case to India in the
1490s – to the eventual establishment of European trade with, and
then colonization of, the East Indies during the following centuries.
A principal spur in these voyages was the determination to find a way
of undercutting the Arab and Venetian grip on the valuable trade in
spices that had hitherto come overland from India or across the Arabian Peninsula – a trade that began with Austronesian inhabitants of
South-East Asia's islands selling cinnamon, cardamom, pepper, gin-
ger, turmeric, nutmeg, and cassia to mainland South-East Asia and
China as early as 1500 BCE, from where – later – they were traded
onwards to the west. The overland route added to the cost of spices,
making Venice exceedingly rich, because it was the point of distribution for the spice trade into Europe, and had been so since at least the
eighth century CE.

Two key technological developments that made long-distance
ocean voyaging possible for European explorers were the lateen sail
and the stern rudder. The lateen sail is triangular in shape; it replaced
the square sail, used on the ancient Fertile Crescent boats and by the
Vikings in medieval times. It enhanced manoeuvrability, for, whereas
a square sail only permits running before the wind, a lateen sail allows
for tacking into the wind. In heavy seas and military operations a
stern rudder is safer and more effective than the older system of a

steering paddle affixed to a ship's side. Together, the innovations made for more efficient and effective sailing.

Such instruments as the astrolabe, cross-staff, compass, quadrant, octant, and sextant came very late to marine navigation in comparison to the methods used by Austronesians long before. These instruments are associated with developments in methods of telling the time – among the earliest being shadow clocks (*gnomons*), their sundial versions, and the water clock (*clepsydra*). Time is not too difficult to determine in a fixed locality, but it becomes more complicated as one constantly changes position on the high seas. And that in turn makes it difficult to fix one's longitude, even if knowing one's latitude – by the height of the Sun and the position of stars – is relatively easy. To determine longitude one needs two clocks aboard: one set to a starting point and the other adjusted daily according to the position of the Sun. Then, together with knowledge of one's latitude, one can fix one's position. Every hour's difference between the two clocks equates to fifteen degrees of longitude; what distance this represents depends on latitude. At the equator it is a thousand miles. At the North Pole and South Pole it is zero miles. Amazingly, it was not until the eighteenth century CE that a self-taught clock-maker, John Harrison, invented a timepiece – the marine chronometer – that would accurately, anywhere in the world, tell the time at the home port of any voyage, thus enabling navigators to know where they were.[12] Until then clocks and watches were unreliable. Pendulum clocks did not function well in the pitching and rolling of ships. Salty sea air, the wet from rainstorms and big waves, and the temperature variations of different climes made timepieces speed up, go slow, or stop. Until Harrison the only supplement to guesswork in navigation was experience – and, even then, guesswork was the chief resource.

A painted stick discovered in China and dating to 2300 BCE is reckoned to be one of the earliest-known versions of a gnomon, a device for telling the time by the shadow it casts. The Greeks invented sundials that apportioned the day into twelve equal parts. Water clocks served at night and in the absence of sunlight, if the weather was not freezing. But the technology of clocks advanced quite rapidly from the medieval period onwards. In the early ninth century CE

Charlemagne was given a clock by the Caliph of Baghdad that, driven by water, activated a chime of bells and made figures of horsemen move. In the eleventh century CE the Chinese engineer Su Sung invented a water clock for his emperor that not only told the time but displayed the movements of the heavenly bodies. These clocks exploited the property of water that made it especially apt for this purpose: its even and easily controllable flow.

Mechanical clocks arrived on the scene in the thirteenth century CE with the invention of the escapement, a device that regularly interrupts a falling weight or (later) the release of tension in a spring, thus producing the characteristic 'tick-tock' of clockwork. Early mechanical clocks were crude and could show only the hour, but that was good enough for a church tower or the great hall of a baronial castle. Development of the mainspring in the sixteenth century led to small accurate portable timepieces – the watch. Clockwork became a metaphor for the machinery of the universe, an influential image in the Enlightenment of the seventeenth and eighteenth centuries.

The invention of all manner of devices of utility occurred alongside the continual improvement and development of existing technologies. Consider, for example, the effect on transport – and thus the economy – of horse collars. The Romans had imposed a legal limit on the weight of wagons to protect horses from choking as they dragged loads. Invention of the padded horse collar made it possible for teams of horses to pull much heavier loads without harm, thus reducing the cost of transport, by the twelfth century CE, to a third of what it had been in the fourth century. The effect on trade and movement of people in Europe was significant.

China, famously, is the home of a number of striking technological innovations. Kites, gunpowder, fireworks, rockets, woodblock printing (dating from the Tang Dynasty 618–907 CE), and an unparalleled mastery of bronze from as early as the middle of the second millennium BCE are among those best known. Ambitious claims are made for many other inventions, some dating to the Neolithic Period; the most secure relate to sericulture (silk production) and canal locks, though ways of raising and lowering water levels existed in pre-Sumerian Mesopotamia. Claims that bells, coffins, horse harnesses, bricks, pottery glazes, seed-drills, arched bridges, stern-mounted rudders, cast iron, steel, wheelbarrows, puppets, smallpox vaccine, even toy helicopters, and much besides, were invented in China over a time span ranging from before the half-legendary Shang Dynasty (second millennium BCE) to the Song Dynasty (960–1279 CE) and later might be true, but these or functionally similar things were invented in other places and times independently. The constrained degree of contact between China and the Near East and Europe until late in history makes it difficult, in regard to many of these innovations, to say what influence occurred, or whether any occurred, and, if so, who influenced whom, in the way of technology transfer. The Silk Roads first began to be mutually traversed quite late in the story so far told, from the second century BCE onwards, but it is hard to imagine that (to take just two of many possible examples) horse domestication and the chariot were not the result of technology transfer across the steppes between east and west, for chariots were in use in China by 1200 BCE and horse bones become frequent in Chinese archaeological sites dated to 2000–1600 BCE – and both horse domestication and the wheel were invented before then on the steppes north of the Caucasus.

Western knowledge of China increased significantly from the seventeenth century CE onwards, by which time the ingenuity displayed in China's earlier history had long been stifled by tradition and ossification. But the roughly contemporaneous appearance of similar technologies in parts of the world that appeared to be not, or not easily, connected to one another is an interesting phenomenon that suggests that human mobility and exchange were more frequent, and more extensive, than we would guess just from looking at the great distances and constrained means of travel in Eurasia in the period from 8 kya onwards. Individuals can carry ideas that, once sown in fertile soil, take on a life of their own; the obstacles to this happening – the inflexibilities of tradition, xenophobic suspicion of foreign ways, and so forth – doubtlessly interfered more often than not; but it must have happened.

A Chinese invention that made its way to the Near East and Europe and profoundly influenced the course of history everywhere thereafter was gunpowder. It is said to have been first discovered by Buddhist monks looking for a potion that would confer immortality, and who mixed charcoal with sulphur and potassium nitrate (salt-petre), creating a fizzing substance that gave off quantities of smoke. This happened in the ninth century CE; intrigued Chinese alchemists continued to experiment with the mixture, not just for entertainment purposes – although that was one motive: fireworks were the outcome – but to make flamethrowers, rocket-propelled arrows, and bombs. By the early thirteenth century the recipe for, and properties of, gunpowder were known in Arabia and Europe. The first cannons were produced in the fourteenth century, the siege guns of the Otto-man Turks were key to the capture of Constantinople in 1453, and by the Thirty Years War in the seventeenth century artillery was an advanced and sophisticated science both in engineering and military terms.[13]

Cannon put an end to the castle and the city wall as defensive technologies. Firearms developed as quickly, and as quickly put an end to the armoured knight. In fact the crossbow had already put this expensive medieval version of a tank at risk, because crossbow darts could penetrate armour; but firearms, even the early unreliable and limited versions, did the same much better.[14] The first firearms were Chinese bamboo tubes stuffed with gunpowder whose ignition expelled a pellet from the tube. From the twelfth century CE metal barrels were developed, and the Chinese called them 'hand cannons'. Arabs and Mameluks had firearms by the thirteenth century, and Europeans by the fourteenth. The earliest versions were muskets, muzzle-loaded smooth-bore long guns, and this basic model survived into the nineteenth century, though better engineering of the parts, rifling of the barrel, and prepared cartridges jointly made the weapons safer to use (for its handler: earlier muskets sometimes blew up in the faces of their firers) and more effective – a trained musketeer at Waterloo could fire six rounds a minute. The very rapid advance in firearms technology in the nineteenth and twentieth centuries, leading via six-shooters, repeater rifles, and machine guns to automatic handguns, powerful assault weapons, and the like – annually a multibillion-dollar industry – has not had a positive effect on peace and stability in the world. A striking claim was made in the closing years of the twentieth century that in the Horn of Africa a Kalashnikov could be acquired in exchange for several small children.

Much more agreeable developments in technology over the same period relate to printing, agriculture, manufacturing, and transport. In agriculture, crop rotation and soil fertility increased yields, thus supporting and prompting population increases. Steam power was harnessed – pumping water from mines and enhancing textile manufacture were among its earliest applications – and it revolutionized transport, the railway being the single biggest development between the invention of the wheel and the near-contemporaneous advent of automobiles and heavier-than-air flight. Both these latter relied on the internal combustion engine, several prototypic versions of which had been invented late in the eighteenth century, though the first patents for the type on which car and aircraft engines are based were not issued until the 1870s.

The internal combustion engine is a striking example of a simple way of controlling natural forces and directing them to effect work, principally in providing mobility, but doing so on scales far more useful than the similar process using steam. When an explosion is created within a cylinder, a piston is made to move. The piston is connected to a system of levers and cogs that transfers motive power to wheels, and at the same time works to reopen an aperture in the cylinder that admits fuel, and a moment later ignites a spark, causing the next explosion – and so on in a cycle, until the fuel supply is stopped. It is breathtakingly simple, and as breathtakingly effective. Metals suitable to the task existed long before the technique itself.

Most of these developments overlap the early history of science, and have increasing connections with it, thus being part of a familiar recent heritage. Accordingly they need no elaboration here. But all of them owe a debt, directly and indirectly, to the new technology mentioned first in the paragraph before last: printing.

The Egyptians had made papyrus from sedge, a flowering grass-like plant that grew in the wetlands of the Nile. Mesopotamians used clay tablets. The Chinese invented paper, pulping mulberry-tree bark, bamboo and rattan into a membrane from which umbrellas, fans, and toilet tissue were made, and when 'sized' with a starch coating was good for writing on. Captured Chinese paper-makers divulged the secret of paper-making to Arabs in the mid eighth century CE, and by the ninth and tenth centuries this had fuelled the growth of a flourishing book trade in the Islamic world, centred on Baghdad, which had over a hundred bookshops, a high point in what was then an open, tolerant, and intellectual period in Islamic culture.

In Europe books were hand-copied on parchment, which is made from sheepskin. Both parchment and copying were expensive. Paper manufacture began in the early twelfth century, in Spain at first because of the Muslim influence, and over the next two centuries spread with increasing rapidity to the rest of Europe, propelled by the increase both of literacy and curiosity characteristic of the Renaissance. Paper was cheaper than parchment, but books were still hand-copied, imposing a limit by quantity because of their production cost. This was changed by the inventions of Johannes Gutenberg.

European textile producers had long used block printing of patterns on cloth, and the same technique was applied to the production of devotional images and playing cards, so the idea of printing was not itself new. But in the years between 1436 and 1453 Gutenberg developed the idea of *movable type*, an innovation perfectly suited to languages written in an alphabet with relatively few symbols that could be combined to represent any word. Whereas Chinese printers and textile-pattern designers alike had to carve individual characters on wood blocks to print from them, movable type meant that the same letters could be used and reused repeatedly, because they can be set in different combinations, and repeatedly inked and pressed to paper leaving iterated copies of legible script upon it.

Gutenberg's innovations touched every part of the process. The type itself he made from an alloy of lead with tin and antimony, which was so durable that even after repeated use it still left a clear impression. The type was cast in a template matrix, resulting in uniformity and ease of production. He invented an oil-based ink that was an advance on the water-based inks used in the manuscript reproduction of texts. Although he experimented briefly with colour printing, it was left to others to carry that development forward.

Gutenberg's press was in Mainz; before the end of the fifteenth century, in less than fifty years, nearly three hundred European cities had presses, and 20 million copies of books had been printed. It is estimated that by the end of the following century that number had increased tenfold, to 200 million copies.

Two examples of the difference made to the world by printing will suffice. Martin Luther, credited with igniting the Reformation, was not the first to voice criticisms of the Church of Rome; indeed they were the very same criticisms as Jan Hus and others had levelled before him. But he lived in the new age of printing, and three hundred thousand copies of his tracts of protest circulated in Europe, most of them being reproduced further by the practice of reading aloud in groups and congregations, and two hundred thousand copies of his German New Testament were in circulation before he died, likewise further disseminated. Both the possibilities and dangers of the new medium were quickly appreciated, with the Church's *Index librorum prohibitorum* coming into existence in 1559, but proving largely ineffectual against the printing-empowered spread of ideas, which fuelled the philosophical and scientific revolutions of the sixteenth and seventeenth centuries and the subsequent Enlightenment of the eighteenth century.[15]

The second example is that the printing and dissemination of such philosophical and scientific classics of antiquity as had survived a thousand years of diminished literacy, neglect, and religious censorship were a spur to the revival of interest in both. This is arguably the more important of these examples, as the survey of the history of science later in these pages shows.

Printing, steam power, and today's digital technologies (especially in the platforms they provide for communication as in social media) are all examples of innovations that were extremely rapid in their widespread adoption and effect on society and history. The production of millions of printed books within fifty years of Gutenberg's inventions is a parallel phenomenon to the hand-held mobile smartphone and what it can do, both in positive and negative ways. Printing destroyed the hegemony of the Church over thought in the sixteenth century and launched a new view of the universe; social media today has created a universal agora whose effects have yet to be fully felt. Luther's tracts and video images shared worldwide of such occurrences as police killings of African-Americans in the United States might turn out to have similarly revolutionary effects.

There are two other technological advances to mention before leaving this sketch of the products of human ingenuity. Each of them played a significant part in the rise of science. One is the telescope; the other is the microscope.

History credits a spectacle-maker called Hans Lipperhey (or Lippershey) of Zeeland in the Netherlands with the invention of the telescope, because he was first to patent a version of one, which he called a *kijker* ('looker'), in 1608. It consisted of a concave eyepiece aligned with a convex lens, giving a magnification of × 3, that is, making an object appear three times nearer. His claim to have originated the telescope was immediately challenged by other eyeglass-makers, and it was said that if he had not taken the idea from them it was because he had got it from seeing children using two aligned lenses to look at a church spire in the distance.

News of the device reached Galileo through his Paris-based friend Jacques Bovedere in 1609. Galileo promptly made his own, eventually achieving magnifications of over twenty times. He presented his device to the Senate of Venice, whose members immediately recognized its maritime and military potential, and who consequently gave him a large reward. Through his telescope in January 1610 Galileo saw the moons of Jupiter. His descriptions of the surface of Earth's Moon, the phases of Venus, sunspots, and the fact that what look like clouds in the Milky Way are actually clusters of stars, revolutionized astronomy and (eventually) humanity's sense of its place in the universe.

Inevitably, the microscope was a companion invention to the telescope, and the same cast of characters – Lipperhey and Galileo included – figure among its early developers. The magnifying properties of lenses were known in antiquity – Aristotle wrote about them – and spectacles had been invented and were in use by the thirteenth century CE. Although the step from a single lens to paired lenses, as in the telescope, must have been taken more than once before, it is a Dutch eyeglass-maker called Zacharias Janssen, experimenting around the year 1590, who is credited with seeing how to make small objects more visible. Before the end of the following century the work of Robert Hooke (whose drawing of a flea in his *Micrographia* (1665) caused a sensation) and Antonie van Leeuwenhoek – the true father of microscopy, with the superb single lenses he ground to yield a resolving power of one millionth of an inch – had matched revelations about the heavens with revelations about the worlds within and beneath the reach of unaided sight.

The foregoing samples some of the more important technological advances of humankind. They are remarkable both in themselves and as a testament to the ingenious mind of *Homo sapiens*. We know that speculation about the principles that underlie natural phenomena, and those that, if understood, explain the effectiveness of the technologies humans developed, abounded in at very least the later stages of *Homo sapiens'* evolution, because the evidence of art and burial practices indicates that our ancestors already surmised that mechanisms of some kind lay behind and beyond the visible surface of things, of which these surfaces are effects or outcomes. What they appear to have thought these mechanisms were is discussed in the next section.

2.   The Rise of Science

The origin of the word 'science' is instructive. By the mid fourteenth century it was used in English to denote 'what is known, what is learned by study, information', borrowed directly from French *science*, which meant the same. In its turn the French word descended from Latin *scientia*, meaning 'knowledge, expertise', formed from *sciens*, 'informed, intelligent', the present participle of the verb *scire*, 'to know' (*scio*, 'I know'). Etymologists surmise that the term is related to *scindere*, 'to cut, divide' – suggesting the connotation 'to distinguish, discern, tell apart' – and that the deep history of this term itself lies in Proto-Indo-European, the common ancient source of the languages of Europe, Iran, and India, in the root *skei-*, 'to cut, to split', appearing in Greek *skhizein*, 'to split', Old English *sceadan*, 'to divide, separate', and in derived words like 'schism', 'schizoid', and 'scatter'.

But, as the word that particularly denotes the sciences as we understand them today, 'science' is in effect a nineteenth-century coining, as is the word 'scientist', invented by the historian of science William Whewell (1794–1866) in response to a challenge by the poet Samuel Taylor Coleridge to find an alternative to the term 'natural philosopher', until then used as the designation for an enquirer into nature. Whewell's coining was based on an analogy with the word 'artist', and for a long time was regarded as a barbarism, gaining acceptance only late in the century. Whewell had a gift for neologism: he also coined the term 'physicist', and in relation to Michael Faraday's discoveries about electricity he invented the words 'anode', cathode', and 'ion'.

Accordingly 'science' no longer denotes knowledge in general but knowledge specifically of the physical world and its basis in material reality, together with the methods of enquiry that yield that knowledge. It is a general label, comprehending a number of enquiries that share methods and assumptions. Its main branches are physics, chemistry, biology, astronomy, geology, and their numerous subdivisions and interconnections, such as astrophysics and biochemistry. The shared methods centrally include empirical techniques of observation and experiment, together with mathematical and statistical techniques of description and measurement. A standard scientific procedure involves making a prediction based on an hypothesis, and then testing it by experiment. These features are distinctive of science; they are what make an enquiry scientific in the modern sense of the word.

Science is to be distinguished from technology, which is the practical application of some of the findings of general experience – and later of science – through the medium of devices, such as machines, constructed to serve specific ends, such as raising water from one level to another, milling corn, harnessing the Sun's energy (for example, by releasing it from its underground stored form in fossils such as coal, oil and gas), moving heavy objects, or enabling communication between people who are located at a distance from one another (*tele* in 'telephone', 'telegram', 'television' means 'far'). Before the rise of modern science in the sixteenth and seventeenth centuries, what we now think of separately as science and technology were often pursued both indistinguishably from each other and in connection with one another, the technological aim generally being the principal one. It was experimental methodology and the application of mathematics that gave modern science its great impetus in the sixteenth and seventeenth centuries and subsequently.

What were the very earliest intimations of scientific curiosity in humankind? The clue lies in the remark made at the end of the last section, about our ancestors coming to understand that there are mechanisms of some kind behind the visible surface of things, of which the phenomena of nature are effects or outcomes. Views of this kind are now identified as 'religion' and 'spirituality', but this is an example of a misleading 'reading-in', for it is more informative to understand them as early attempts at frameworks of explanation: as *proto-science* in fact.[1] Correlatively therefore we must see such observances as prayer, ritual, and sacrifice as efforts to engage with these forces and to influence them – which is precisely what any piece of technology does, whether it is a *shaduf* for irrigating crops, a steam engine, or an airplane.

Efforts to disentangle enquiry into the principles of nature from accretions of mythology and superstition must often have been made by reflective individuals, but the first such efforts we know of were made by the Greeks of the sixth and fifth centuries BCE. These efforts were the harbingers of science. Although some of the mathematical and scientific work of the period from then until the second century CE was significant – say, from Pythagoras to Ptolemy's astronomy – it was not until the beginning of the modern era, in the sixteenth century CE, that science, as we understand this word today, can be said properly to have begun.[2]

A reconstruction of the early history of ideas suggests how the nascent scientific impulse detoured into religion, as religion is understood today. No account of knowledge and the pursuit of knowledge can ignore the human propensity for belief and superstition, readily understood in psychological terms as the desire for explanatory closure in the face of uncertainty, and the fact that true enquirers constitute a minority of any population.

First one has to define 'religion'. This copious word has vague boundaries of application, demonstrated by the first ambiguity-embodying definitions one encounters when seeking them by, for example, today's commonest method: online. Here are two examples: 'the belief in and worship of a superhuman controlling power, especially a personal God or gods; a particular system of faith and worship'; 'a set of beliefs concerning the cause, nature and purpose of the universe, especially when considered as the creation of a superhuman agency or agencies'.

The separate parts of these definitions are (a) the central place of deity or superhuman agency of some kind; and (b) a system of thought about the cause and nature of the universe. Label these respectively as (a) the 'god part' and (b) the 'explanation part'. The 'purpose' aspect alluded to in the second definition might, in connection with the (a) god part, relate to a deity's supposed intentions and aims for humanity, or in the (b) explanation part can be regarded as relating to the question of the meaning and value of human existence quite independently of, and therefore without belief in, the existence of deity.

Running (a) and (b) together blurs the distinction between a *religion*, focally considered in (a) terms as focusing on the concept of deity, and a *philosophy*, understood in the most general (b) sense of 'a system of thought about the universe'. It is important to notice this, because certain systems of thought, such as Buddhism, Jainism, and Confucianism, are not predicated on belief in the existence and activities of a god or gods, and are therefore more accurately understood in (b) terms as philosophies rather than religions. The world-view implied by natural science, taken as a whole, is a (b) philosophy in this sense too. As constituting world-views, religions might accordingly be regarded as a subset of philosophies, but, given that what differentiates them is belief in supernatural agency, while philosophies are based on broadly naturalistic premises, it better clarifies matters to regard religion and philosophy (and therefore science) as different species.

The ground for thus distinguishing them is captured in the definitions given by somewhat more authoritative sources: 'Religion: Belief in or acknowledgement of some superhuman power or powers (esp. a god or gods)' (*OED*); 'Religion: The outward act or form by which men indicate their recognition of the existence of a god or of gods having power over their destiny . . . the feeling or expression of human love, fear, or awe of some superhuman and overruling power' (*Merriam Webster*). These identifications of what is *focal* to the concept of religion anchors it to the (a) god aspect, belief in the existence of a deity, and requires us to view non-theistic ways of thinking about the world as *essentially* different from them.

Some will point to the temples, rituals, and other observances found in Buddhism, Jainism, and Confucianism and claim that these distinguish them as 'religions'. Since almost all regular purposive activities of human beings, such as a daily exercise routine, a dietary regimen, a pattern of work, provide value-endowing structure to life, on this account everything is religion; and if everything is, nothing is, for the word is thereby stretched into uselessness. As it happens, metaphoric uses of the term, as in 'football is his religion', borrow from that aspect of religious observance that it shares with any form of systematic behaviour regarded as value-producing, but with special relevance to the consuming character of especially fervent religious belief. Yet no one takes the use of the word 'religion' in this context literally.

In what follows the centrality to religion of belief in supernatural agency – a god or gods – is key to reconstructing religion's origin and evolution, and to the way in which the ancestral human observation that there are mechanisms working within natural phenomena gave rise to such belief as – so to speak – the first approximation to science.

The evidence from mythology is that our ancestors, those in the Upper Palaeolithic or Mesolithic periods of the Stone Age being those most commonly nominated, sought to make sense of the world around them by projecting their experience of *agency* on to it. That is, they understood their own responsibility for the splash that occurred when they threw a stone into water, or a branch breaking away from a tree when they pulled it with sufficient force. The word 'agent' comes from Latin *ago, agere, egi, actus*, 'to drive, do', which we hereby see gives us the word 'actor' also.[3] In contemplating the wind, thunder, lightning, and other phenomena, the most direct step for our forebears was to go from recognizing their own agency in causing things to happen or change, to the idea of there being an agent likewise causally at work behind the wind or thunder, obviously bigger and more powerful than themselves, and often invisible – though the Sun, Moon, stars, mists, whirlwinds, bushfires, and the like might readily have been thought of as visible such agents.

We see this idea represented directly in the mythologies of our more recent ancestors. Consider Greek mythology, for example – the dryads in the trees and nymphs in the streams, Zeus hurling thunderbolts, Poseidon causing earthquakes, the smoke from volcanoes evidence of Hephaestus's forge at work, the rainbow as a path from sky to Earth for Iris to descend. The imputation of agency to what happens in the world is a form of explanation – an explanatory framework. It organizes and makes sense of things. Moreover it extends into narratives about the origin and course of the universe, as well as

explaining highly individual occurrences construed as the intervention of a deity – Athene deflecting the spear Hector had hurled at
Achilles on the plain of Troy, for example. From the most general
to the most particular it therefore offers a complete explanatory
framework.

The mythology of the Greeks was probably not believed *literally* by
everyone; many might instead have regarded the stories instrumen-
tally, as a figurative way of representing matters otherwise not
understood. What it does is to embody the kind of explanation that
came most readily to hand, an inference from observation of causal
connections involving ourselves to events not involving ourselves.

It is easy to see how speculation about underlying principles
evolved into myths and religions, and how efforts to influence nature
turned into superstition. In brief it is that the forces of nature were
personified, and stories grew around them – a process known as
mythopoeia; a reading of the Greek myths with a view to understanding their sources demonstrates the process.

As an explanatory framework, the idea that what happens in the
universe is agent-caused is in itself a form of proto-science. And this
equally readily suggests the proto-technology that goes with it:
namely, engagement with the agencies in question, to secure their aid
or intervention, and to prevent them from undertaking activities
harmful to human interests. The idea that there can be communication with the agencies could readily have been inferred from the
experience of dreams, hallucinations when fevered, exhausted, or in
a trance induced by repetitive movement, the effects of mind-altering
substances in (for example) mushrooms eaten or stored foods that had
fermented, and so on. Ritual, sacrifice, prayer and incantation,
taboos, and treating certain numinous places – a forest glade, a hill-
top where the wind sounds like voices – as sites where contact could
most easily occur are psychologically natural adjuncts of this.

But continued observation and inference, and experiment with
different ways of influencing what happens in the world – this time
by acting directly on it: diverting a watercourse to irrigate a patch of
land, planting seeds for crops, living in proximity to the young of
animals to domesticate them – would gradually have begun to effect

a separation between ways of thinking about nature and how to interact with it, and the view that there are directly involved agents at work in the processes in question. One might continue to pray to the gods for a good crop, but one would do the actual work of planting, watering, hoeing, and harvesting it anyway.

It is plausible to see the conjunction of two processes as eventually fileting away the characteristically *religious* aspect of the beliefs that had constituted these first efforts at explanation and interaction. One is that efforts to influence the agencies behind phenomena took the form of practices that we now recognize as taboos, and beliefs that we now recognize as motivating moral restrictions. To have the agencies on one's side, there are things one must do and not do, ways one must be and not be, to secure their goodwill. To put it graphically, taboo and morality are the joint analogue of not stepping on the cracks in the pavement in order to avoid something bad happening.

The other process is that those who were in charge of communicating with the agencies – and communicating the agencies' messages back to the rest of the tribe: a significant role indeed – came to have great power in the community. They either combined temporal and religious authority in their own persons, or were seen by temporal authorities as highly useful allies in the maintenance of social and political order. Again, to put matters graphically: the idea that there is an agency that always observes everything people do, even in private, and who will reward or punish them accordingly – an invisible, ubiquitous and omnipotent policeman relentlessly intent on everyone's activities – is a useful instrument of social management.

Thus, as naturalistic explanations cumulatively replaced agency explanations, the agencies increasingly became *super*natural, because no longer individually and causally operative in nature – in the streams (nymphs) and woods (dryads) or at the source of lightning bolts (Zeus). The agencies thus became detached from their original explanatory role – except, typically, in the most general and abstract sense of being credited with creating the universe in the first place – while retaining the features that had been implicit in the proto-technology of influencing them, such as supplication, prayer, and sacrifice. As the proto-technology of humanity's proto-science, such activities could be expected to achieve the desired manipulation of natural events; but by the time that our ancestors knew that (for example) storm clouds presage thunder and lightning in a predictably normal way, prayer and supplication had come to be requests for a miracle, a shortcut to an end other than the one nature would arrive at.

At the same time we notice that the residue of our ancestors' proto-science continued to provide a god-of-the-gaps resource for explaining what is inexplicable or not yet understood, and it still does this; it persists, in generally inexplicit and allusive form, in the background of increasingly secure naturalistic understandings of the world, because, even where there are in fact no gaps, generalized ignorance of science creates them adventitiously, and our ancestors' proto-science comes to the rescue.

Modern science began – so most agree to date it – in 1543, the year in which Nicolaus Copernicus's *De revolutionibus orbium coelestium* (*On the Revolution of the Celestial Spheres*) was published. It was also the year in which Andreas Vesalius published his treatise correcting the anatomical errors of ancient physicians, *De humani corporis fabrica* (*On the Structure of the Human Body*), thereby revolutionizing the understanding of anatomy. Until then most proto-scientific and pre-scientific ideas about nature were derived from the often inaccurate authority of the ancients as contained in the works of Aristotle, Galen, Pliny the Elder (*Naturalis historia* (*Natural History*)), and others. Application of empirical methods and the use of the quantitative techniques of mathematics allowed the enquirers of the later Renaissance to challenge the hegemony over thought both of these ancient thinkers and of religious orthodoxy, and to begin the process of understanding the world more deeply and systematically. The success of science from that day to this has been, as noted and without hyperbole, the greatest of human achievements – and this despite the regrettable applications (guns and bombs) to which politics has too often put science via technology.

Regardless of the fact that there had been several millennia of astronomical observation already, together with much pre- and

proto-science in the centuries beforehand – some of the most important in India, the Near East, and China – choosing the year 1543 in
Europe is not arbitrary, either as a start-date or a location. This is
because the time and place in which Copernicus and Vesalius lived
was one that at last had proper standards of measurement, employing
a number system derived from India, paper and printing that made
the communication of ideas more rapid and general, and Latin as the
common language of research and scholarship. This state of affairs
was soon further improved by the arrival of devices – the telescope
and microscope first among them – that were seized upon by serious
enquirers, such as Galileo, who recognized their potential as instruments of discovery. Not least among the enabling factors was the new
inefficacy of religious prohibitions against scientific enquiry, at least
in the Protestant states of Europe, where theological authorities did
not have the power to inhibit research and publication.[4]

The authoritative-seeming systems of thought of classical authors
had figured among the obstacles to progress for their later successors,
because these latter had been reluctant to challenge them. Their
reluctance was a function of the long-held belief that humanity and
society had deteriorated since a Golden Age in the distant past, implying that the ancestors knew better, and were better, than all their
descendants. In the new mood of the sixteenth and seventeenth centuries these pieties, as with those of orthodox religion, ceased to be a
barrier. The quality and quantity of scientific advance from the mid
sixteenth century onwards is of a different order from anything
before it, and its results lie before our eyes: the transformation of the
world and human experience. If anything deserves the name 'revolution', it is this; and, true to the nature of science, it is the work of
many hands – not just Copernicus, Galileo, Newton, Priestley, Faraday, Maxwell, Marie Curie, Einstein, Bohr, Heisenberg, Rosalind
Franklin, Crick, and others whose names are salient in the history of
science, but the collegial, critical, collaborative and competitive host
of other talented individuals who built and continue to build the
house of science, brick by brick.[5]

It has to be acknowledged, though, that the makers of the scientific
revolution did not see themselves as complete repudiators of the

ancients. On the contrary, they saw themselves as resuming an enterprise that had been interrupted for more than a millennium. When the new printing presses began issuing the texts of ancient authors, enquirers in the Renaissance saw themselves as picking up where the ancients left off. As the historian of Greek science Benjamin Farrington wrote,

> The old Greek books, which the invention of printing and the birth of modern scholarship were putting into their hands, were the best available, were, in fact, the most up-to-date books in the various departments of knowledge. For Vesalius and Stevin in the sixteenth century the works of Galen and Archimedes were not historical curiosities. They were the best anatomical and mechanical treatises in existence. Even in the eighteenth century for Ramazzini, the founder of industrial medicine, Hippocratic medicine was still a living tradition . . . A generation ago Euclid and geometry were still synonymous terms in English schools.[6]

Not long before the scientific revolution began, European nations had started to send ships across the oceans on voyages of discovery. Although the motivation was principally economic, naturalists and artists accompanied the venturers and brought home news of wonders and portable treasuries of flora and fauna from diverse climes. 'Cabinets of curiosities' assembled by collectors anticipated the first museums and provided prompts to yet further speculation about nature's extent and variety.

It will doubtless surprise some to learn that one of the proximate roots of modern science is alchemy. This is because alchemy has come to be identified in the popular imagination with science's least successful and most notorious aspects, given that it was not primarily a search for knowledge for its own sake but rather a practical effort to transform base metals into gold, to find the elixir of immortality or perpetual youth (or at least longevity), to discover magical means to wealth, power, influence, health, and love, to poison people undetectably, or to foresee the future.

Although all this is indeed true of alchemy, its efforts to understand nature and control parts of it also had good aims, as in its application to medicine. In the absence of scientific method and suitable accompanying mathematical tools, alchemy was haphazard and disorganized, and therefore serious and quack investigators were indistinguishable as to which of them was the real savant.

The two most celebrated aims of alchemy were to change common metals into precious metals, and to find a cure-all for disease and even death. It was believed by many that both aims would be realized by discovery of the 'philosopher's stone', a supposed potent substance that would make this possible.

That the alchemists' efforts were not entirely arbitrary can be seen from the fact that they assumed that the objects of familiar experience, such as trees, people, and rocks, are made from different mixtures of elements, and they followed the ancients in thinking these were four in number – earth, air, fire, and water – each with one or more of the four properties hot, cold, wet, and dry (thus yielding such compounds as hot dry air, cold wet air, hot wet air, and so forth). If lead and gold differ just in the mixture of the elements constituting them, why cannot one rearrange the elements in lead to turn it into gold? In this respect the alchemists were, as it happens, right in basic principle, though wrong about the constitutive elements.

As noted in the section on pre-science technology above, work with natural substances (including not just stone, wood, water, bone, animal skins and furs, metals, and fire but also the many varieties of each of these, with each variety exhibiting different properties – and other stuffs such as dyes, herbs, and gems besides) had been continuous since before the Neolithic Period, and it is from these that the various aspects of alchemy derived. The earliest-known alchemical text is the *Physika kai mystika* (*On Physical and Mystical Matters*), attributed to the fifth-century BCE Greek philosopher Democritus, though more probably written by Bolos of Mendes in the third century BCE. Together with a few Egyptian papyri this early record shows that efforts to make gold or to turn small quantities of gold into larger quantities, and to produce other valuable substances, had already long been made, as had discussion of the utensils, stills, and furnaces necessary for the task. Here too the thinking is not arbitrary: if you can make a number of plants eventually proliferate from a single seed, why not see if you can make a lot of gold proliferate from a single nugget?

To the story of alchemy as an immediate precursor of modern science might be added the long story of astrology, because observation of the heavens had always had both astrological and astronomical interest for those who engaged in it. In Egypt and Mesopotamia one of the non-astrological uses was supervision of the calendar and thus dating the onset of changing seasons; in Egypt, given the regularity of the Nile floods, timings structured the agricultural year. Predictive deductions such as these from the movement of the heavenly bodies were taken as confirmation that they could be employed to predict other things – whether a battle would be successful, for example.

This was still the case in early-modern times. Take, for example, in 1572, the occurrence of 'Tycho's Nova', supernova SN 1572, and, some years later, in 1577, the appearance of the Great Comet (now known as C/1577 VI). Late in the year 1572 a supernova became visible in the constellation Cassiopeia, glowing more brightly than Venus in the night sky, and therefore a striking sight. It was observed by many people around the world, and recorded in detail by Tycho Brahe, hence its colloquial name.[7] It had a seminal effect on astronomy and science, not least by disproving Aristotle's view that the heavens are unchanging – if a new star comes into existence among the constellations, how could the heavens be immutable? For astrologers and astronomers this meant that a new model of the universe was needed. More significantly and generally, sixteenth-century beliefs were presented with a crisis by the appearance of the astonishing new star: for, as just noted, it implied either that the received cosmology both of religion and the ancient thinkers was wrong, or that, if the received cosmology were right after all, the new star's appearance meant that something enormous and unthinkable, and probably disastrous, was about to happen. Popular opinion inclined to the latter view – and not just popular opinion: Queen Elizabeth I consulted the astrologer Thomas Allen about the implications. The nova remained visible until 1574.

In 1577 a more terrifying phenomenon occurred: not a star appearing in one place and then vanishing eighteen or so months later, but a star moving across the sky with a fiery tail. This was the Great Comet C/1577/VI. Its appearance was ominous: wildly speculative pamphlets were published, saying that it was shaped like a Turkish scimitar, which meant that the Ottoman hordes were about to ravage Europe; that it had traversed the Seventh House of marriage and partnerships, which implied that religious disunities would deepen; or that the then proposed marriage of Elizabeth I to the Holy Roman Emperor Rudolf II would not take place. The comet had appeared in the west, suggesting something significant was about to happen in the New World, though, as Brahe observed, its tail pointed east, which suggested that it would instead scatter poison, plague, and division among the Russians and Chinese.

That nothing extraordinary (in the literal sense of this word) followed these portents is doubtless part of the growing sense among a number of thinkers that disciplined methods of enquiry were needed to separate sense from nonsense, with the aim of attaining genuine understanding of the world. Two significant names in this connection are Francis Bacon and René Descartes. They each in their different ways recognized that, to sort the sheep from the goats among methods of enquiry – so to say: to ensure that each in the pairings of chemistry and alchemy, astronomy and astrology, medicine and magic, is kept apart from the other – responsible methods of enquiry had to be identified. In such works as *The Advancement of Learning* and *The Great Instauration* Bacon promoted empirical and inductive methods and urged collaborative endeavour as the way to accumulate and mutually check information (until then alchemists and others had jealously guarded their researches in order to keep the results to themselves). Bacon's advocacy of a 'House of Solomon' – a scientific research institute – was one of the inspirations for the Royal Society, which received its charter from Charles II in 1662. In his *Discourse on Method* and in the classic *Meditations on First Philosophy* Descartes described the methodological task as one that must start 'from the foundations' if one wished 'to establish anything at all in the sciences that is stable and likely to last', and must proceed by careful, small, constantly reviewed steps to exclude error.

The key point here is that as more responsible and disciplined approaches to the investigation of nature began to sift the genuine from the spurious, and as techniques and instruments progressed – not least empirical and quantitative techniques and the improvement of telescopes and microscopes – so the separation of real science from spurious science allowed the former to develop, with extraordinary rapidity in historical terms, into the set of disciplines we know today. It did not happen immediately; Isaac Newton spent more time on alchemy, numerology, and efforts to crack what he thought was a secret code in the bible than on the science that made him famous. But it is the science that endured; in his *Principia mathematica* and in almost all (apart from the last section) of his *Optics* these other, more speculative researches played no part.

Newton is indeed the colossus at the gateway to modern science. Kepler and Galileo were among the 'giants' on whose shoulders he said he had stood in arriving at his theory of gravity and his Laws of Motion, but, unlike their work, which addressed particular phenomena, his contained an explanatory framework that brought together the concepts of gravity, mass, force, and motion, and that proved applicable to other areas of science, for example, to the kinetic theory of gases.[8]

Newton's Three Laws of Motion explain everything about the relationship between bodies like apples and planets and the forces acting on them. The First Law says that an object will either remain at rest, or, if moving, will continue to move, unless a force acts upon it. The Second Law says that the force acting on an object is equal to the mass of the object multiplied by its acceleration.[9] The Third Law says that when a body exerts a force on a second body, the second body will simultaneously exert an equal and opposite force on the first body. These laws, constituting 'classical mechanics', apply to anything and everything, provided that they are slow-moving relative to the speed of light and large relative to the size of an atom.

Newton's theory of gravity was inspired by thinking about falling bodies considered in light of the Laws of Motion. Why would an apple, stationary in its place on a twig, ever fall from the tree unless a force were applied to it? He knew from Galileo's work that falling bodies accelerate at a constant rate, entailing that the force making them fall is likewise constant. He knew from Kepler's Laws of Planetary Motion that the amount of time it takes for a planet to orbit the Sun is precisely related to its distance from the Sun. Newton's Laws of Motion are generalizations of Kepler's Laws of Planetary Motion; conjoining the insights of his more general laws with the theory of falling bodies was the inspiration for his Law of Universal Gravitation. This law states that every mass attracts every other mass with a force that is proportional to the product of their masses (the mass of one multiplied by the mass of the other), and inversely proportional to the square of the distance between them (the distance between them multiplied by itself). Gravity always attracts and never repels, acts instantaneously, and is independent of any other property of objects, such as their electric charge or chemical composition. The strength of the gravitational force weakens rapidly as the distance between masses increases.

One way of understanding this last point is to recall that Kepler had shown that the orbits of planets around the Sun are ellipses, not circles. It had been a vexation to him that he could not make the observational data fit with the supposition that, because circles are perfect figures, planetary orbits in a divinely created heaven should be circular. But the data showed that the Sun sits at one focus of an ellipse, and that the line between the planet and the Sun sweeps out equal areas in equal times, meaning that the planet speeds up as it swings round the Sun at the closest part of its orbit to it (the perihelion arc of its journey), and travels more slowly the further away it is (the aphelion arc). Treating Sun and planet as a single system, we see that its centre of mass lies not at the centre of the Sun but at a point between the Sun and the planet, though very much closer to the Sun as the more massive object. Sun and planet swing round this centre of mass analogously to a seesaw whose fulcrum has to be much closer to the heavier end, so that the whole can be balanced. The 'equal areas equal times' point is Kepler's Second Law of Planetary Motion, and leads to the Third Law: that the time it takes for a planet to complete one orbit is related to the average distance from the Sun in this ratio – the square of the time is proportional to the cube of the average distance. It was this law that directly led to Newton's insight that the gravitational force must be proportional to the product of the masses of the mutually attractive bodies.

A thought-experiment, devised by Newton himself, exemplifies the relationship between gravity and the Laws of Motion. Imagine a cannonball fired from the top of a mountain. Observation shows that the faster the cannonball is propelled, the further it flies. If there were no force of gravity acting on it, then, according to the First Law of Motion, the cannonball should fly in a straight line away from the cannon's mouth forever. In the presence of gravity the trajectory of the cannonball will depend on its momentum. If it is moving slowly, it will fall to the ground. If it is going too fast for the Earth's gravitational field to hold it, it will escape into space. If its momentum balances gravity, it will orbit the Earth.

Newton's work was a brilliant synthesis, and it constituted a picture of a deterministic universe in which, if you knew all the positions of things and all the forces acting on those things at any given point, you would be able to infer all past and future dispositions of things. This was, in essentials, physics until the nineteenth century, even though scientists progressively found inconsistencies with the picture thus drawn. It was a problem from the outset that the Newtonian conception required 'action at a distance', which raises the question of how the gravitational force is mediated. What transmits it across the space between masses? Newton himself refused to hypothesize; in the *Principia* he wrote, 'I have not as yet been able to discover the reason for these properties of gravity from phenomena, and I do not frame hypotheses' – his words were *hypotheses non fingo*. But he recognized the difficulty: 'It is inconceivable that inanimate Matter should, without the Mediation of something else, which is not material, operate upon, and affect other matter without mutual Contact . . . Gravity must be caused by an Agent acting constantly according to certain laws; but whether this Agent be material or immaterial, I have left to the Consideration of my readers.'

Increasingly it became apparent that Newton's theory is inconsistent with observation. One example is the orbit of Mercury around the sun, which does not follow exactly the same path each time, with the result that the point at which it is closest to the sun – the perihelion – is different on each orbit (the 'perihelion advance of Mercury'). This calls into question the concept of invariance, the assumption that physical descriptions are unaffected by differences in the frame of reference from the standpoint of which they are made. Maxwell's discoveries about electromagnetism in the nineteenth century called this assumption into question, and subsequent investigation, most significantly Einstein's discoveries, showed why.

In Newton's visualization of the universe as a frame of absolute space and absolute time, with the Laws of Motion and gravity working within them like clockwork, a significant problem is that it gives no explanation of 'time's arrow': the apparently irreversible passage of time from past to future. For consider: Newtonian mechanics describes both the forward and the reverse of a sequence of events with equal adequacy. Think of the motion of billiard balls: the cue ball strikes an assembly of coloured balls and they ricochet off each other and the edge of the table in obedience to the Laws of Motion. Run the sequence backwards and the laws equally well describe the balls' motions as they reassemble into an orderly triangle. The idea that time's arrow flies only in one direction had to wait for advances in understanding gases, which in Newton's picture consist of particles like little billiard balls bouncing off each other, their interactions explaining the relationship between pressure and volume. The seventeenth-century chemist Robert Boyle had shown that increasing the pressure on a gas decreases its volume. On the Newtonian view, what is happening is that increasing pressure is diminishing the spaces between the gas particles; they are being squeezed together. But this does not account for the fact that another way of reducing the volume of a gas is to cool it. The significance of temperature, which Newton ignored, was noted by Daniel Bernoulli, who proposed that the particles constituting gases move about rapidly, the temperature of the gas being directly related to the rapidity of their motion: the faster they go, the higher the temperature. And, as temperature rises, the pressure exerted by the gas – the pressure of particles hitting the walls of the container – rises with it. In a liquid this effect takes the form of an increase in volume.

The next step towards an understanding of time's arrow was taken by correlating the concepts of 'work' and heat. 'Work' is a measure of the amount of force applied over a given distance, or, alternatively put, the amount of energy expended in doing so – think of someone pushing a wheelbarrow, or a piston forcing liquid along a tube. James Joule showed that a given quantity of work will always produce the same quantity of heat; work and heat are thus shown to be forms of energy. The First Law of Thermodynamics captures the fact that, whereas energy can be transformed either into work or heat, the overall amount of energy stays the same: it is 'conserved'. But there is an important asymmetry between the transformations: in principle, work can be transformed wholly into heat, but heat cannot be transformed into work without some loss. Friction, or dissipation of heat into the surrounding environment, carries off part of the energy being applied. The total energy is the same, split into both work and heat. This is captured in the Second Law of Thermodynamics.

That heat loss cannot be reversed is the key to the arrow of time. The concept introduced to explain this is *entropy*. In any irreversible process the degree of entropy increases. One way to understand entropy is as disorder: to impose order on a system, one has to apply work to it, as with the efforts that continually have to be made to keep one's house tidy. Time's arrow flies from more order towards less order; the passage from past to future is the increase in disorder in any closed system disconnected from a source of work that would counteract entropy. The entropy of such a system can increase to the maximum possible point of complete randomness, a state known as 'thermodynamic equilibrium'. The opposite state of minimum entropy is only attainable at the lowest possible temperature, zero on the Kelvin Scale (–273 degrees Celsius).

Because heat had been recognized as motion – the faster the motion of particles in a gas, the hotter it is – and because Newton had identified the Laws of Motion, the result appears to be a satisfying and inclusive theory of physical reality at the level of ordinary experience – at the level of 'medium-sized dry goods', as the philosopher J. L. Austin put it, referring to that slice of reality that lies approximately midway between the very small (atomic) and very large (cosmological) scales of things. Note that Newton's physics had no need of a theory of the atom, and it does not incorporate an account of the phenomena of magnetism, electricity, and light, which in the nineteenth century were only beginning to be explored. It was these latter explorations that laid the foundations for Einstein's theories and for quantum mechanics.

Both electricity and magnetism had been long known; Thales, in the sixth century BCE, discussed the properties of lodestones and the electrical phenomena produced when amber is rubbed against another substance such as wool (the *triboelectric effect*, generating static). The Greek for 'amber' is *elektron*. In the nineteenth-century exploration of these phenomena the concept of a 'field' was invoked; both electricity and magnetism were conceived as fields that mediate the forces associated with each, with the strength of the forces varying in just the same way as gravity, namely, inversely as the square of the distance between charges. The difference is that, whereas the force of gravity is always attractive, electricity has both a negative and a positive aspect. It was known that like charges repel each other, and unlike charges attract each other.

Observation and experiment established that electric currents generate magnetic fields. Michael Faraday saw that magnets, conversely, can create electric fields, if the magnet is accelerated, for example by being rotated. Producing electric current by moving a magnet is known as 'induction'. James Clerk Maxwell formulated a theory combining electricity and magnetism into the single force of electromagnetism, in the process demonstrating that light is a wave in the electromagnetic field. This was shown by combining two observations. One is that changes in the electric field produce changes in the complementary magnetic field and vice versa, in such a way that if the electric field oscillates, its frequency of oscillation will be matched by the oscillation frequency of the magnetic field. Faraday had measured the resistance offered by a vacuum to the rate of oscillation of electric and magnetic fields; when this measurement was entered into Maxwell's equations, they showed that the electromagnetic wave moves at 186,000 miles per second – the speed of light.

There is, however, a problem with Maxwell's theory. The temperature of a body is the result of vibration of the atoms in it. The hotter a body grows, the more its constituent atoms oscillate, the frequency of the electromagnetic radiation they emit rising through the spectrum of colours to red, yellow, and eventually white. On Maxwell's theory, the intensity of electromagnetic frequencies above violet grows unlimitedly, independently of temperature. This problem came to be known as the 'Ultraviolet Catastrophe'.

The solution was found by Max Planck, who proposed – as an heuristic merely: like Maxwell he thought nature was a continuum – that the radiation emitted by a hot body comes in discrete jumps he called *quanta*, enabling him to write down a formula describing the relationship between the intensity of radiation and its temperature and frequency. This relationship is a constant, denoted h: the 'Planck Constant'.

At the very end of the nineteenth century Maxwell's theories about light and Planck's introduction of the idea of quanta were joined by J. J. Thomson's discovery of the electron to put physics on the brink of an era of rapid and dramatic scientific progress. In one of the great ironies of history, some scientists thought that their pursuit was over, and only details remained to be filled in; that view is apocryphally ascribed to Lord Kelvin, but it was certainly true of Planck's teacher Philipp von Jolly, who advised him to find a career outside physics because physics was complete. Before the first decade of the twentieth century was over, Einstein had published his Special Theory of Relativity, and the experimental and theoretical work of the next thirty years produced general relativity and quantum theory.

As this shows, the story of twentieth-century physics is the story of contemporary physics.[10]

3.   The Scientific World Picture

Observation tells us that the objects of familiar experience are complex, made of smaller parts, and many of them appear, even to the unaided eye, to have structure. It is a natural inference to think that the parts constituting a complex whole might themselves have parts – when you take a clod of earth in your hand and rub it, it disintegrates into smaller and smaller pieces – such that eventually you must arrive either at nothing or at the smallest possible pieces. The idea that a space-occupying solid thing could ultimately be made out of nothing seems contrary to common sense, so the most natural inference is to the idea that there are 'smallest possible things' out of which other things are made.

Or rather: that might seem to be the natural inference. But the idea of a granular ultimate structure to things – of things that cannot themselves be made of parts, or be any smaller: 'uncuttable' or 'indivisible' things (for which the ancient Greek word is *atomos*) – was not what was hypothesized by those who made the first attempt at systematic science, namely, the ancient Greek philosophers. Their first inference was that the universe must be a single pervasive stuff capable of taking on a multiplicity of forms, themselves capable of a multiplicity of effects (Thales's 'water') or a combination of basic stuffs (air, earth, fire, and water). It was a sophistication of such views that led to the idea that physical reality consists of ultimate constituents that are *atomos*. It is of great interest to note that the arc of modern scientific enquiry has gone from thinking of nature in the latter *atomic* sense to conceiving of it in a way that is reminiscent of the earlier notion of something pervasive or the interplay of pervasive continua, as in talk of 'fields' in contemporary physics. This is a crude and remote analogy, of course; but it is worth pausing for a moment to recall some earlier ideas about the underlying form of physical reality, because certain of the assumptions still made about what nature must be like, and what can count as a good theory of nature – notice the strong modalities of the words 'must' and 'can' – have their source there, and are at work in science today.

Three of the first 'physicists' (*phusikoi*), Anaxagoras of Clazomenae and a teacher–pupil pair, Leucippus and Democritus – these two known as 'the Atomists' – hypothesized that everything is made of fundamental elements, which Anaxagoras called 'seeds' and the Atomists called 'atoms'. All three were Presocratic philosophers: Anaxagoras flourished in the first half of the fifth century BCE, the Atomists in its second half.

Anaxagoras held that 'coming to be' and 'passing away' are not creation from nothing and destruction into nothing, but instead are rearrangements, effected by mixings and separations of eternally existing elements. What exists is conserved; physical occurrences are changes caused by interactions. He also thought that the 'seeds' of all things, *panspermia*, are always all present together in everything, individual things being differentiated only by the preponderance of one type of seed over the others, not by the absence of other types of seeds.

What originally existed before 'the worlds' came into existence, Anaxagoras held, is an undifferentiated and unlimited mass of stuff, consisting of an indiscriminate mixture of the seeds of things. Individual things are the result of aggregations of seeds, while separation of them causes their demise. Aggregation and separation are effected by some active principle or force for which he used the analogy of *nous*, which in its literal sense means 'mind' but by which Anaxagoras did not mean a thinking mind or god, but rather an active power such as one observes in a magnet that draws iron to it. There is no void, no 'nothingness'; the universe is everything there is, namely, the totality of seeds; and in support of this he gave, as Empedocles did before him, experimental demonstrations of the real corporeal existence of air to show that it is not the 'nothing' that the senses seem to suggest.

It is not clear whether the idea of 'seeds' in Anaxagoras's theory had any influence on the Atomism of Democritus and Leucippus, but there is at least a superficial similarity in the basic conception.

Atomism is the theory that everything is composed of tiny imperceptible objects, each of which is 'uncuttable'. Aristotle, impressed, even though he disagreed with their idea, felt obliged to study Atomism in detail. He therefore wrote a work in several volumes, all now lost, about the Atomists' views.

The nub of the atomic theory is that there is an infinite number of indivisible fundamental entities that are eternal and unchangeable in every respect but position. In addition there is 'the void,' nothingness – but nothingness is real; it is the space that separates atoms, which are therefore able to move in the void and to collide with one another, their various shapes making it possible for them to link together into larger agglomerations when they collide, and for these agglomerations to break apart again later, thus giving rise to all the phenomena of things and their changes in the sensible world. This captures the idea, found also in Anaxagoras, that 'coming-to-be' and 'passing-away' are just changes, not actual creations and destructions of what exists. The totality of what exists is conserved.

The Atomists called atoms 'what-is' and the void 'what-is-not'. Aristotle in the *Metaphysics* describes the Atomists' account of how atoms constitute things as follows:

> They declare that the differences [between atoms, 'what-is'] are three: shape, arrangement and position. They say that what-is differs only in 'rhythm', 'touching' and 'turning' – 'rhythm' is shape, 'touching' is arrangement, and 'turning' is position. Thus A differs from N in shape, AN from NA in arrangement, and Z from N in position.[1]

A thread running through these ideas has its source in the thinking of Parmenides, arguably the most powerful and influential philosopher in the Presocratic tradition. Parmenides argued that reality must be a single unchanging thing, because anything complex and mutable is unstable and therefore unable to persist in existence over the long term. Combine the two ideas that the concept of 'nothingness' (not the vacuum of space full of quantum fluctuations but *nothingness*) is meaningless because unthinkable, with the observation that the world does indeed exist, and you are logically committed to thinking that the world is eternal – it cannot come into existence from a previous nothingness, nor vanish into a future nothingness, since nothingness does not exist – and it must be impervious to change and variety, because either, and especially both, threaten its ultimate decay. This motivates Parmenides's reductionism of all things to a single thing, a One, and his description of this One as *necessarily* unchanging and permanent.

Now note that, in their thinking about the ultimate constituents of reality, Anaxagoras and the Atomists respect the Parmenidean requirement that what exists must be eternal and in itself unchanging, but at the same time seek to account for the observed plurality of things, and the movement and change they engage in, by postulating that there are (in effect) many Parmenidean 'ones': the seeds or atoms.

The reductionist impulse to find an ultimate unity in what underlies reality is a theme in science that, as we here see, has ancient roots. The question to be asked is: is this impulse driven by what observation and successful theory tell us must be so, or is it a deep function of the way we think things *must* be, because of something other than observation and theory – for example, something to do with the cognitive architecture we bring to thinking about what the world must be like?

These early ideas of an underlying structure to things whose combinations and interactions give rise to all perceivable phenomena prompt us to ask: why did the first scientists, given that they were thinking along these suggestive lines, not go further with their enquiries? Why did they not experiment? Why did science remain in this merely speculative state, and have to wait another two thousand years – until the era beginning with Copernicus and Galileo – to get going properly?

The answer is complex. One part of it is that these thinkers were beginning from scratch, working in a small community of thinkers dispersed both in geography and date, with little in the way of a context that would cumulatively support building on ideas and finding ways of testing them. It has been observed that cultural evolution requires a minimum density of population as its environment for ideas and practices to take hold, providing a basis for development.

The early scientists relied on observation and reason but rarely on experiment. Social conditions in the Greece of the sixth to fourth centuries BCE nevertheless were conducive to philosophy, and the first flickerings of science, even in this limited way. But the degree of robustness and inventiveness of thought evident in this period did not continue in the same degree in the Roman Period. By the first centuries CE one of the dominant strands of philosophy, Neoplatonism, had begun to assume the lineaments of religion; by the time that the Emperor Justinian in 529 CE closed the School of Athens, banning philosophy and banishing the philosophers on the grounds that philosophy conflicted with Christian doctrine, the intellectual circumstances were inimical to further enquiry. Not until the Reformation of the sixteenth century was enquiry again sufficiently free of the need to comply with religious doctrine for science to resume.

Nevertheless, the surmises of Anaxagoras and the Atomists proved to be intimations of a powerful way of thinking of the structure and properties of physical reality. The 'corpuscular' theory ('corpuscle' means 'little body') of the seventeenth century's natural philosophers was a reprise of the Atomist views of antiquity, but it was not until the nineteenth and especially twentieth centuries that a fully scientific and experimentally supported theory of the atom was formulated.

Notice one thing: the speculations of Parmenides and others were prompted by empirical observation and resulted in inference to theories that would explain it. Parmenides saw decay and change, wondered how a world could continue to exist in their presence, and hypothesized that human perceptual awareness does not reveal what logic requires in regard to the underlying nature of reality. Anaxagoras and the Atomists observed that wholes have parts, that nothing can come from nothing (*ex nihilo nihil fit*), and that there must be ultimate units of reality whose relationships give rise to observed phenomena. The inferences made by today's 'string theorists' from the current anomalies of fundamental physics have the same *form* as these speculations, though with a vastly richer apparatus of background theory and mathematical power in which to articulate them.

Since the late 1960s the best and most widely accepted scientific theory of the fundamental structure of matter and radiation has been the 'Standard Model' describing the elementary particles and the forces that bind them together into atoms.[2] It was formulated by physicists Steven Weinberg, Abdus Salam and Sheldon Glashow after the last of these had proposed a way of unifying the *electromagnetic* and *weak nuclear forces* to constitute the *electroweak force*, and further combining this unification with a theory (most closely associated with the work of Peter Higgs and François Englert, hypothesizers of the 'Higgs boson') explaining how the elementary particles acquire mass. For this seminal work Weinberg, Salam and Glashow jointly received the Nobel Prize in Physics in 1979, and Peter Higgs and François Englert jointly received the Nobel Prize in Physics in 2013.

The importance of the Standard Model is that it offers a relatively simple and economical description of the basic constitution of matter. A powerful way of thinking of the structure of matter is to see it as built out of elementary particles interacting and combining according to laws expressible in mathematical terms. Particles of matter interact by the mediation of force-bearing particles. Matter particles are called *fermions* and consist of two groups: *quarks*, which constitute the protons and neutrons forming the atomic nucleus, and *leptons*, which include electrons and neutrinos. When quarks are bound together, they are called *hadrons*.[3] A simple (but beware: a misleading – see below) way of picturing an atom is as a little solar system, with the nucleus at the centre and electrons orbiting it. If this were actually what an atom was like, then – to get a sense of its inner scale – if you expanded it to the size of the Royal Albert Hall in London, the nucleus would be the size of a fly at the centre of the hall, and the outermost shell of electrons would be the walls.

The matter particles interact by passing force particles – the *bosons* – between them. The bosons are *photons*, *gluons*, and *W* and *Z particles*. The electromagnetic force, mediated by photons, binds electrically charged particles together; the massive W and Z bosons mediate the weak force in the nucleus; and the massless gluons constitute the *strong nuclear force* that binds quarks into protons and neutrons – the matter *hadrons*.

Among the charged leptons in the model, only electrons are stable, that is, retain the same charge; the others (the *muon* and *tau* leptons) decay extremely quickly, in fractions of a second.

A key element in the whole structure of matter thus depicted is the need for a particle that imparts mass to the other particles apart from the photon and gluon. All the other elementary particles of the Standard Model had been experimentally observed, but this particle, the *Higgs boson*, was not observed until the Large Hadron Collider at CERN in Switzerland reached the high levels of energy required to produce a Higgs boson in laboratory conditions. The discovery was announced in 2012.

The all-important role for the Higgs boson is that it answers the question of where leptons and quarks get their masses, and how the difference arises between photons, which have no mass, and the massive W and Z bosons. How this works can be explained by noting that talk of 'particles' does not adequately describe phenomena at the subatomic level, which for certain purposes are better represented as 'fields'. Thus an electron does not orbit an atomic nucleus as a planet orbits the Sun, but rather is a field in the neighbourhood of the nucleus. Now visualize the Higgs field as a swimming pool into which objects representing leptons and quarks are thrown: the water slows down the objects by dragging on them: the drag effect is the mass of the object. Photons have no mass; an object representing a photon would pass through the Higgs swimming pool without pause, uninterrupted in its passage.

Although discovery of the Higgs boson provides further confirmation of it, the Standard Model is still not complete. It does not offer a way of combining the electroweak and strong forces with the other fundamental force of nature, namely gravity; and the numerical parameters of the theory, of which there are about twenty, are not derived from physical principles but were ascertained experimentally. The effort to combine gravity with the atomic forces (in other words, to find a quantum theory of gravity) has been given some support by 'String Theory', which remains controversial: see below.

The hope that the atomic forces operative at the microscopic level of nature can be unified with gravity, which operates at the large – indeed universal – scale of nature, turns on the assumption that nature is at its profoundest level simple, entailing that the four forces of nature – the electromagnetic, the weak, the strong and gravity – are in fact just versions of a single underlying force that we cannot yet grasp. The strong reductionist impulse of Parmenides is fully at work in this assumption.

The theory that provides insight into the subatomic realm is quantum theory. It is a powerful and successful theory that underwrites many practical applications, even though it appears to give a radically unintuitive picture of the deep levels of reality. The success of quantum theory is demonstrated by its great precision, yielding exquisite accuracies to many decimal places.

To understand quantum theory, it helps to state the points just made about the development of the Standard Model from a slightly different direction, as follows.

Quantum theory arose out of the inadequacy of classical physics, derived and developed jointly from the work of Newton and eighteenth-century chemistry, to explain a number of increasingly apparent anomalies. When, in the mid nineteenth century, James Clerk Maxwell successfully gave a statistical description of the behaviour of gases, he was able to assume that gas consists of tiny featureless atoms interacting as if they were miniature billiard balls. But then experimental discoveries later in that century showed that atoms have inner structure, for particles were observed with mass less than the atom of least mass, the hydrogen atom, and moreover atoms were displaying what came to be called radioactivity (the natural disintegration over time of atomic nuclei), accepting and giving electric charge and changing into other atoms. The crucial discovery was made by J. J. Thomson in 1897 when investigating cathode rays: he discovered that the 'rays' were in fact streams of negatively charged particles, to which he gave the name 'electrons'.

This led to speculation about the distribution of positive charge in the atom. Thomson proposed a 'plum pudding' model in which the negatively charged electrons were 'plums' in a mass of positively charged 'dough'. Ernest Rutherford and his colleagues discovered, by means of 'scattering' experiments, that the greatest part of an atom's mass and its positive charge are concentrated in a tiny percentage of its volume. Rutherford hypothesized that the mass and positive charge were focused at the volume's centre, thus constituting the atom's nucleus. The planetary model that results posits three kinds of subatomic particles: electrons, protons, and neutrons, with the electrons 'orbiting' the protons and neutrons forming the nucleus. (This way of describing the model owes itself chiefly to Niels Bohr in the early twentieth century.) Protons are positively charged particles, neutrons have, as the name suggests, no charge; together they carry most of the atom's mass. Electrons are very 'light' particles, having a tiny fraction of the mass of a proton, and they are negatively charged. The difference in charge between electron and nucleus is what keeps the former 'in orbit' round the latter.

Protons and electrons are always equal in number in a non-ionized atom; when ionized an atom contains either fewer or more electrons than protons. Atoms differ from one another according to the number of protons in the nucleus, and this determines what element they are; thus hydrogen is number one on the periodic table of elements because it has one proton; oxygen is number eight because it has eight protons; and so on.

This classical picture of the atom soon proved inadequate. One simple reason is that, on classical principles, electrons should lose energy as they fly around their nuclei, and therefore spiral into them. But they do not; atoms are relatively stable structures. Moreover it turned out that atoms do indeed absorb and radiate energy but only at specific wavelengths. Such puzzles as the problem of black body radiation and the photoelectric effect showed that a new way of thinking about the elementary structure of matter was required. It came when Max Planck showed in 1900 that he could solve some of these problems by assuming that energy comes in discrete packets, which he called quanta (Latin for 'amounts').

Planck himself regarded this as a purely heuristic notion, a mathematical sleight of hand for solving a puzzle, and did not accept his own suggestion literally until much later. But in the third of his four famously seminal 1905 papers, Albert Einstein used Planck's idea to explain the photoelectric effect (why metals produce electricity when light of a high enough frequency is shone on to them). The strange thing, as it seemed to Einstein and his contemporaries, is that, whereas there was very good experimental evidence that light is a wave – and by the end of the nineteenth century it was assumed in physics that light is indeed a wave – it behaves as a particle in Planck's solution to the black body problem and Einstein's explanation of the photo-electric effect. Which is it: wave or particle? The answer, by now familiarly, came to be that it should be treated as both – or, rather, as either, depending upon which is most convenient for a given purpose.

At the level of ordinary experience we are familiar with the phe-nomena of waves and particles, such as waves at sea and individual pebbles on a beach respectively. When a particle moves it carries both mass and energy from one locality to another, whereas waves carry only energy, and are spread out in space. But at the scale of things described by quantum theory this intuitive wave–particle distinction breaks down, and in its place there is the *wave–particle duality* that makes quantum theory seem so unintuitive.

Louis de Broglie (the only duke – or, as he was French, duc – to win a Nobel Prize) extended the idea of wave–particle duality from photons to other particles in 1912, in the face of much initial scepti-cism, but the work of Niels Bohr had already confirmed that the interpretation suggested by Planck – that energy comes in packets at discrete intervals on the energy gradient, not continuously but with jumps between – is right. Bohr was therefore able to explain how atoms are structurally stable, while being able to absorb and radiate energy; it is that the wavelength of an electron always has to have a whole-number value, so that if an electron emits or absorbs energy, it has to do so by jumping to another whole-number wavelength, there being no resting place between.

With these developments in place, physicists (chief among them Erwin Schrödinger and Werner Heisenberg) were able to work out the mathematics required for the theory. Among the important results is a radically different view of electrons – not as particles but in effect as probability smears around nuclei. When electrons absorb or lose energy, they vanish from one 'position' in the vicinity of the nucleus and instantaneously reappear at another.

Heisenberg postulated the 'Uncertainty Principle', which states that one cannot simultaneously measure both the position and the momentum of a subatomic particle (where 'momentum' is the mass of a particle multiplied by its velocity). This has profound implications for causality and predictions concerning the future behaviour of particles. On the classical view, if one knows everything about the current state of a physical system, together with the laws governing it, one can predict precisely what its future states will be. But Heisenberg remarked, 'In the sharp formulation of the law of causality, "if we know the present exactly, we can calculate the future", it is not the conclusion but the premise which is wrong.'

Quantum theory has equally profound implications for an understanding of reality itself. It appears to state that how things are in the microscopic world is the result of measurement – that is, a quantum state does not have a definite character until it is measured (for example, a particle does not have a definite path until the path is calculated). This is because, prior to the measurement being made, the quantum situation consists of a range of possibilities, and measurement settles which of them counts as actual. This raises a vexing question: must there be a measurer to make a measurement before reality can have a determinate character? How can the nature of reality depend upon a measurement being made of it? And, in any case, how does reality 'decide', when it is measured, which determinate state it will adopt? This is known as the 'Measurement Problem', and it is the focus of much discussion and disagreement about the right way to *interpret* quantum theory, that is, of how to make sense of it as a picture of reality.

In what has become known as the 'Copenhagen Interpretation' of quantum theory (because it was advanced by Niels Bohr and Werner Heisenberg when working together in that city, where Bohr's institute was located), the amazing-seeming fact that quantum systems are indeterminate until measured is simply accepted as given; and, when it is, all the rest follows happily. But what the interpretation itself asks us to accept is so philosophically and scientifically puzzling that it has remained a major point of debate, with a variety of sometimes exotic theories aimed at making sense of the puzzle.

One such is the 'Many Worlds Theory', which proposes that all possibilities are actualized by the splitting into different worlds of the world in which the possibilities arose, each of the new worlds in turn splitting into new actual worlds from the possibilities in them, with none of the new worlds able to communicate with the others after the split. The 'Consistent Histories Theory', also known as 'Decoherent Histories', in essence asserts that the environment of a quantum event acts as the observer of it, causing all such events to assume a classical actualized state. The Copenhagen Interpretation itself is a version of *instrumentalism*, which comes down to saying 'The theory does not describe reality, but somehow it works, so let's just get on and use it': as the dictum attributed to Feynman and various other physicists has it, 'Shut up and calculate!'

And what was long the least popular alternative, more recently making a comeback, is the theory advanced by David Bohm: that the universe is an 'implicate order', in the sense that there is an underlying 'quantum potential' in the universe that connects all quantum phenomena, whose character is the result of the unfolding of the deterministic structure of that underlying reality.

Einstein himself was never convinced by interpretations of quantum theory that made reality look indeterminate or probabilistic. In what is called the 'Einstein–Podolsky–Rosen Thought-experiment' (EPR), which Einstein used to demonstrate that quantum theory is incomplete, the following happens: suppose you fire a pair of just-interacted particles in exactly opposite 'spin' states away from each other in a straight line, then measure the spin state of one, thus (according to the Copenhagen Interpretation) fixing which state it is in. It follows that instantaneously the other particle is fixed in the opposite spin state, without anything being done to it. But for this to happen the information about the spin state of the first particle, as fixed by observation, would by hypothesis have to be transmitted to the second particle at a speed greater than the speed of light ('superluminal transference of information'). Yet a fundamental commitment of physics makes this impossible: nothing can travel faster than the speed of light. Therefore there is something wrong with quantum theory. Einstein took this as vindication of his view that the universe obeys the principle of locality.

Alas for Einstein, and to the added amazement of all, the EPR effect was experimentally shown to be correct by the work of Alain Aspect in experiments at the University of Paris in 1982. This has proved one of the motivations for continued debate about, and development of, theories such as those mentioned to find other ways of accounting for the apparent mysteries of quantum theory.

A major puzzle is that the other powerful theory in physics, Einstein's General Theory of Relativity describing the nature of gravity, space, and time, seems not to be consistent with quantum theory. This really matters, because another whole realm of science – cosmology – rests upon it.

The phrase 'the theory of relativity' denotes two related theories developed by Einstein in the first two decades of the twentieth century. In 1905 he published the paper 'On the Electrodynamics of Moving Bodies', which set out the basic postulates of the Special Theory of Relativity, where 'special' means 'restricted', because it applies only to objects moving at constant velocities. In 1916 he published his General Theory of Relativity, which takes account of accelerating bodies also ('acceleration' means 'change of velocity', either increasing or decreasing). The General Theory therefore deals with gravity, because it shows that acceleration and gravity have identical effects; this is known as the Equivalence Principle.

Einstein's work grew out of the fact that, by the late nineteenth century, it was apparent that there was a serious problem in the classical physics based on Newton's work. This was that the Newtonian account of the relations between moving bodies was not consistent with the equations of electromagnetism discovered by Maxwell. The Newtonian view is in effect a common-sense theory about how moving bodies behave; for example, if you are in a vehicle moving at 20 miles an hour, and throw a ball ahead of you at 10 miles an hour, the ball's speed will be the sum of these two speeds – 30 miles an hour. In the background of such obvious-seeming facts is the idea that the laws of physics are the same for everyone in uniform motion relative to one another, a notion developed in seventeenth-century physics to provide frames of reference in which to describe and apply the laws of nature. In the ball-throwing case, the vehicle's speed is relative to something stationary outside itself, and the speed of the ball is computed in the same frame of reference when it is added to the speed of the vehicle itself. An important assumption in this is that time is absolute – it ticks away regularly no matter what, the same in all frames of reference.

But Maxwell's equations of electromagnetism give the speed of light, *c*, a constant value of about 186,000 miles an hour (more accurately, 299,792,458 metres per second in a vacuum; a photon will orbit the Earth seven times a second), independently of whether a light source is stationary or moving, and independently also of the velocity of an observer of the light. If one is sitting in a vehicle moving at 100,000 miles an hour and switches on a forward-shining beam of light, its speed will not be *c* + 100,000 miles an hour but just *c*. It seems paradoxical, but, whereas two balls thrown towards each other approach at the sum of their individual speeds, two beams of light shone towards each other approach at the speed of light, not at twice the speed of light.

Einstein's contribution was to show that it is not contradictory to postulate that the laws of physics are indeed the same for all observers in the same frame of reference, yet also to postulate that the speed of light is constant irrespective of frame. Accepting both (with a suitable mathematical adjustment to the former[4]) involves accepting some astonishing new ways of thinking about nature. For one thing, time no longer appears absolute and invariant, but is slower when measured on a moving vehicle than at a place stationary with respect to that vehicle. Also, objects grow shorter in the direction in which they are moving, as measured by an observer. And the faster an object moves, the greater its mass becomes. This is why *c* is an absolute speed-limit for anything with mass; reaching it requires an infinite amount of energy. There is no longer any such thing as absolute simultaneity; two events that appear to happen at the same time for one observer can appear to happen at different times to a different observer. And, perhaps most famously, Einstein showed that mass and energy are equivalent and interconvertible: this is expressed by the famous formula $E = mc^2$ where E stands for energy, m for mass and c for the speed of light.

Special Relativity agrees with Newtonian physics at speeds that are low in comparison to the speed of light, but diverges from it increasingly as speeds become significantly larger fractions of the speed of light. This has been repeatedly tested and confirmed, and applying it experimentally and technologically is found to yield greatly more accurate results than can be achieved by applying Newtonian principles. Physicists use the Special Theory of Relativity in their work all the time, where it serves as – so to speak – the wallpaper of their thought.

Einstein was helped in developing the General Theory by something that he at first found an annoyance: namely, the demonstration by Hermann Minkowski (who had been Einstein's physics teacher and who thought him a lazy student) that his Special Theory of Relativity could best be expressed geometrically in terms of a four-dimensional space–time. Einstein had already worked out the 'Equivalence Principle', stating that the effects of a gravitational field are exactly the same as the effects of acceleration. Among other things this explains weightlessness as experienced in a manned rocketship. When its engines are switched off and therefore the rocket is not being accelerated, its occupants will float about inside in 'free fall'.

An immediate consequence of the Equivalence Principle is that light is bent by gravity. This was not a novel idea, because it was accepted in Newtonian physics that light, thought of as streams of particles – photons – will feel the gravitational force. Einstein's work showed that the effect of gravity on light is twice as strong as Newtonian physics predicts, as a result of the new model of space–time entailed by his theory. The theory states that space–time is itself distorted by the presence of matter in it, just as a stretched-out sheet will be pulled down into a dip if a heavy object is placed in the middle of it. Think of light moving through space–time as behaving like a ball being rolled across a stretched sheet with a weight in the middle; the ball's trajectory will be affected by the slope of the dip, being drawn down it towards the weight in the middle. Rolled with sufficient force, it will go from one edge of the sheet to the other, swerving round the contour of the slope yet not ending up in the dip with the object – a bit like a golfer's putt that skirts but does not fall into the hole. This is how light travels through space, occasionally being bent by massive objects such as the Sun (as demonstrated in a critical observation supporting Einstein's theory in 1919). This phenomenon is known as *gravitational lensing.*

In addition to reconceiving space–time as curved and gravitation as the effect of this curvature, the General Theory also predicts the *gravitational redshift of light, gravitational waves,* and the existence of *black holes.* Its effectiveness makes it the basis of the standard cosmological model of the universe.

Gravitational redshift is the lengthening of the wavelength of electromagnetic radiation as it escapes from a gravity well – a 'gravity well' being the gravitational pull of any massive body. Because a photon must expend energy to escape the well, but not by slowing down – light, of course, always travels at the speed of light – the decrease in its energy has to take the form of a decrease in its frequency, thus increasing its wavelength in the direction of the red end of the spectrum. This prediction of the General Theory was experimentally confirmed in the 1960s.

Gravitational waves are disturbances in the curvature of space–time. Like electromagnetic radiation, gravitational waves are a form of energy propagation, caused by the movement in space–time of massive objects that disturb the curvature of space–time in that region. As a rough analogy, think of a stone falling into a pond and sending out ripples. Gravitational waves move at the speed of light, squeezing and stretching the space–time through which they pass, so that if one were observing the effects of a wave passing in (say) a region of space occupied by two stars, one would see them bob backwards and forwards towards and then away from each other alternately as the space–time they occupy narrows and expands. Direct observation of gravitational waves was achieved by the scientists of the Laser Interferometer Gravitational-Wave Observatory (LIGO) in 2015, earning a Nobel Prize for Kip Thorne and his colleagues there.

Black holes are astronomical entities whose gravitational field is so strong that nothing, not even light (hence their blackness), can escape from them. Einstein's General Theory of Relativity entails them in its description of the curvature of space–time, for this tells us that, when space–time is curved round upon itself tightly, the result will be so compact that nothing can escape the strength of gravity thus generated. As it happens, the existence of black holes was anticipated as early as 1795 on the basis of Newton's account of gravity; Pierre-Simon Laplace (1749–1827) worked out that if an object were compacted into a small enough radius, its 'escape velocity' (the velocity required to overcome the gravitational force it 'feels') would have to be greater than the speed of light – which is impossible.

The standard account in contemporary cosmology locates the origin of black holes in the death of stars (specifically, stars at least four times larger than our own Sun). Stars are gigantic fusion reactors that exist as long as the forces fuelling them outweigh the large gravitational forces their size generates. When a star of the appropriate size begins to exhaust its fuel, gravity pulls it in upon itself, collapsing it inwards. At a certain point the degree of compression of its core, and the heat thus generated, becomes too great, and the star explodes as a supernova. What remains is a highly dense remnant, whose gravitational field is so great that nothing can attain an escape velocity sufficient to leave it – not even light.

The first person to recognize the implication of Einstein's account of gravity in this connection was Karl Schwarzschild, who worked it out while serving on the Russian Front in the First World War, just months before he died. He gave a description of the geometry of space–time around a spherical mass, showing that for any such mass there is a critical radius within which any matter would be compacted to such a degree that it would in effect seal itself off from the rest of the universe. He sent his paper with these calculations to Einstein, who presented it to the Prussian Academy of Sciences in 1916. That critical radius is now called the 'Schwarzschild Radius'.

What the Schwarzschild Radius measures is the 'event horizon' of a black hole, that is, the boundary line from inside which nothing can escape. At the centre of a black hole is a 'singularity', the name physicists give to an entity to which the standard laws of physics do not apply.

There are two types of black holes, distinguished by whether or not they are rotating. The non-rotating kind are called 'Schwarz-schild Black Holes', with just a singularity as their core and an event horizon. If the core is rotating – and most black holes will be of this type, because the stars from which they formed rotated – there will be two further features. One is an *ergosphere*, an egg-shaped region of space outside the event horizon, so shaped because it has been distorted by the black hole's gravity dragging on space–time in the neighbourhood. The second additional feature is the *static limit*, the boundary between the ergosphere and normal space. If by mistake one flew one's spacecraft into the ergosphere, there would still be a chance of escaping across the static limit by exploiting the energy of the black hole's rotation. But once across the event horizon there would be no going back.

Although nothing is known about the singularity that lies at the core of a black hole, the black hole's mass, electric charge, and angular momentum (rate of rotation) can be calculated. This is not done directly but by means of the behaviour of objects in the black hole's vicinity: the wobbling or spinning of a nearby star; gravitational lensing effects, occurring when light is bent by the black hole's gravity, as predicted in the General Theory of Relativity; X-ray emissions caused by a black hole drawing material from a neighbouring star and heating it up so greatly by compressing it in its gravitational field ('superheating' it) that the material emits X-rays; and more. It is also hypothesized that supermassive black holes (those with masses billions of times greater than the Sun) can throw off high-speed jets of matter, presumably from the ergosphere, and emit powerful radio signals.

Black holes are such exotic entities that they suggest other ideas. One was put forward in 1963 by the mathematician Roy Kerr, who suggested a way that black holes might be formed without a singularity at their cores, and, if so, flying into one would not result in being crushed to an infinitesimal point but might result in being spewed out on the other side into a different time or even a different universe, through a 'white hole' – the reverse or other side of a black hole, which ejects matter rather than sucks it in.

This was one suggestion for time-travel or (if there is more than one universe existing in parallel or honeycomb fashion) for visiting other universes. Other suggestions – for example, the existence of 'wormholes' in the fabric of curved space–time that would allow one to shortcut through time – do not hypothesize the absence of singularities at the core of black holes.

The standard cosmological model of the universe has at its base the Big Bang Theory of the universe's origin. A composite summary of it, combining the idea of an expanding universe with influential ideas about how the first moment of expansion occurred, yields a picture that has the universe coming into existence 13.72 billion years ago in a 'singularity' that, from an initial immensely rapid expansion in the first infinitesimal fractions of a second, set the universe on the course that yields its present condition.

The origin of the Big Bang Theory lies in the observation made in 1929 by astronomer Edwin Hubble that the universe is expanding. A few years earlier he had established that the universe is more than just the Milky Way Galaxy in which our solar system is located, and is vastly larger than it, containing very many galaxies like our own – current estimates have it that there are *two trillion* galaxies altogether. The observation that this vastness is, in addition, expanding in all directions was a yet greater surprise.

That the universe is expanding implies that at earlier points in its history everything was closer together; running the clock back eventually gives us everything crunched together at a starting point. This idea had been suggested in 1927 by the physicist Georges Lemaître (a priest who taught science at the Catholic University of Louvain), so when, two years later, Hubble noted that the further away galaxies are, the faster they are travelling – as shown by the fact that the light they emit is shifted further towards the red end of the electromagnetic spectrum – the entailed idea of a birth event was inevitable.

As it happens, the name 'Big Bang' was coined as a joke by proponents of the then rival 'Steady State Theory', which has it that the universe exists eternally, with matter spontaneously coming into existence in the vacuum of space. But the joke name stuck and ceased to be a joke.

Subsequent investigation of the possibilities thus suggested included, as the current front-runner hypothesis, the idea already mentioned: that the first instant of the universe's existence consisted of a singularity that extremely quickly – in infinitesimal fractions of the first second – 'inflated' into the universe's earliest primordial state, from which it has continued to expand and develop. Not unsurprisingly this is known as the 'Inflationary Model'.

At the very beginning of the universe's history, says this theory, what had a moment before been nothing was an enormously hot plasma, which, as it cooled after 10^{-43} seconds into its history (the 'Planck Time'), came to consist of almost equal numbers of matter and antimatter particles annihilating each other as they collided. Because of a tiny asymmetry in favour of matter over antimatter particles to the initial order of about one part per billion, the dominance of matter increased as the universe matured, so that matter particles could interact and decay in the way current theories of the structure of matter describe. As the initial 'quark soup' cooled to about 3,000 billion kelvins, a 'phase transition' led to the formation of the heavy particles, the protons and neutrons, and then the lighter particles, the photons, electrons, and neutrinos. (A more familiar example of a phase transition is the transformation of water into ice when the temperature of the water reaches zero degrees Celsius.)

Between one and three minutes into the universe's history the formation of hydrogen and helium began, these being the commonest elements in the universe, in a ratio of about one helium atom to every ten hydrogen atoms. Another element formed in the early process of nucleosynthesis was lithium. As the universe continued its expansion, gravity started to operate on the matter present in it, triggering the process of star and galaxy formation.

Hubble's observation of the universe's expansion consisted in his noting that, in every direction in which one looks in the sky, (almost) every galaxy is travelling away from us, and that the rate at which they are receding is proportional to their distance from us. This is 'Hubble's Law': the further away galaxies are, the faster they are receding. To understand this, imagine that our galaxy is a raisin in a lump of dough swelling up in a hot oven; from the point of view of our raisin, all the other raisins will be seen to be getting further and further away as the dough expands in all directions, and, the further away they are, the faster they will be receding, just as Hubble's Law states.

The speed and distance of galaxies can be calculated by measuring the degree to which the light coming from them is shifted towards the red end of the spectrum. Light behaves in a way analogous to the 'Doppler Effect' in sound, familiarly illustrated by the way the noise of a car drops in pitch as it moves away from an observer. In a similar way, as a source of light moves away from an observer, so the light streaming from it increasingly displays redshift. If a light source is coming towards an observer, its light is shifted towards the blue end of the spectrum. Therefore the greater the redshift, the faster the source of light is moving away, and the further away it is.

The Big Bang Theory received powerful support from observations of cosmic background microwave radiation, left over from the universe's earliest history. This observation won the 1978 Nobel Prize in Physics for the two astronomers who made it, Arno Penzias and Robert Woodrow Wilson – who had at first thought that the phenomenon they were observing was caused by bird droppings on their equipment. It is also supported by the observation that the most abundant elements in the universe are helium and hydrogen, just as the Big Bang Theory predicts.

The standard version of the Big Bang Theory requires a consistent mathematical account of the universe's large-scale properties, and the foundation of such an account is a theory of gravity that explains how large structures in the universe interact. This is Einstein's General Theory of Relativity, which gives a description of the curved geometry of space–time and supplies equations describing how gravity works.

The standard cosmological model premises that the universe is homogeneous and isotropic, meaning that the same laws operate everywhere and that we (the observers) do not occupy a special spatial location in it – which entails that the universe looks the same to observers anywhere in it. These assumptions, jointly known as the 'Cosmological Principle', are just that: assumptions, and are of course challengeable – and there are indeed questions about them, not least one that asks how the universe's properties had sufficient time to evolve (especially in the very early history of the universe) to be as they now are. This is known as the 'Horizon Problem'. The Inflationary Model is one answer to this problem, and it has the advantage that it appeals to known laws of physics to explain how the universe's properties arose after the first infinitesimal moments. There are other, less conservative answers, some requiring adjustments to Einstein's equations, or more generally requiring acceptance of the idea that the values of what we now think of as the constants of nature (such as the speed of light) might have been different in the early universe. These less conservative answers are prompted by puzzles that current theory wrestles with. The greatest puzzles concern dark matter and dark energy, of which more below. But even if these puzzles did not exist, others have vexed cosmology since Hubble's observations.

One such puzzle concerns whether the universe will continue to expand forever, or whether gravity will eventually slow down its expansion and then pull it back into an eventual 'Big Crunch' – which perhaps, if the cycle repeats itself endlessly, will be a new Big Bang starting everything over again. The answer depends on the density of the universe. This is estimated by working out the density of our own and nearby galaxies, and extrapolating the figure to the whole universe – which involves the assumption that the universe is everywhere homogeneous, something there might be reason to doubt. This is the *observed density*. The ratio of this density to the *critical density* – the density of the universe that, so it can be calculated, would eventually stop it expanding – is known as *omega**. If omega* is less than or equal to 1, then the universe will expand until it cools to the point of extinction (a *cold death*). If it is greater than 1, it will stop expanding and begin to contract, suffering a catastrophically explosive death in a Big Crunch.

For reasons of theoretical convenience, omega* is assigned the value 1. Measurements achieved by observation suggest that it is about 0.1, which, if right, predicts the continual expansion to a cold death scenario.

Although the Big Bang Theory is the one most widely held by cosmologists, and the one best attested to observationally, it is not uncontroversial. One historical rival to the theory, already mentioned, is the Steady State Theory put forward by Fred Hoyle, Hermann Bondi, and others, which hypothesizes that the universe exists infinitely in the same average density, with new matter being spontaneously generated in galaxies at a rate that equals the rate at which distant objects become unobservable at the edge of the expanding universe. Hoyle and Bondi accepted that the universe must be expanding, because in a static universe stellar energy could not be dispersed and would heat up, eventually destroying the universe. The rate of appearance of new matter required for the steady state need only be very small – just one nucleon per cubic kilometre per year.

Apart from the discovery of the cosmic background radiation, which powerfully supports the Big Bang model, another reason for scepticism about the Steady State Theory is that its assumption that the universe does not change appears to be refuted by the existence of *quasars* (quasi-stellar objects) and radio galaxies only in distant regions of the universe, showing that the earlier universe was different from how it is today. Distance in space is equal to remoteness in past time; to look at far objects in space is to see into the history of the universe; if it was different in the past, it is not in a steady state.

There are a number of other rivals to the Big Bang Theory, in the form of alternative models: 'Plasma Cosmology', the 'Ekpyrotic Model', 'Subquantum Kinetics Cosmology', and others. These proposed alternatives have different degrees of plausibility.

Plasma Cosmology was suggested in the 1960s by the physicist Hannes Alfvén, who won the Nobel Prize for his work on plasmas. He suggested that electromagnetism is as important as gravity at cosmological scales, and that galaxies are formed by its effect on plasmas. This idea has been revived in connection with the dark matter problem.

The Ekpyrotic (from the Greek 'out of fire') Model is suggested by String Theory and supports the idea of an endlessly cyclical universe. It hypothesizes that the original hot expansive beginning of the universe was caused by the collision of two three-dimensional precursor universes propagating along an additional dimension. The two universes mingle, their energy converting to the particles (quarks and leptons) of the three-dimensional present universe. The two precursor universes collide at every point almost simultaneously, but the occasional point of non-simultaneity gives rise to the temperature variations in the background microwave radiation and the formation of galaxies.

Subquantum Kinetics Cosmology offers a unified field theory drawn from the way non-equilibrium reaction systems generate self-organizing wave patterns, hypothesized as giving rise to matter continuously in the universe – a sophisticated version of the Steady State Theory.

These competitors of the Big Bang model are motivated by efforts to address its problems or answer criticisms of it. Among the criticisms it faces are these. It has to adjust parameters, such as the cosmic deceleration parameter or those that relate to the relative abundance of elements in the universe, to conform to observation. It has to explain why the cosmic microwave background radiation temperature is the residuum of the heat of the Big Bang rather than the warming of space effected by radiation from stars. It has to account for the fact that the universe has too much large-scale structure to have formed in just 13 to 14 billion years, thus needing the Inflationary Model to render consistent, ad hoc and untestably, the apparent age of the universe and the greater age needed for the formation of its structures. A particular example is that the age of some globular clusters appears to be greater than the calculated age of the universe. Some observers claim that the most distant – and by hypothesis therefore the oldest – galaxies in the universe, those in the 'Hubble Deep Field', show a level of evolution out of keeping with their supposed age.

And, perhaps most puzzling of all, the Big Bang Theory requires that we accept that we know nothing about 95 per cent of the universe, which has to take the form of dark matter and dark energy; dark matter accounting for the distribution and relationships of observed galaxies and galaxy clusters, and the other mysterious ingredient – dark energy – pushing the universe apart at an increasing rate, the speeding-up only seeming to have begun about halfway through the universe's known life to date.

Nevertheless: as the foregoing shows, both quantum theory and cosmology individually have, despite the open questions and puzzles that attend them, strong foundations and powerful support by experiment and application. It is therefore a vexing matter that they appear to be irreconcilable, thus galvanizing the search for a unifying theory that combines an understanding of the large-scale phenomena of gravity with theories about the structure and properties of the quantum level. This has proved exceedingly difficult to do, but one of the most promising-seeming ways of achieving it is String Theory, a first version of which was mooted in the early 1980s.

String Theory postulates the existence of minuscule vibrating string-like strands and loops from whose vibrations the phenomena of gravity and the elementary particles alike arise. String Theory succeeds in this remarkable unification, first, by proposing that there are nine spatial dimensions, six of them curled up so minutely that they are undetectable; and, second, by bringing certain other assumptions to bear, among them that there is an unchanging background geometry, and that the cosmological constant – the degree of energy in the universe hypothesized by Einstein as counteracting the gravitational pull of the universe's mass – is zero.

The mathematics describing strings and their behaviour is beautiful, and the laws required to govern string behaviour are elegant and simple. These facts, together with the power of the theory to achieve the grail of unification (in 'Supersymmetric' versions – see below – the theory unifies all the matter and force particles, the fermions and bosons), are attractive reasons for thinking, as some of its proponents do, that 'it *must* be true'.

A little more detail explains why. Relativity theory offers powerful insights into the universe, and other theories – about the Big Bang, the evolution of galaxies, stars and planets, black holes, gravitational lensing effects, planetary orbits, and much besides – depend upon it. Quantum theory plays no part in this; the universe on the large scale is regarded as a purely classical domain. By the same token, quantum theory works very well as a description of the microscopic realm, where gravity is ignored. If a way were found of connecting the two theories, almost certainly into a third theory that embraces both, it would require that there be a particle to carry the gravitational force, a *graviton*, with a particular property, namely, zero mass and two units of spin. This idea was common property in physics for quite some time but only as a proposal, on the grounds that the mathematics of adding gravitons into the quantum mix simply did not work. Particles can interact at zero distance, but the effort to make gravitons do so – as they would need to – gives rise to mathematical nonsense; infinities appear in the calculations.

Then in the early 1980s a happy coincidence suggested a way forward. Strings were proposed initially in an effort to explain the relationship between spin and mass in hadrons, the composite elementary particles – protons and neutrons – consisting of combinations of quarks. The idea did not work well, and an alternative called 'Quantum Chromodynamics' proved more successful. But viewing particles as excitations of strings allowed for a particle with zero mass and two units of spin, and further allowed that interactions between particles could be spread out in a way that unscrambled the mathematics of interactions between such a particle and others. This was a hallelujah discovery, and the accompanying hallelujah thought was that String Theory might constitute the long-sought and vastly desired theory of quantum gravity at last.

Some theorists, however, voice deep concerns about String Theory, chief among the first being that there is no complete formulation of it, and that no one has proposed its basic principles or specified what its main equations should be. Worst of all for a scientific theory, given that science lives and dies on this crucial point, String Theory makes no directly testable predictions, because the number of possible interpretations of it is so large. Indeed string theorists talk of an indefinitely large 'landscape' of many possible string theories. To the dismay of its critics this last fact has led some of String Theory's senior proponents to claim that experimental verification of the theory is not necessary – the sheer beauty of the mathematics in which the theory is expressed, they say, is convincing enough by itself.

Other defenders of String Theory also appeal to the 'Anthropic Principle' – the brute fact that the fundamental constants of physics and chemistry are fine-tuned in just such a way as to produce and sustain the life that exists on our planet – as a way out of the difficulty that otherwise no single version of the theory's many possible versions presents itself as uniquely right.

Because of the importance of fundamental physics and the apparent power of String Theory within it, it matters that the question of testability should lie at the centre of debate about it. This is what gives critics their greatest concern, because it reaches beyond questions about the theory itself to the very basis of scientific culture. And this is so even if one accepts that there is no single correct methodology that applies across all branches of science, for the one thing that binds them together as sciences is answerability to test and conformity to nature. Together these are assumed as the *sine qua non* that anything properly describable as science must observe.

This is where the Criteria Problem bites. Are simplicity, elegance and beauty sufficient as criteria to justify adoption of a theory? When two or more empirically adequate theories conflict, such 'extra-theoretical criteria' might be invoked to decide between them. Doing so is controversial enough in such a case; but if 'beauty' and the others are the sole reason for choosing a theory, what justification can be offered? For recall that the judgement of beauty will be made on the basis of our human reactions to how things appear at the pinhole through which we try to see nature.

If there is something that might count as a test of String Theory, it is the possibility that empirical work might show that the speed of light has varied during the universe's history. Anything that shows that the General Theory of Relativity might need adjustment would call String Theory into question, for the theory assumes that the General Theory is correct.

Another possibility is that very high-energy particle colliders, such as the one at CERN in Switzerland, might discover the *super-symmetric* partners of currently known particles, which would provide experimental evidence for the family of string theories that hypothesize that every boson has a fermion partner (this pairing is known as supersymmetry). The proposed supersymmetric partners of known particles have not been detected by current experimental means; very high-energy colliders – reaching energies that CERN cannot yet attain, and perhaps might not be able to attain – are needed to produce them if they exist.

So String Theory could be indirectly testable after all, in that it could be undermined or supported by these means, even though by itself the theory says little that is subject to direct experimental scrutiny.

But to sceptics about String Theory, it also matters that physics should welcome and encourage a variety of other approaches to the five fundamental problems facing physics, only one of which – the unification problem – is addressed by String Theory. The other problems are: first, the need to make sense of quantum mechanics itself, which is full of unresolved puzzles and anomalies; the quantum world appears to us a strange place, and its oddity suggests that something more fundamental waits to be discovered. The second, related, problem is the need to determine whether all the particles and forces of the Standard Model can be understood in terms of a more inclusive theory that describes them as manifestations of that hypothesized deeper reality. The third is to explain why the values of the free constants of nature – the numbers describing (for example) the masses of quarks and the strengths of the forces binding the atom – are as they are. And the fourth is to come up with an account of two profoundly puzzling phenomena that recent astronomical observations seem to reveal: the already mentioned dark matter and dark energy.

The approaches that might help with some or all five of these major challenges include 'Loop Quantum Gravity' (LQG), 'Doubly Special Relativity' (DSR), and 'Modified Newtonian Dynamics' (MOND). Unlike String Theory, they all make directly testable predictions, and if wrong can be shown to be so.

LQG suggests quantizing gravitational fields (*quantizing* means restricting the permissible number of values a variable can take) at the Planck Scale into *spin networks*, thus reconciling gravity with the fact that the structure of matter is itself quantized. One advantage of LQG over String Theory is that it does not require extra spatial dimensions.

DSR adds to the invariability of the speed of light the concept of the invariability of the values of 'Planck Length' and 'Planck Energy'.

This helps to deal with the problem that different observers can disagree about the value of the Planck Energy in their respective inertial frames, making it hard to write 'transformation laws' between their frames, and thus hard to reconcile gravity with quantum theory.

MOND offers an alternative explanation of why stars in galaxies have higher velocities than Newtonian mechanics would predict. It is an alternative to dark matter as an explanation of this phenomenon, and involves adjusting the relationship between a star and its galaxy's centre of mass, for example, by saying that the gravitational force varies in strength with the radius, rather than the square of the radius, of its orbit.

The testability matter is the key question for critics of String Theory, and it is why they contrast it unfavourably with these other possible approaches, and characterize String Theory itself as metaphysics (in the pejorative sense of this term) rather than physics.

It is a correlative of the tenet that experiment and observation lie at the heart of science that ideas that are consistent with everything – positive and negative – that nature says about them are empty; ideas that cannot be tested against nature lie under suspicion until some way of subjecting them to empirical interrogation is found. This commitment is the bottom line, the stubborn requirement, in science: that there has to be anchorage in repeatable experimental data.

A moment's reflection prompts a question: what does this commitment itself mean? This question presses throughout the next section.

4.   Through the Pinhole

The question asked at the end of the preceding section concerns what we are to say when enquiry in science has gone beyond the current limits of experiment, and is afloat in an ocean of possibilities too deep for such anchorage. String Theory and the Many Worlds Theory are typically cited as examples of speculation in physics, and with them such ideas as that the universe is a hologram, or consists of information, or exists in the precise form it takes so that intelligent beings can exist (the 'Anthropic Cosmological Principle'), which cannot be subjected to experimental test. A hard-line experimentalist might ask: are these speculations science at all? Are they not something different altogether – namely, metaphysics?

A first step to considering this question is to draw a distinction between hypotheses for which, as a practical matter, experiment cannot (yet) be done, because of cost or size constraints, and those speculations for which it is in principle impossible to devise a test. As a matter of scientific policy it is questionable whether we should ever say that experiment is 'in principle impossible'; human ingenuity, discoveries elsewhere in science, empirical demolition of parts of currently accepted theory that underlie the speculations in question, and more, could change things. And, in any case, to say that it is *in principle impossible* to do such and such is a form of defeatism, to which aspiration is a far better alternative.

A second step is to note that the key to worthwhile speculation – using this word to denote the formulation of adventurous or apparently wild-seeming hypotheses – that cannot (or cannot yet) be subjected to experimental test, is the nature of its connection to already tested theory. On this basis an argument can be constructed as follows.

Disciplined inference from current theory into the landscape of possibilities that it suggests is more than acceptable; it is necessary.

That is how new lines of enquiry are opened. When a field of enquiry has reached an experimental limit, the number of possible ways forward increases, and with them the role of imagination – which is what is happening in the fields mentioned above. But if the speculation is to some degree constrained by keeping in sight of current theory, if the imagination is disciplined, if as far as possible it 'saves the appearances' of current theory as Aristotle long ago demanded of all hypothesis-formulation, and does not premise itself on notions that have no plausibility from current perspectives, then it merits the name of science.

A proponent of such an argument can acknowledge that new departures might end by revising current science, or might arrive at a point very far from present theory, perhaps overthrowing it altogether. The idea that there should always be a plausibility-endowing link to current understanding is not intended to make any of this impossible; it is rather that the conceptual journey to a point beyond current theory has to be linked to the latter by explicable steps. We see (in the usual case) how theory evolves and adapts, or why new evidence and better reasons may require replacing an existing paradigm with a new one.

If the experiments at CERN had not detected a particle at the predicted Higgs energy level, thus completing the current Standard Model of the atom, an exciting new quest to understand how elementary particles acquire mass would have begun. There were various speculations on this head before empirical confirmation of the Higgs happened. But none of them involved ideas wholly disconnected from what experiment had so far revealed about atomic structure. No one was tempted to explain why electrons are far less massive than nuclear particles by speculating that (say) large gnomes sit on protons and neutrons and little gnomes sit on electrons.

Again, the beautiful mathematics that suggests a vast landscape of ideas about how the four forces of nature might be reducible to a single force – the string theories – is not mere fancy; the speculations have their roots (somewhat distantly, it is true) in the Standard Model. Likewise puzzles about dark matter and dark energy, why the constants of nature have the values they do, what sense to make of the collapse of the wave function that transforms a range of probabilities into a definite value – consider the Many Worlds Theory's answer to this – themselves arise from theories that have been tested to exquisite levels of precision, and that underlie so many of the technologies we use constantly: the transistors in our smartphones, their LED screens, the electricity supply to our homes from nuclear power stations, and much more. Most proponents of these views are committed to holding that experiment and observation, even if indirect and requiring extremely subtle detective work, are not ruled out *in principle*. It is this all-important fact that would prevent them from being mere science fiction or fantasy but instead legitimate parts of science proper.

This argument has much to recommend it. But it does not lay to rest the various thoughts prompted by the problems that beset the acquisition of knowledge on its frontiers: the Ptolemy, Pinhole, Metaphor, Map, Criteria, and Truth problems.

The Ptolemy Problem concerns the fact that the efficacy of a given theory for a given range of purposes is not a guarantee of its truth. The Ptolemaic view of the universe has Earth at its centre and is ingeniously devised to explain the motion of the planets – *planetoi* is Greek for 'wanderers' – which appear to divagate across the face of the fixed stars in apparently inconstant ways, sometimes going backwards (retrograde motion) and sometimes appearing nearer and sometimes further away. Ptolemy's model is an elaborate scheme in which the planets move in epicycles, or small orbits, around a deferent or larger orbit centred upon a point, the 'eccentric', lying between the planet in question and the Earth. The fixed stars are attached to a celestial sphere beyond the planetary sphere. This scheme works pretty well for predicting positions of the planets and eclipses, and for navigating the oceans; but it is purely instrumental within its limited range, and does not provide what current astronomy would regard as a description of the 'true' positions and movements of planets and stars.

The moral is that just because a theory 'works' does not mean that it is true. We generally assume otherwise: if a vaccine protects against infection, the theory underlying vaccination – that an inert version of a pathogen can stimulate the immune system with readiness to deal with a live version of it – seems thereby proved right. But it remains that, although empirical support for a theory raises the probability that it is correct, even a very high level of probability leaves in place a possibility that it is wrong. This is one of the reasons why science is regarded, as a methodological principle, as defeasible, and why the degree of probability that the outcomes of experiment are not the result of error or some other factor has to be very high.

What is the status of the Ptolemy Problem as a potential subverter of our aspirations to know? The most convincing answer is that, while it may strike a cautionary note, it does not come near to derailing the quest for knowledge. This applies without qualification to *knowledge how*, where in any case the aim is practicality, not explanation. The great body of technology is indifferent to the Ptolemy Problem, because the effectiveness of a technique or device is indifferent to what choice is made of explanations for it. Of course, understanding the reason why something works is both interesting in itself and a potential source of improvements and analogous applications; but the lever and fulcrum was being used to move heavy objects long before Archimedes explained the underlying principles.

However, the significance of the Ptolemy consideration for the point at issue – whether nothing can be science that does not admit of experimental test – is that outcomes of experimental tests that corroborate hypotheses do not *settle* matters either way. The philosopher of science Karl Popper introduced the notion of 'falsification' in response to the observation that a positive experimental outcome is not proof of truth, to argue that the best one can aim for is refutation of an hypothesis by showing that what it predicts does not happen. But, in response, it was argued that a result contrary to prediction no more refutes an hypothesis than a result in line with prediction confirms it. The respective cases add to or subtract from confidence in the hypothesis, yes; but they do not settle matters outright. Often it is the case that other factors – internal coherence; consistency with existing theory; simplicity; the beauty of the mathematics in play – enhance the degree of plausibility of a positive experimental outcome, or, if they are lacking alongside a negative experimental outcome, tell against the unlucky hypothesis the more.

What does this mean for hypotheses for which no experimental investigation is possible? Does the fact that such investigation would anyway not settle matters mean that the other factors – simplicity, coherence, consistency with already accepted theory, beautiful mathematics – by themselves might be enough? To say this is to decouple science from the *absolute* requirement of empirical test.

What is at issue here is highly relevant to the Pinhole Problem: namely, that the starting point in all our enquiries is the limited and highly circumscribed nature of information available to us locally in space and time and from our finite point of view, allowing us a view of the universe (and the past) as if through a pinhole positioned at just our restricted scale. The question asked in the Introduction is: do our methods successfully carry us through and beyond the pinhole?

It is helpful to have a sense of scale to understand the nature of the problem. A first shot at nominating the smallest and largest scales of the universe would respectively cite the Planck Length, 1.6×10^{-35} metres, and the distance from Earth to the edge of the observable universe, 4.4×10^{26} metres (46.5 billion light years = 14.26 gigaparsecs). Some like to point out that this puts us on planet Earth about halfway between the size of an atom and the size of the universe, and some of those some like to find a special significance in that thought; but the more interesting consideration is to speculate whether these smallest and largest measures are (so to speak) 'as far as we can reach' with theory and observation, given our limitations as we squint through the pinhole. That is, might whatever is truly fundamental to reality be vastly smaller than the Planck Length, or is 'length' an inapplicable, inappropriate concept altogether in that connection? And, on the other hand, might the universe be vastly larger than what we observe, despite the mathematics of the Big Bang Theory, for example, by being one bubble in a multiverse, with other bubble universes varying in the physical laws that operate within them?

The motivation for such questions lies in part in the fact that what we strive for in the way of testable, empirical support for theory in science consists in what we can achieve in extending the reach of our powers of enquiry – and ultimately that means our powers of observation. In its turn, observation is effected by means of instruments:

telescopes of various kinds (gathering light, radio waves, gamma rays; on mountains, in space), microscopes (optical, electron, and others), oscilloscopes and other measuring and detecting devices all the way up to the Large Hadron Collider at CERN. They are connected to our unaided powers of empirical access to nature – our five senses, and chiefly vision, hearing, and touch – and our mental powers of inference, comparison, analysis, and comprehension, as in effect their amplifiers and potentiators; but they are connected to them, in the final analysis inescapably so. Is this ultimate anchorage in our cognitive endowment of the demand for empirical check on theory a limiting factor, even a distorting one, because we only allow ourselves on principle to accept what our highly local and constrained powers permit us to investigate?

A thought might be that mathematics is another instrument – this time the instrument of reason, which extends the reach of enquiry into realms that physical instruments of observation cannot reach. Just this thought underlies the idea that the beauty of the mathematics in which (for a significant example) String Theory is expressed is such that experimental verification employing the other 'ways of seeing' is not required; on this view, mathematical beauty is physical truth.

The striving to overcome limits imposed by our starting point at the pinhole results in great ingenuity in experiments and the design of instruments that carry them out. This reduces one aspect of the problem at least, for, although the nature of experience and how it is conceptualized are inescapably human constructs – that is, functions of human cognitive capacities – it does not follow that science is a matter of mere subjectivity, because intersubjective and cooperative endeavour eliminates the *merely* subjective, as it were by triangulation. The mark of successful enquiry and its theoretical and experimental potentiations is convergence: whenever an experiment can be replicated, the hypothesis it tests – as acknowledged above – thereby gains support. But one line of thought about this will insist that this is still in the end an outcome of the human mind applying itself to understand what it encounters, and the human mind is bringing to the task its character and its limitations.

To dramatize this, consider the following two sets of considerations side by side: how the Planck quantities are arrived at, and the question of the relation between cognition and its objects.

The Planck Length (*Pl*) is found by multiplying the constant of gravitation (*G*) by the reduced Planck Constant (*ℏ*), and dividing the result by the cube of the speed of light. This latter value is very large, which is why *Pl* is very small – the smallest length that can be determined by combining the constants of nature in this way. Note that 'constant' really means what it says: the speed of light, *c*, is fixed in all circumstances. Because *c* is constant, other quantities have to change in relation to it: space contracts and time expands for someone travelling at large fractions of the speed of light; the faster you go, the more you shrink and the more slowly your clock ticks. This is the theory of relativity in action.

Divide *Pl* by *c* and you get Planck Time, 5.4×10^{-44} seconds. There is also the 'Planck Mass', which by the same kind of algebra turns out to be very large, 10^{19} times greater than the mass of a proton. Recall, as a parallel, Einstein's famous formula $E = mc^2$; if you multiply the tiny-seeming mass of a subatomic particle by c^2, you get an enormous amount of energy. Hence atom bombs.[1]

Now, our sceptic about the concepts we use to squeeze our understanding through the pinhole might ask, 'The Planck dimensions are derived from the constants of nature; what happens to Planck Length and Planck Time at large fractions of the speed of light – do they contract and dilate respectively? If, as products of the interrelationship of the constants, they do not, how is this consistent with what is happening at larger scales?' We are very far indeed from being able to observe effects on the Planck dimensions by experimental means, but in any case it is a question whether, at the ultra-extreme scales involved, the concepts in play even make sense. At these scales, quantum theory and general relativity meet (a motivating factor for Loop Quantum Gravity (LQG) Theory, as mentioned above). A point to note is that the values of the constants, as parameters in descriptions of the properties of elementary particles, yield very precise results in the theory known as 'Quantum Electrodynamics' (QED). The degree of precision was once illustrated by Richard Feynman, when he said that if he were asked how far it is to the Moon, he could answer, 'Do you mean from my head or from my feet?' and in his book *QED: The Strange Theory of Light and Matter* he remarks that by its means one could arrive at a measurement of the distance between New York and Los Angeles 'that would be exact to the thickness of a human hair'. He won the Nobel Prize for his work on Quantum Electrodynamics.[2]

Now put alongside these points the following considerations about human cognitive capacities. To give an initial intuitive clarification of questions about the relation between cognition and its objects, note something instructive about investigation in the social sciences. Research in these fields has to be conscious of spoilers such as the 'Hawthorne Effect', in which subjects who are being studied alter their behaviour because they know that they are being studied, or the 'Observer-expectancy Effect', in which the researcher, usually without realizing it, influences the behaviour of the subjects being studied. The most familiar problem is the researcher's own unconscious biases in interpreting what he sees in the data. Here the researcher, or the research situation, interferes with what is being studied, so that the phenomena are not being seen for what they are in themselves, but instead are phenomena as they are when being studied. Think of observing chimpanzees in the jungle: is one observing chimpanzees, or is one observing chimpanzees that are under observation? Would they behave that way if you were not looking at them?

In the natural sciences analogous effects can occur. The act of observation, and the presence and use of observational equipment, cannot be excluded from consideration of the phenomena observed. These are the so-called Observer and Probe effects respectively. To what extent does the preparation of a specimen for microscopic analysis – freezing, slicing, fixing, staining, mounting, squashing, smearing, whatever methods are used – interfere with what is seen? Does the difference between a cell and a cell prepared for microscopy admit of successful discounting? The answer is, on the whole, yes, though doubtless persistent sceptics will seek to persist in their scepticism.

In the theory of knowledge the question of the nature and structure of cognition and its relationship with its targets is an important and much discussed matter. Take the simplest case: looking around one's room. One can describe this familiar proceeding by invoking the neurology of visual perception – light reflected from the surfaces of objects falls on the lens of the eye, which focuses it on the retina, whose constitutive rods and cones are activated and stimulate impulses along the optic nerves to the primary visual cortex in the occipital region of the brain. One can also describe this in terms of the psychology of perception, in which one does not simply see but always *sees as* – that is, the act of seeing is intrinsically interpretative, because the incoming sensory stimulation is, in the very act of arriving, subsumed under conceptual categories that tell the seer (whether consciously or unconsciously) what it is that is being seen, or that at least offer a theory of what is being seen. This involves memory, a grid of sortal concepts about the kinds of things encountered in visual experience from which selections are made to classify the thing in question, and a significant amount of computation about (variously) the status, behaviour, qualities and their significance, intentions and motives if the thing seen is an agent of some kind – and so on. An ordinary act of seeing is, in this way, very complex, and it involves a great deal of mental activity, consisting in the application of a rich apparatus of concepts and computational capacities. 'Seeing' is thus something far beyond the mere irradiation of the retina and stimulation of the visual pathway.

Not only is perception intrinsically interpretative; it is also selective in its focus. This is what earns magicians their living. It functions to provide data useful in the given circumstances (at the bottom of this explanatory pile one would say, 'Useful for purposes of survival and reproduction'). The interpretations made of the data chosen as significant, excluding other possible data, might be thought of as coming in two broad types. One is the evolutionary endowment of software for the perceptual and reasoning systems; the other is culturally and experientially acquired software. Between them they have a *constitutive* function: they build a world with the perceiver at its spatial, temporal, and meaning-giving point of origin. Indeed it is literally correct to say that they build a *virtual reality* with the perceiver at its point of origin. It is easy to demonstrate this. Vision consists in electrochemical patterns of activity in parts of the occipital and temporal regions of the brain. These activations give the perceiver the illusion of looking out through her eyes at a world arrayed in a three-dimensional space beyond her head. Correlation of this with the proprioceptive and tactual experiences generated by reaching out to touch an object in that space reinforces the illusion. But in another sense it is not an illusion: it *is* the reality of perceived space and movement. A good analogy might be the relation between the icons on the screen of one's laptop and the activity in the interior of the laptop that actually instantiates the pattern of information that the icon represents. The world of perceptual experience is, so to speak, iconic.

Vary the language of this account – now standard in cognitive neuroscience – only somewhat, and you have almost exactly the view expressed in Kant's *Critique of Pure Reason* about our experience of the apparent world. His argument likewise was that how the world appears to us is a construct of our cognitive capacities.[3] Though this is now a commonplace of thinking about perceptual cognition, his view carries an additional philosophical punch – which is that the world we perceive is inextricably conditioned by the manner of our perceiving it. This remains so even when we extend the power and range of inputs by means of instruments: the ultimate destination of the inputs gathered by the instruments is the cognitive reception we give them. Incoming information activates cognitive structures that organize and interpret what – so those structures have it – this input is telling us about what is out there, whether in the room, in intergalactic space, or in the inner structure of the atom.

Of course, the structures in question are not just the basic infant's visual software. Far from it: there is a sophisticated acquired apparatus of cognitive structures, now additionally consisting in the theories of physics and competency with mathematics. An experimental physicist at CERN's Compact Muon Solenoid experiment 'sees' something much more than an array of traces on a screen, because her interpretative equipment is richly conformed to recognize what they mean. But note that: *what they mean*. Interpretation *is* experience. In the case of the CERN physicist, the innate endowment of cognitive structure is enhanced by acquired structure.

The deeper aspect of the philosophical punch in Kant's view is that if we were not equipped with the cognitive apparatus that obliges us to interpret incoming data as arrayed in space and sequenced in time – with space and time implicitly regarded, in everyday experience, as Newtonian absolutes – we would simply not have experience of the ordinary everyday kind. In Kant's terms, possession of this cognitive architecture is required for the very possibility of experience, because it is its constitutive structural framework. And this in turn means that the world of our pre-primed expectations is a classical world from which it is very hard to escape – a factor in the perplexity aroused by the Measurement Problem.

This perplexity can be summarized as the apparently insurmountable difficulty of interpreting quantum reality in terms of classical reality. The efforts to make sense of quantum reality – many worlds, hidden variables, Copenhagen instrumentalism – are in whole or part motivated by the desire to have quantum theory make sense to classically conditioned thought. The assumption that the world, the classical world, of familiar experience is the *real* world – or the upper surface, so to speak, of the real world, continuous with whatever its underlying microstructure is – is shown by the Kant-style entailments of the psychology of perception to be cognitively inescapable. And in this lies what could be a major clue.

It should be mentioned that the Many Worlds Theory is one solution to the Measurement Problem in quantum physics and not to be confused with the idea of the 'multiverse', which denotes the proposal in cosmology that the universe could consist of many regions, perhaps an infinite number, with different physical laws at work in them. Multiverse theories are developments of Alan Guth's inflationary model of the universe's very early history, postulated to deal with the 'Horizon Problem' in cosmology: the problem of how the universe could contain what it does, given that there appears to have been too little time since the Big Bang for it to get that way.

The difficulty experienced in efforts to interpret quantum theory from the standpoint of what makes classical sense to us could be – I suggest – the result of having matters the wrong way round. It is not the world hinted at by science's efforts to see through the pinhole that is a 'merely theoretical' world, but rather it is the classical world we occupy that is the theoretical world – a world configured by us for our own convenience, given the scale at which we exist, in a very narrow band at a point on the gradient of scales between the very small and the very large. We organize the data that our sensory systems are equipped to detect – a section of the electromagnetic spectrum – into a world of causally interactive objects, such as apples and trees, human bodies, and trucks. These are collections of quantum events that interact with the quantum events constituting what we describe as our nervous systems, and our nervous systems organize what they are undergoing into experience of a world according to the innate and acquired organizational principles we bring to bear in interpreting the excitations of our nervous systems.[4]

A good analogy is provided by socially constructed institutions, such as 'the government', 'parliament', 'the health service', 'the army'. Although there are physical entities associated with each of these constructions, such as people and buildings, filing cabinets and rocket-launchers, they are themselves agreed-upon ideal ('made of ideas') entities, such that if the great majority of us decided to stop thinking and acting as if there were such a thing as (say) parliament, it would simply cease to be. While it exists, it is as 'real' as a mountain and can affect our lives; we cannot individually wish it away or change it without taking significant action; but it is a construct nevertheless, based upon our agreements, both tacit and explicit, to treat it as existing and causally efficacious.

There is of course a difference between a mountain and a social institution, though both are ideal – that is, constructs – in the intended sense. But a mountain is constructed by organizing input data, while a social institution is a projection, in essence, though not in effect, as fictive as a character in a novel.

Another way to put the point is to say that we carve up the fundamental data, conceived in the most general terms as events or interactions between events, according to our needs and interests at our scale. Imagine someone brought up in a jungle who had never been in a library, suddenly teletransported into the stacks of the Library of Congress. Why would she see the books as individual entities rather than a whole shelf of them as a single, multicoloured, snake-like thing? How she carves up the domain might change as she interacts with it, finding it more convenient to treat books as individual things rather than treating whole, book-filled bookshelves as a single thing. But until needs and interests introduce refinements into how she determines the optimum degree of pluralism for her ontology (theory of what exists), the events that we think of as interactions between those constituting her nervous system and those impinging on it have no *necessary* division into books and bookshelves, indeed no *necessary* division even into her and what is not her. The sorting principle, from the point of view of a given system of events capable of representing other events to itself, is, in short, *utility* – the convenience of creatures of our size and constitution. The question of what the representing system of events itself is – fundamentally, a brain and its functions – is another and further question: see Part III below.

This suggestion amounts to saying that the perplexity induced by the Measurement Problem is an artefact of the Pinhole Problem – the fact that we are the kind of creature we are, constituted to deal with existence at the scale we occupy relative to the rest of the universe. We are not natively endowed with the cognitive structures that make superpositions of quantum states seem natural. Our cognitive capacities organize the world into causally interactive spatio-temporal individuals with determinate properties. To think that *this* familiar world we see around us *is reality*, that *this* is the benchmark for what reality is really like in itself, is the source of the mystery. Of course, this world is indeed part of, or at least connected to, reality; in some cases it is an aspect or sliver of reality, as a single molecule of H_2O is part of the ocean, and in other cases it is the ideal constructs we devise to help us to manage our intercourse with reality, rather as we use the idealized lines of latitude and longitude to map and navigate the planet.

There are broadly two ways that our being confined to squinting through the pinhole might make quantum phenomena look weird, as our classical preconceptions lead us to calibrate weirdness. Either we are nowhere near really grasping what the fundamental nature of reality is – our theories are incomplete, perhaps even wrong; unlikely, given the success of their applications, *pace* the Ptolemy Problem – or the cognitive architecture of our thought imposes conceptions of order, causality, linearity, consistency, monotonicity, uniformity, predictability, and so on, which are as useful to conceptualizing quantum phenomena as it would be to apply dog-grooming techniques to solving quadratic equations.

This differs from an instrumentalist view in that the latter is in effect a version of 'just calculate', leaving the question of what is real to one side. The Pinhole Problem identifies something definite and factual: that we occupy a very narrow band of the scale of things, and our cognitive equipment has evolved to deal with that narrow band effectively; and that therefore our cognitive equipment – by means of what it forces us to think and makes us wish to believe – interferes with our reception of, and ability to organize, data unconformable to it. That is why the Measurement Problem appears so problematic: we are not configured to think that way naturally.

But we *are* able to think that way mathematically. This brings into view another large and interesting question: what Eugene Wigner called 'the unreasonable effectiveness of mathematics', 'the miracle of the appropriateness of the language of mathematics for the formulation of the laws of physics'.[5] The mathematics of theoretical physics and cosmology often predicts the existence and nature of phenomena that experiment then proceeds to find. A classic case is Paul Dirac's postulation of the existence of antimatter particles (see below). A familiar example of mathematical properties evident in nature is the 'Fibonacci Sequence' and the related 'Golden Ratio', 'Golden Spiral', and 'Golden Angle' phenomena, seen for example in the arrangements of the flowers, leaves, and fruits of many plants. The sequence is named after Leonardo Fibonacci – 'son of Bonacci' – of Pisa, who introduced Europe to Indo-Arabic numerals in his book *Liber abaci* (*Book of Calculation*, 1202), written for merchants to help them with profits, prices, and currency exchange. In it he showed that the pattern exemplified by the way rabbit populations increase is a sequence in which the successive numbers are the sums of their two preceding numbers: 1, 1, 2, 3, 5, 8, 13, 21, 34, 55, etc., thus: 1 + 1 = 2, 1 + 2 = 3, 2 + 3 = 5, 3 + 5 = 8, etc.

The flowers of sunflowers, daisies, cauliflowers, and broccoli exemplify Fibonacci patterns; the number of petals in various flower types come in Fibonacci quantities: lilies and irises 3, some daisies 13, chicory 21, Asteraceae (sunflowers, other daisies, asters, many others) 34, 55, or 89. More generally, natural forms in plants, animals, and minerals display symmetries and patterns, many of which are close to exemplifying the Golden Ratio or the geometries and spirals one finds in snail shells, beehives, the skin of a pineapple, even the diving swoop of a bird of prey. In the case of snail and seashells the spiral forms are not pure Golden Spirals but close; symmetries are found in crystals such as snowflakes, and five-pointed-star shapes abound in nature (the pericarp of an apple core, the arms of most types of star-fish), constructed by nature on the model of the regular pentagon, which exemplifies the Golden Ratio in the intersection of its diag-onals. Biomathematics, geology, physics, and astrophysics provide many such examples.

The Golden Ratio, symbolized as ϕ (phi) is found by cutting a line into two unequal lengths such that the result of dividing the longer segment by the shorter segment is the same as dividing the whole line by the longer segment. Thus: in the line AB there is a point Z closer to B such that, when AZ is divided by ZB, the result is equal to AB divided by AZ. The ratio of $AZ:ZB$ is 1.61803398874 . . . (the dots showing that this is an irrational number like pi; it does not termin-ate or recur). The arithmetic underlying this ratio and its siblings – the 'Golden' figures mentioned – or versions or approximations of them, is promiscuously littered about nature, and we often find their mani-festations very beautiful.

The equations of physics appear to represent mathematics in the secret heart of nature that is even more precise and fundamental than many of these manifest examples in biology and mineralogy. This is why it seems so 'unreasonably effective' or even 'miraculous' in its aptness for describing nature and predicting its properties and behav-iour. Are there ways to address this apparent 'unreasonable effectiveness' without resorting to some form of magical thinking – perhaps by the unexplanatory move of appealing to a deity as the mathematical author of nature (unexplanatory because doing this simply pushes the 'mystery', if it is one, into the deeper darkness of another and even less well defined 'mystery')? Several have been offered, well summarized by R. W. Hamming.[6]

One is the familiar point captured in the saying 'If your tool is a hammer, everything looks like a nail.' Hamming puts it by saying, 'We see what we look for.' The suggestion is that use of mathematical techniques shapes, rather than reflects, what we take nature to be, giving as examples the inverse square law – 'If you believe in anything like the conservation of energy and think that we live in a three-dimensional Euclidean space, then how else could a symmetric central-force field fall off?' – and the Uncertainty Principle, which prompts the question 'Why should I do all of the analysis in terms of Fourier integrals? Why are they the natural tools for the problem?'[7] He also cites the distribution of the physical constants, 60 per cent of which have the numbers 1, 2, and 3 as their leading digit, all the remainder up to 9 occurring only 40 per cent of the time, a phenomenon described in 'Benford's Law', which applies most accurately when data range over several orders of magnitude.[8]

The second reason he identifies is that 'We select the kind of mathematics we use. Mathematics does not always work. When we found that scalars did not work for forces, we invented a new mathematics, vectors. And going further we invented tensors . . . we select the mathematics to fit the situation, and it is simply not true that the same mathematics works every place.'[9] A *scalar* is a number identifying a point on a scale, such as temperature. A *vector* is a magnitude and direction combined, for example, velocity, which is speed plus direction. *Tensors* are generalizations of vectors; roughly put, they describe transformations that occur when the vector bases change.

The third reason Hamming considers is that mathematics is not a magic wand for explaining everything, for there is much it does not, and almost certainly cannot, explain; he gives 'truth', 'beauty', and 'justice' as examples. And finally he cites the idea that perhaps there are some things we are simply not equipped to understand, given the limitations on our cognitive powers: 'Just as there are odors that dogs can smell and we cannot, as well as sounds that dogs can hear and we cannot, so too there are wavelengths of light we cannot see and flavors we cannot taste. Why then, given our brains [are] wired the way they are, does the remark, "Perhaps there are thoughts we cannot think," surprise you?'[10]

But he is not in fact persuaded by these considerations. 'I am forced to conclude both that mathematics is unreasonably effective and that all of the explanations I have given when added together simply are not enough to explain what I set out to account for . . . The logical side of the nature of the universe requires further exploration.'[11]

Without question, the efficacy of mathematics in science is a striking phenomenon. To make sense of this, first note some of the reasons why nature itself so often exemplifies symmetries and regularities, for a clue lies here.

A symmetrical structure is stable, iterative, reproducible, homogeneous, and its parts are interchangeable. Certain geometries have great utility: in living organisms a spherical shape promotes conservation of body temperature, given that heat loss varies with the relation of a body's volume to its surface area. In the human embryo, development occurs evenly in all directions until bilateral symmetry is broken to allow the development of organs that exist in a single copy, such as the liver and heart; but even here development is orderly, the organs taking positions lateral to the torso's central axis to save space. Patterns, in short, are economical and efficient. It is no surprise that, as natural structures become more complex, they should so often consist in patterns built upon patterns. Think of a drawing made by a pendulum with a pencil attached to its end, so that as it swings it leaves marks on a piece of paper that is attached to a slowly rotating disc. As the paper revolves, the resulting pattern will be a marvel to behold. That in essence is how patterns are built up in nature. Arguably, symmetry is nature's default, and departures from it have sufficient reasons – in the case of biological entities, mostly adaptive ones, including sometimes *spandrel* effects, the result of the accidental conjunctions of other adaptations.

Now consider mathematics. Mathematics concerns patterns, dimensions, and relationships. The two latter are in fact themselves patterns, mappable deviations from patterns, or approximable by patterns. As a way of capturing and organizing patterns, mathematics allows long chains of pursuit in following the unfolding of patterns and discerning their relations to other patterns. A simplified genealogy of a familiar concept such as number might look like this: we intuitively grasp that some collections or sets of things are bigger than others; some look about the same.[12] It does not matter what the things are; we wish to compare them; I want to see if you have more of those things than I do. We lay them out side by side in a line, until one of us has none left to match to the remainder possessed by the other. From doing that with different kinds of things we realize that number, as an adjective of collections, can be considered independently of the collections, and their relations to each other investigated. We can then, in turn, abstract from number, and consider just the relations, as in algebra; we can represent algebra geometrically, and vice versa; and the patterns and relationships explored can have no relationship to anything relatable to empirical experience – think of the topology of higher-dimensional spaces. The entities of mathematics are abstract structures. Most of mathematics is not arithmetic, but even arithmetic is in essence about patterns. In the exploration of patterns yet further patterns can be generated by manipulation and combination of existing patterns: in quantum theory in physics, *complex numbers* (combinations of real numbers with numbers multiplied by the imaginary number i, the square root of -1) are useful.

From these thoughts a – yes – pattern emerges. Mathematics is the investigation and manipulation of patterns. Nature is full of patterns. Consider some analogies of how the connection between the two can be made.

It is reasonable to assume that the weight of a lump of iron varies with the size of the lump. The bigger the lump, the heavier it is. Now suppose that, in addition to that fact, you know the dimensions of two differently sized lumps of iron, and the weight of one of them. Obviously, you would be able to deduce the weight of the other. Now suppose that the two lumps of iron are hidden in a complicated structure whose activity is being disturbed by an imbalance between their weights. You know the dimensions of one, and you know the degree of disturbance, so you can calculate the dimensions of the other – and can tell an investigator what to look for. Suppose your calculations say the lump is about the size of a cannon ball 12 centimetres in diameter. When the investigator finds just such a lump deep in the machinery, it will seem amazing to him that it so accurately fits the mathematics describing it.

In this case there is no surprise for the mathematician. But now suppose that the various values of the system are coming to light partially from experiment and partially from trying out different values for some of what cannot be measured. Some of the unmeasured values might be constrained in regard to what range they can occupy by the measured values. The values are put into the equations and the handle turned – and a suggestion emerges that there is an as yet-undetected entity with such-and-such properties that must be there if the measured values are correct. An experiment is designed that, if the entity exists, will detect it – and the experiment either detects it or does not. Failure to detect it might be attributable to a number of factors, only one of which is that the entity does not exist. But if it is found, the mathematics that described it is vindicated. This is how the planet Neptune was discovered in 1846: by working from the known values for gravity together with the orbit and eccentricities of Uranus. It is how the Higgs boson was hypothesized, sought, and found. Reality seems to be mathematical or to obey mathematics, whereas what mathematics is doing is identifying a place to look on the assumption that certain already established or given values are correct. In these examples, it identifies and fills a gap in a pattern. It is very apt for doing this. Suppose you and someone far away both have the same aerial photograph of a given landscape, and you wish to identify a point in it for that person. You both lay over it a transparency marked out as a numbered grid; you locate the point for them by giving the point's coordinates. Mathematics serves figuratively as a grid, except that the underlying photograph has blanks, and the grid directs attention to what might possibly fill them.

A close analogy of this is the childhood game of Battleships. You and your opponent each draw a grid of squares, and dispose your battleships in different but contiguous configurations about the grid.

Suppose the grid is 12 × 12, each square identified by a number and letter combination; each battleship occupies four squares, and you each have three of them. Whoever sinks all the other's battleships first is the winner. You fire your shots by turns, recording hits and misses and progressively refining your search according to the emerging pattern. Each hit immediately constrains the possibilities for neighbouring squares; at least one of the eight must have more of the battleship. To begin with your shots will be random; you are feeling your way blindly; but even empty squares are information. Now add a feature: misses close to an occupied square have a value – say, half for a contiguous square, one quarter for a square next to a contiguous square. Now mathematics can considerably reduce the degree of (literally) hit and miss in the search. A pattern of values of ones, halves, and quarters will sharply raise the probability that a given square contains a battleship part; they might uniquely identify the square. This too analogizes the way reality seems to *be* mathematical in its nature, whereas what the mathematics is doing is disciplining and sharpening the focus of enquiry. It is helping to suggest where the missing bit of a pattern might be found.

If you see the sequence 1 3 5 7 9 13 15 . . . and you ask what is missing, it must be '11'. This is merely discerning the pattern. In essence, seeking equations in physics – looking for the ways in which phenomena can be represented in terms of, or as functions of, or as combinations of, other phenomena – is likewise looking for the pattern. Dirac's famous equation is a picture of the pattern of behaviour of an electron in an electromagnetic field. The mathematical pattern was such as to entail that the electron could exist in a positively charged form as well as its known negatively charged form. To begin with Dirac thought there was something wrong with the pattern suggested by the mathematics, but eventually accepted it as compelling, even though it implied the existence of antimatter that would annihilate when interacting with matter. The anti-electron – the positron (named for its positive charge) – was experimentally observed in 1932 in a cloud chamber experiment at Caltech by Carl Anderson, at a stroke doubling the number of known particles – and incidentally creating the puzzle of why there is a universe at all, given that if a consequence of the Big Bang was the production of equal amounts of matter and antimatter, there should not be a universe at all.

The Neptune example illuminates why there is a 'Dark Matter' hypothesis. The speed of rotation of stars in the outer arms of a disc galaxy should be greater than that of stars closer to the galactic centre, but observation – made by astronomer Vera Rubin – shows that there is not much difference. Combine this with the known value of gravity, and the conclusion seems to be that there is a large amount of undetected mass in and around the visible galaxy. Other observations – for example, clusters of galaxies that should fly apart unless something is holding them together, and gravitational lensing effects in which light coming from further away than a given galaxy is bent to a degree greater than is explicable on the visible galaxy's mass – add to the force of the implication that a whopping 95 per cent of the matter in the universe is dark energy and dark matter, stuff that does not interact with the electromagnetic field and therefore cannot be seen or otherwise detected.

Consider now a disanalogy between the effectiveness of mathematics as a language for describing nature and a language such as English. The categories of a natural language like English, namely nouns, verbs, adjectives, and adverbs, enable reference to things and descriptions of – respectively – their activities, properties including relational ones, and modes, in response to the way our cognitive capacities organize experience into things with properties involved in events themselves having properties. Thus: in 'the red ball rolls slowly' we speak of the red (adjective: property) ball (noun: thing) rolling (verb: action, event) slowly (adverb: manner or mode of the action or event) – and we can therefore think of natural languages as reflecting the way the world appears to us. But the ontology of mathematical domains – things such as sets and functions – are not encountered in ordinary experience but are abstractions, and the language constructed to talk about them, the language of mathematics, is devised by us to do this.[13] Recall Alice in Wonderland checking whether she is still herself, and not some other person, after eating the cake that makes her grow very tall, by doing her multiplication tables: 'Four times five is twelve, and four times six is thirteen, and

four times seven – oh dear! I shall never get to twenty at that rate,'
she says – correctly, if multiplying in the bases 18, 21 (then 24, 27 . . .);
and in this sequence she never will get to 20 because, although 4 × 12
= 19 in the base 39, the next step – 4 × 13 in the base 42 – does not get
you to 20.[14] Change the base – change the definitions, the axioms, the
rules – of any formal system and you get different results.[15]

Yet, as Sundar Sarukkai observes, 'the surprise is all the more
exaggerated when it is found that mathematical objects, which presumably exist independently of our physical world, are very apt in
describing our physical world . . . And ironically, it seems to be doing
it "better" than natural language.'[16] The 'better' is explained by
mathematics' precision and predictive power in physics. He attributes
this to the utility of mathematics in representing *form* in nature, a way
of analogizing or picturing structure. This is in essentials the idea
mooted above about pattern: nature is highly patterned and pattern-
related, and mathematics as systems of abstract patterns is very apt for
describing it. The cognitive architecture of some minds – the minds
of those who find mathematics easy, attractive, or both – might find
recognizing and manipulating patterns as pleasing as it is useful.

On this view there is a natural fit between mathematics and the
world that makes the former a powerful instrument for advancing
beyond the pinhole into regions where perception and imagination
cannot reach. It is, in effect, another eye – the eye that can see the
patterns in and underlying nature.

If one were to take this idea seriously, it would support those
who say that it is restrictive to seek experimental verification of
mathematics-generated theories such as String Theory because doing
so is asking for the primitive resources available at the pinhole to
check what can only be seen with mathematics. But to repeat: such a
break with the principle of *empirical* control on theory would be very
radical. The empirical principle seems to have so much to recommend it; surely, we think, it is not arbitrary to leave the decision to
nature itself as to whether some hypothesis is correct. But then we
recall that empirically accessible nature is itself an ideal reality – a
virtual reality – predicated on the cognitive capacities we possess to
deal with the highly local scale at which we exist; and the question

therefore arises: can the relationship between this ideal virtual reality be sufficient to serve as a check on whatever we say about all other scales of reality?

There is a reprise here of an old and long-standing philosophical debate between the outlooks of empiricism and 'rationalism', this latter label understood as denoting the epistemological view that, whereas empirical enquiry can only arrive at local, partial, and temporary opinion about the world because of the limits on our perceptual powers, rational thought can arrive at eternal and immutable truths such as those of mathematics. Parmenides and Plato are the great ancestors of the rationalist tradition, whose distinguished later members include Descartes, Spinoza, and Leibniz. As an epistemological outlook rationalism is naturally allied to realism about abstract entities – for example, Platonism in mathematics, which holds that mathematical entities are genuinely existing things in a non-spatio-temporal realm. The eternity and immutability of their truth is taken as a mark of greater reality than the perishing and imperfect things encountered in empirical experience. By extension a commitment to mathematical Platonism motivates realism about theoretical entities; empiricists such as the physicist Ernst Mach were inclined to be instrumentalist (sometimes called 'anti-realist') about the entities postulated in physics. His view has not withstood the experimental observation of entities he thought were mere conveniences of theory: at CERN the protons collided together in high-energy experiments are very real.

As noted, Kant attempted a rapprochement between the empiricist and rationalist viewpoints by arguing that both are partially right: the world presses upon our sensory surfaces (the empirical part) and the mind organizes (the rational part) these stimulations into a version of the world suitable for our activity. He thought that the barrier of how the world seems – the phenomena – is too great for us to penetrate, and that therefore we can never know anything about the world beyond our experience, the 'things as they are in themselves', which he called *noumenal* reality. Note that in his view both phenomenal and noumenal reality are *reality*, the reverse and obverse of the same coin; it is, he thought, just that we cannot turn the coin over.

Science seeks to show that he is wrong about that. Science is the endeavour of peering through the pinhole in the barrier between phenomena and noumena with the expectation that it will connect the two realities; not only will phenomenal reality serve *via* experiment to check on what we find or say about noumenal reality, but the two realities will be continuous; how things are in the noumena will explain why things seem as they do in the phenomena.

But this is where we encounter the problem of the conflict of understanding regarding quantum and classical realities. Do these thoughts show that we mix together hopes and assumptions in a way that explains why the quantum world seems so puzzling? The thought earlier was that the puzzlement arises because we try to find an *interpretation* of quantum theory that makes sense in classical terms. This, in alternative language, is what the expectation that there will be a mapping from noumenal to phenomenal reality means. After all, the expectation is realized in significant cases – colour perception is well explained by our sensitivity to a certain range of frequencies in the electromagnetic spectrum; indeed all seeing and hearing consists in this. That is what makes it credible to think that squinting through the pinhole and pushing our probes through it – including the probe of mathematics, which is the furthest-reaching probe of all – will reveal the underlying pattern of all things eventually.

But, in turn, this lends credibility to the idea that when we can *only* see with mathematics because no empirical experiment can reach that far beyond the pinhole, then the mathematics should itself count as the test. If it is an extension of, or derived from, what has empirical support – as String Theory is an inference from the Standard Model but going beyond it and providing ways to resolve some of its problems – then on this view the theory has as much scientific legitimacy as results in a laboratory can confer.

What can decide between the empiricist constraint that no scientific theory can pass muster unless it is mapped back through the pinhole into our classical experience, and the claim that the rationalist is in effect making, that mathematics is a way of seeing that can confirm itself?

At this point it might seem that at the frontier of knowledge the same conflict of view will always arise, unfolding even as the frontier itself advances and greater terrains of ignorance come into view: for the frontier is defined as the point where the reach of reason outstrips experience as the means to seek new insight.

Two sidebar thoughts suggest themselves at this juncture. One is that there is no way of dispensing with imagination, speculation, and surmise at the frontier of knowledge. The other is that it would be a mistake to think that experimental scientists cannot think up ways to test even the wildest-seeming of such surmises. They have done it often enough. Who would have thought that the leftover radiation from the Big Bang could be detected today, buzzing faintly across the universe; that it could be mapped with great precision, and that from its configuration so much could be deduced about the universe's history? And that is just one example.

A different application of the Pinhole Problem, this time coupled with the Reading-in Problem, seeks some significance or relevance *to us*, we humans who are squinting through the pinhole, of what we see through it. A prime example is the Anthropic Principle. This states that if the universe were different from how it is, there could be no life in it – and, in particular, we would not exist to observe and study the universe. Put another way, it states that the 'constants' of nature (the speed of light, the electric charge, the Planck Constant) are fine-tuned for the production of life, making it appear that we occupy a preferred time and place in a preferred universe.

The implications of this view, if crudely drawn, are pleasing to those who think the universe exists so that we can exist – indeed that it was expressly designed for that purpose. This was the prevailing view for many centuries before Copernicus showed that planet Earth is not the centre of the solar system, from which it follows that the solar system is not the centre of the universe but something relatively insignificant in the grand scheme of things – a cluster of bodies around an ordinary star, itself one of a very large number of stars. Less than four centuries later Hubble showed that the extent of that insignificance is vastly greater still, because we occupy a corner of one galaxy among many billions of galaxies.

But the impulse to self-congratulatory admiration of the way the universe appears to be arranged for the express purpose of producing humanity (which, remember, counts among its members Adolf Hitler and Pol Pot as well as Leonardo da Vinci, Johannes Brahms, and Albert Einstein) has been given a boost by the observation that the basic constants of physics and chemistry seem to be fine-tuned to give rise to exactly a universe that can produce and sustain life and therefore us, and which would not, because it could not, exist if the constants did not have just the values they do.

For example: if electrons and protons did not have equal and opposite charges, chemistry would be radically different, and life as we know it improbable if not impossible. Likewise there would be no water if the weak nuclear force were any weaker than it is, for then all the universe's hydrogen would turn to helium, and there can be no water without hydrogen. In any case the properties of water are miraculously suitable too: alone among molecules, because of the properties of the hydrogen atom, it is lighter in its solid than in its liquid form, meaning that ice floats. If this were not so, the oceans would freeze and planet Earth would be an ice ball, inimical to life.

Equally marvellous is how carbon synthesis happens. Carbon is the key component of all organic molecules, and therefore underlies the very possibility of life – at least, again, of life as we understand it. If the ratio of the strong nuclear force to the electromagnetic force (which keeps electrons in nuclear 'orbit' in atoms) were different, the processes in the heart of stars that synthesize carbon could not happen. Moreover the window of opportunity for carbon synthesis to happen is a very small one, requiring precise energy levels and temperatures and tiny timescales. The age of the universe is crucial too: at 13.72 billion years old the universe is ripe for the production of carbon. If it were ten times younger, there would not have been enough time for carbon synthesis; if it were ten times older, the main sequence stars in which carbon is produced would have passed their sell-by date for the process.

So this must be what has been called the 'Golden Age' of the universe for life. Add to this the fact that gravity is 10^{39} times weaker than electromagnetism; were it not this much weaker, stars would be far more massive and would therefore burn up more quickly. The value of the strong nuclear force is also crucial, for if it were even just a little stronger atoms could not form, and if it were just a little weaker neither could stars.

The term 'Anthropic Principle' was coined by astrophysicist Brandon Carter in his 1973 contribution to a symposium in honour of Nicolaus Copernicus's five-hundredth birthday. As a corrective to the view that Copernicus had shown we occupy no special place in the universe, Carter sought to show that there is a relation between the constants of nature and our existence. He defined two forms of the Anthropic Principle: a 'Weak' one, explaining the striking relations between some of the constants of nature defining 'here and now' as the point and place in the universe's history that permits life of our kind to exist; and a 'Strong' one, suggesting that we can infer what the constants should be from facts about carbon-based life of our kind, or alternatively that we occupy one of a number of universes in which the constants are as we observe them to be.

John Barrow and Frank Tipler provide the following statement of what is respectively involved in weak and strong versions of the Anthropic Principle:

(1) The observed values of all physical and cosmological quantities are not equally probable, but they take on values restricted by the requirement that there exist sites where carbon-based life can evolve, and by the requirement that the universe be old enough for it to have already done so.

(2) The universe must have those properties that allow life to develop within it at some stage of its history either because (i) there exists one possible universe 'designed' with the goal of generating and sustaining observers; or (ii) observers are necessary to bringing the universe into existence; or (iii) an ensemble of other universes is necessary for the existence of our universe (a multiplicity of coexisting universes as envisaged in 'multiverse' theory).[17]

(1) describes a 'Weak Anthropic Principle' and (2) describes versions of a 'Strong Anthropic Principle'. In effect (1) can be glossed as an

iteration of the point that the constants are as they are because we, made possible by them being so, now happen to observe and measure them. (2) is more controversial under any of its readings, the first of which has it that the universe is a consciously designed entity, the second that its existence depends upon our observation of it, and the third that there are many universes whose parameters differ from one to another, and with us occupying the one, or one of the ones, suitable for life.

Familiarly, some choose the version that has the universe existing expressly, by design, in order for us to exist. But any version that makes it seem surprising, wondrous, or improbable that the universe is so apt for our existence rests upon a simple mistake. As with finding the nature of quantum reality puzzling from the classical perspective, the mistake is to have matters the wrong way round. For the fact that we exist means that of course the physical constants of the universe have to be such that we can exist, for if they were not such that we can exist, we would not be here to measure them. It is like thinking that one's own existence must be miraculous, given the additive happenstances of all our personal ancestors meeting and mating. So to take the values of the constants as proving that they were fitted to produce us is to put the cart before the horse – or better: to agree with Voltaire's Dr Pangloss that because some of us wear spectacles, noses came specifically into existence to support them. A more accurate perspective on the matter is to recognize that with the value of the constants as they are, in the conditions prevailing on this planet over the course of the last 4 billion years, it would have been more surprising had life *not* evolved. That perspective will gain much plausibility if life is found elsewhere in the solar system or in space.

Might life be discovered elsewhere? The physics and chemistry of matter on this planet are common to the observable universe, removing any *prima facie* barrier to an affirmative answer. Statistics alone suggests that life is abundant in our galaxy, and, if so, then superabundant in the observable universe. 'Life' is one thing, 'intelligent life' another; naturally many people are interested in the possibility of the latter, but both are fascinating prospects. Moreover, there is no reason to expect that life would be like life as we know it – life requiring water, the presence of which is standardly taken as a key sign for the possibility of life by astrobiologists – for it involves a big assumption to think that the only kind of life there can be is carbon-based life, or that different universes could not produce life and even intelligence on the basis of physical arrangements other than those in the observable universe. The restriction to water-involving possibilities for life beyond Earth is another product of the Pinhole Problem, taking what we know to constrain what we can, or even what we are conceptually permitted to, imagine.

A driving ambition in fundamental physics is to arrive at a single, comprehensive, unifying Theory of Everything. An assumption that attends this ambition is that such a theory will show that there is ultimately just one kind of thing, out of which the apparent variety of less fundamental things arises. To unify the four known forces of nature – the strong and weak nuclear forces, the electromagnetic force, and gravity – and to show that the quarks, leptons, and bosons of the Standard Model are at base one kind of thing – indeed to show that there just is *one* thing underlying all forces, fields, and particles – is the grail of physics.

The reductive assumption that reality must ultimately be a single (or at least a single kind of) thing harks back to the earliest Greek philosophers, the 'Presocratic' thinkers who hypothesized a unitary underlying reality out of which the apparent diversity of nature comes. As noted in the Introduction, the Milesian thinker Thales, flourishing in the mid sixth century BCE, took the view that the one underlying stuff, or *arche*, is water. His pupil Anaximander called the underlying reality *apeiron*, by which he meant an undifferentiated, infinite, primordial something from which the variety of things endlessly arises and is renewed. He was alone among the Presocratic thinkers in nominating an unknown underlying substrate; his successor Anaximenes chose air, and later Heraclitus chose fire as the *arche*.

Recall that the most significant argument for the *necessary* unity, eternity, and unchangeability of the fundamental nature of the universe was given by Parmenides, who said that if the underlying basis of reality were plural, changeable, and temporary, it would be unstable and therefore unable to remain in existence. That is not of course the argument in the present drive for unification, though an aspect of it might be: the idea that it is more intellectually satisfying to arrive at a single, simple, unitary x for which the explanation itself reflects the simplicity and comprehensiveness of what it explains. The assumption that fundamental reality and the truth about it must be simple, rational, and comprehensive is one that has played a dominant role in guiding the direction of enquiry in the philosophy of nature and its heir, science, from the outset. What should we think about this assumption itself?

Suppose you ask why there might not be seventeen and a half forces and eight and three-quarter fundamental entities or kinds of thing, or any other arbitrary number of them – why the apparent fetish for one of each, and even for just one that is both? Why the drive for ultimate unicity and simplicity? One thought might be that by looking for the simplest ultimate explanation, even if you do not find such a thing, you will have expunged the ad hoc and the conceptually adventitious from your theories, and got closer to the truth thereby. This is a persuasive point. It is allied to another idea: that the simplest theory is the most likely to be true (everything else being equal). This is not such a good point, however; it is certainly not so in its application elsewhere in enquiry, for example, in the study of history, or of morality, or politics. This consideration ought to generate a degree of unease, because, in suggesting that the truth might be complicated rather than simple, it hints that in the journey to reductive explanations the reason for the 'appearance' of complexity might be overlooked.

A large part of the ambition for a Theory of Everything lies in the history of science itself, in which successive advances have tended to embrace and include earlier advances as special cases of themselves, and to exhibit what the physicist David Deutsch calls 'reach', that is, applicability to more than the original target of enquiry.[18] Newton's theory of gravity is included in, and superseded by, Einstein's theory of gravity, and Einstein's theories explain more than gravity: they 'reach out' to generate or inform theories about black holes, gravitational lensing, and more. Maxwell had unified the phenomena of electricity and magnetism; Einstein had unified space, time, and gravity; atomic phenomena (that is, above the level of the nucleus and its inner structure) were unified into the theory of the atom by Schrödinger, Heisenberg, and Dirac; the Standard Model with Quantum Chromodynamics, electrodynamics, and the Higgs field provide a unified account of the subatomic level. The Standard Model, with its multiple particles of very different masses (or, for photons and gluons, no mass) and forces of different strengths, looks – for all the extreme precision it affords and the unequivocal experimental support it has – messy and unfinished. It works: inviting Ptolemy Problem thoughts. It is profoundly unintuitive and weird; inviting Pinhole Problem thoughts. It cannot be rendered consistent with the General Theory of Relativity without much speculative hypothesizing. Efforts to clean it up and suggest underlying simpler and more unitary entities and mechanisms – as in 'String Theory' – go beyond the reach of experiment, and so another aspect of the Pinhole Problem arises, discussed above. The irresolvable-seeming difficulty of putting quantum phenomena and gravity into a single framework prompts the thought that if they are both expressions of something far more fundamental, all the familiar concepts, including space and time themselves, might need to go.

There is a pincer movement, as it were, at work in current science; there are questions prompted by the Standard Model and there are questions at the other end of the scale, in cosmology, concerning the phenomena of dark matter and dark energy. The quantity of observable matter in the universe is too small to account for the gravity at work in it; that is the dark matter conundrum. The rate of expansion of the universe has been increasing over the last half of its existence since the Big Bang; that is the dark energy conundrum. As noted earlier, before the Hubble Space Telescope revealed, from observations of very distant and therefore very ancient supernovae (distance is time in the universe), the increase in the rate of the universe's expansion, it was thought that the universe might have sufficient energy density eventually to stop expanding and begin collapsing in on itself. If the universe had too little energy density, it would continue to expand but more and more slowly because of gravity's drag effect. The Hubble Space Telescope observations suggest two alternative possibilities: one is that a previously discarded aspect of Einstein's theory of gravity might have been right after all, or that the theory itself is wrong; the other is that there is a form of energy at work in the universe that is not understood, though it is known from the universe's rate of expansion how much of it there has to be – namely, about 70 per cent of the universe. Given that a further 25 per cent of the universe consists of dark matter, it follows that 95 per cent is made of unknown stuff. The spectacularly successful science we have addresses just 5 per cent of the universe.

Recall that Georges Lemaître showed in 1927 that on Einstein's theory the universe must be expanding. As the accepted view then was that the universe is static, Einstein invoked an idea he had put forward in a 1917 paper: that the universe has a vacuum energy density that acts to counteract the effect of gravity, the value of which is given by the 'Cosmological Constant', or lambda (λ), which has to be just right to achieve its effect; for if it were even a little higher or lower, the universe would expand or contract accordingly. When Edwin Hubble showed that the universe is expanding, Einstein was very pleased; it meant he could abandon his ad hoc device invented to keep the universe still. Indeed he described the introduction of the Cosmological Constant as 'the greatest blunder of his life'.

To abandon the Cosmological Constant is the same as saying that its value is zero. In his early effort to keep the universe static Einstein had proposed that its value is small but negative. The Hubble Space Telescope observations suggest that the constant's value is not zero but positive. The General Theory of Relativity plus a Cosmological Constant with a small positive value accommodate dark energy, so dark energy by itself does not require abandoning General Relativity; but it does nothing to resolve the problem of how to render the theory consistent with quantum theory. If anything, it makes it harder to see how to achieve a quantum theory of gravity. This is because the properties of an hypothesized massless particle to carry the gravitational force, the graviton, produce infinities in the mathematics describing their properties and interactions. In the case of the force particles described in Quantum Chromodynamics, which explains the activity of gluons binding quarks together into nuclear particles, infinities also occur, but can be disposed of by 'renormalization', because gluons (like photons) are 'Spin 1' bosons. The graviton has to be a 'Spin 2' particle if it exists. The difference is significant: Spin 1 bosons are the entities of a quantum theory for a vector field, but a quantum theory of gravity is a theory for a much more complicated thing, namely a tensor field, and it is this that requires massless Spin 2 bosons, and blocks renormalization of the mathematics describing their interactions. When infinities appear in the equations, something has gone wrong.

Moreover, paradox results from the effort to think of gravity's effects on the particle and antiparticle pairs that flash in and out of existence as quantum fluctuations in the vacuum of space. This is because such fluctuations at the Planck Scale generate enormous energies, making black holes likewise flash in and out of existence, gobbling up all intervals of space at the Planck Length. The result would be what the physicist John Wheeler described as a constantly bubbling space–time 'foam', in which energy, distance, and time all become subject to the Uncertainty Principle. The amount of Planck Scale vacuum energy is far too great to be the dark energy pushing the universe apart; it has to be explained away somehow. If the grail of a unification between quantum and relativity theories is to be achieved, a solution to this problem, among others, is required – and yet looks insurmountable without the more adventurous proposals that currently fall outside the possibility of experimental test.[19]

The Pinhole Problem appears to be at work here. The effort to observe and describe the phenomena of nature at the furthest current extent of science's reach to the smallest and largest scales respectively – yielding pictures accurate to the sixth decimal place and more, and in impressive ways applicable via advanced technologies – ends in paradox when the attempt is made to combine them. Whatever inference one draws – that human cognitive powers are insufficient (too pessimistic), that we are inhibiting ourselves by requiring that everything detected beyond the pinhole be dragged back through it so that we can understand it classically, that nature is intrinsically inconsistent or incoherent (unlikely), that we are still at a primitive stage of science (this and the second suggestion are the most likely) – there is still the question of why the goal should be a single, simple, comprehensive account, and whether it is right to seek it. To ask this is to ask whether we are under a Parmenides-like spell, and what it would be like to think outside the paradigm it implicitly represents.

PART II
History

In the Introduction it was pointed out that, until the nineteenth century, history – in the sense of a past known about with some degree of reliability – stretched back less than three thousand years, to a period somewhat approximately and patchily remembered in the Hebrew bible and the archaic period of Greece. Herodotus, although alluding to Min, the priapic fertility god of Egypt, as that country's 'first king', begins his account of Egyptian history with a pharaoh he calls Sesostris, a name shared by several pharaohs, none of whom did what Herodotus says his Sesostris did, and who is probably a composite of several pharaohs dating from Egypt's Nineteenth Dynasty (twelfth century BCE).[1] Most of what Herodotus learned from priests and scribes during his visit to Egypt is mixed with legend and filled with references designed for a Greek audience, and he gives no dates; but he knows of the great monuments, and compares them and Egypt's cities to Nineveh and Babylon.

The Hellenization of Egypt after the conquests of Alexander the Great, and its subsequent inclusion in the Roman Empire, hastened the dimming of historical interest and memory, so that the Greek and Latin authors who wrote about it knew little more than Herodotus did. A parallel loss of influence by Persia and the states of Mesopotamia had a similar effect: Greek historians remembered Xerxes and before him Darius and Cyrus because of their culture's struggle for survival against the then mighty Persian Empire, just as the editors of the Hebrew bible's first five books – the Torah or Pentateuch – remembered King Nebuchadnezzar because he took their leaders into exile in Babylon. The Torah speaks of Pharaonic Egypt, Babylon and Nineveh, Kish and Ur of the Chaldees where Abraham was born, because, relative to its own authors' patch of the Levant, these were mighty places of note.

Until recently, then, relatively secure history (as viewed from

Europe) extended back only as far as the first half of the first millennium BCE – say, not much before 750 BCE – and what was known of it, and even more so of periods preceding it, was only hazily remembered. The beginning of globalization in the Early-Modern Period CE brought the monuments of Egypt and the traditions of other peoples and places, not least China, into focus from the Eurocentric perspective; in the *Records of the Grand Historian (Shiji)* of Sima Tan and Sima Qian, written in the late second and early first centuries BCE, the history of China putatively reaches back to the Yellow Emperor in the mid third millennium BCE, though the 'Doubting Antiquity' school of historians in China in the 1920s argued that the Yellow Emperor was a legendary figure only, that anything before the Shang Period of the second millennium has to be ascribed to myth – China's earliest written records date from the reign of King Wu Ding, said to be the twenty-first king of the Shang Dynasty, in the mid thirteenth century BCE – but that anything before the 'Spring and Autumn' Period (*c.* 770–475 BCE) was mainly suppositious.

What was not known about origins was indeed generally supplied by myth everywhere, and in the European and Near Eastern case the monuments surviving from much earlier times into the classical period served chiefly as hooks upon which the myths and legends could be hung.

Thus it was that the history of times before the composition of the Homeric poems and the Pentateuch – both occurring, roughly speaking, between the ninth and seventh centuries BCE – was only impressionistically known until very recently indeed: in fact until the nineteenth century CE.

1. The Beginning of History

The lost world recovered by nineteenth-century archaeologists and historians is the world of the first civilizations, which arose in the Near East's Fertile Crescent.[1] Discoveries about the Bronze Age there and in Europe, *c.* 3000–1200 BCE, necessarily prompted interest in what preceded it – the Chalcolithic, or Copper Age, roughly 6.5 to 3.5 kya, and the period from the end of the Younger Dryas climate change event around 12 kya, taken as the beginning of the Neolithic Period. These dates and period labels give the misleading impression of an orthodoxy in archaeology, predicated on the idea of a 'Neolithic Revolution', to use the archaeologist Gordon Childe's phrase, in which there was a change from nomadic hunter-gathering to settled living accompanied by agriculture, followed after several thousand years by the rise of towns and cities – marking a sharp break from the preceding fifty thousand years of occupancy of the Near East and western Eurasia by anatomically modern humans, *Homo sapiens*. Moreover this orthodoxy is of a piece with thinking that the development of new technologies and styles of life was a linear and progressive one – 'progressive' in the sense of development from less to more sophistication in the means and manner of life – save for periods when a catastrophe of some kind intermitted that development, for example, whatever it was that precipitated the Bronze Age Collapse around 1200 BCE.

But the neatness of this picture is a much challenged one. There is evidence that some people were living in houses as early as 20,000 BCE; that some hunter-gatherers were not nomadic; that harvesting and cooking grains, and perhaps even planting and cultivating them, began before the Neolithic; that animal-herding long preceded agriculture; even that settled living might in some cases have given way to a resumption of nomadic life after climate changes or other disruptive events. Cave art, ornaments, and burial practices, some of them stretching back over a hundred thousand years, suggest a more complicated picture of prehistory than the one presented by the linear story.

These points have to be remembered. At the same time it is true that the very recent discoveries in Neolithic and especially Chalcolithic and Bronze Age history show a remarkable change in the conditions of human social and economic life. Given that what we know, or at least think, about the past has as much effect on us as how our current conceptions and interests colour our interpretations of the past, that this elaborate chunk of history rose up so recently and suddenly, like a great populous island emerging from the sea – the sea of time – before our amazed eyes, is a striking fact. Leaving historiographical controversies aside for the moment, the story of this recovered world, together with the story of its recovery, can be sketched as follows.

The curiosity of collectors at the beginning of modern times was the first scratching at the surface of a deeper past. There had been surprisingly little interest in pre-classical history before then. The many and complex effects on Europe of what medieval Crusaders brought back from the Near East, from the late eleventh century onwards, did not include deep history; for them, as for everyone, the past was what the Old Testament said it was. European traders had been active in the Levant since the days of the Roman Empire, but a large increase in demand for spices during the early Renaissance made the merchants of Venice and Genoa rich. Indeed the increasing wealth from trade in the late medieval period onwards is one of the causes of the Renaissance, placing into secular hands opportunities for leisure and the commissioning of art. This in turn was followed by competition between French and English merchants for a share of the Levant trade, beginning in the sixteenth century.[2] But none of this led to the discovery of a past much earlier than the classical past.

Things began to change in the late sixteenth century. 'Cabinets of curiosities' became fashionable, and, although at first they consisted of mere miscellanies of marvels – including objects mistaken for mermaids' tails and unicorns' horns, and with dinosaur bones interpreted as giants' thigh-bones – they were the origins of museums and spurs to scientific investigation.[3] Wealthy participants in the Grand Tour who went beyond the Roman world to the Greek, and the few who penetrated yet further into the Levant – named for the *rising* of the Sun – were often most interested in Christian antiquities and in the purchase of memorabilia associated with them, or, like Johann Wansleben before them in the 1670s, in finding manuscripts.[4] There were some others who, as early as Wansleben or Jean Chardin (whose 1686 book on his travels in Persia contains the first suggestion that wedge-shaped marks found on clay plaques were writing, not arbitrary decoration), guessed that much earlier history existed. Also in the seventeenth century CE the Jesuit polymath Athanasius Kircher published a grammar of the Coptic language, and John Greaves, an Oxford mathematician and linguist, travelled in the Levant and Egypt and measured the height of the Great Pyramid of Giza. In the eighteenth century a boost was given to interest in pre-classical antiquity in general, and Egypt in particular, by the French scholar Charles Rollin in his twelve-volume *Ancient History of the Egyptians, Carthaginians, Assyrians, Babylonians, Medes and Persians, Macedonians and Grecians* (1730–38).[5] Far more popular than accurate, this work was the spark that led eventually to a scholar-army accompanying the fighting army in Napoleon's invasion of Egypt in 1798. This in turn unleashed the nineteenth century's interest in the possibility of deeper history, and a quest for it. When it happened, Assyria and Babylonia were the first to be found, thereby extending the reach of history back into the late third millennium BCE. The first half of the twentieth century (after some mid-nineteenth-century glimpses) saw this reach extend even further, to Sumer in the fourth millennium BCE.

The discoveries thus made are still far from complete, even as these words are written. Millions of texts from those periods have been recovered, but not much more than a tenth have been read, and there are many hundreds of sites yet to be explored – these being the *tells*, or mounds, dotted numerously over the landscape of today's Iraq and neighbouring regions, which are the layers of ruins and debris of occupation accumulated over thousands of years (*tell* is Arabic; variants are *tel*, *tall*, *tal*; the sites are called *tepe* in Persian and *huyuk* in

Turkish). At time of writing the *tells* are inaccessible because of war and civil war, politics and diplomatic tensions, but the vast collections of clay tablets inscribed with information about economic and military affairs, government activity, buildings, trade, money owed and received, medicine and science, literary works, temple activity, and more, are available in museum collections, constituting a rich source of information about a world completely lost until recently.[6]

We call 'ancient' the time we contemplate, across two and a half thousand years, to Greece's classical period of the fifth century BCE – that is, halfway through the first millennium BCE. A Greek living then would have looked back, had he known of it, even further: across three thousand years before his own day to the rise of Sumer from the middle of the fourth millennium. By Greece's classical age even the history of its relatively recent Bronze Age past, captured as legend rather than history in the amber of Homeric epic, was forgotten; if there was a siege of Troy by 'Achaeans' and 'Argives' in the late thirteenth century BCE, around the period of the Bronze Age Collapse, it had become a much embroidered memory by the time Peisistratus, tyrant of Athens in the sixth century, ordered the hitherto orally transmitted poems to be written down. The Greeks' mythologized version of history reached therefore barely five hundred years before their own time. Yet themes and beliefs persisted from the far deeper past: the story of a great flood, told in Greek mythology as Deucalion's flood and in the Hebrew bible as Noah's flood, was related in *The Epic of Gilgamesh*, originating in the third millennium BCE at the latest. The Venus of the Romans was the Aphrodite of the Greeks, who was the Ishtar of the Babylonians, who was the Innana of the Sumerians – the same goddess with the same attributes governing the same set of human concerns about love, sex, beauty, fertility, and strife, stretching back more than four thousand years before Augustus became the first emperor of Rome.

Napoleon's scientific commission for the study of Egypt was in part a result of France's long-nourished interest in the eastern Mediterranean and the Levant, the French having established trading relations with the Ottoman Porte in the sixteenth century, before the English but after the Venetians and Genoese. It was also, as mentioned, a result of the interest roused by Charles Rollin's popular history of the region. Prompted by the celebrity of the Napoleonic researches in Egypt, a form of trophy-tourism began. Archaeology indeed started as the search for plunder, both by trophy-tourists themselves and by locals keen to supply the tourists' desires for mementos and collectables.

The desire for plunder, though in slightly more sober and scholarly dress, was also felt by such august institutions as the British Museum and the museums of Berlin and Paris. Along with it came a dawning sense of academic responsibility; but the desire for displayable ancient wonders was at first the major motive. In 1854 the British Consul at Basra, J. G. Taylor, was asked by the British Museum to investigate some of the *tells* in southern Mesopotamia. He chose the *tell* of Pitch, first noticed as significant by a briefly visiting fellow British official, William Kennett Loftus. Taylor found inscriptions identifying the site as the bible's 'Ur of the Chaldees'. Taylor dug there for two seasons only, followed at the end of the nineteenth century by an expedition from the University of Pennsylvania. It was not until a full-scale British Museum and University of Pennsylvania joint archaeological effort was mounted in the 1930s, under Leonard Woolley's direction, that Sumer came properly into view.

Apart from Abraham's connection with Ur, the only biblical reference to Sumer occurs as 'the land of Shinar' in Genesis.[7] In the nineteenth century French, German, British, and some American archaeologists, looking for biblical sites – following in the footsteps of the curiosity collectors and plunderers already mentioned – had investigated Assyrian *tells* concealing Nineveh and Assur in the northern part of Mesopotamia, and Babylonian sites in central Mesopotamia. Assyriology, the general name for study of the region, flourished as a result. Nineveh's discovery by Austen Layard in the 1840s was a key moment in this development. It was an Assyriologist, the French–German Jules Oppert, who realized that the 'Shinar' reference in Genesis meant Sumer in southern Mesopotamia, and he further recognized that cuneiform writing originated there.

The archaeology of Mesopotamia developed with increasing rapidity after the mid century's discoveries. The aforementioned

William Loftus, a young geologist working for the British government's Turco-Persian Boundary Commission, was seconded to the
British Museum's investigations in Mesopotamia, and was among
those who worked at Nineveh and Nimrud, at which latter site he
discovered the palace of the Assyrian King Ashurnasirpal II and a
trove of carved ivories dating to the ninth century BCE. But it was his
earlier brief excavations in southern Mesopotamia that proved most
significant for Sumerian history. He found Uruk and the Ziggurat
of Ur in 1850, the discoveries of which were the British Museum's
prompt for sending Taylor there. Meanwhile brilliant work on the
cuneiform texts was being done by Sir Henry Rawlinson and George
Smith of the British Museum, one great result of which was the latter's discovery of *The Epic of Gilgamesh* among the tens of thousands of
clay tablets and fragments that had been brought to London from the
Assyrian excavations.

Smith's story is a moral tale in its own right. Born into a working-
class family in London in 1840, he left school at fourteen to be
apprenticed as an engraver to a printing firm. His private passion was
Assyriology, the new and exciting discoveries from the Near East
filling his imagination. He spent his lunch hours in the British
Museum, reading reports about what was being found at the excavations and teaching himself to read cuneiform. He eventually came
to the attention of the Museum authorities, who, impressed by his
expertise, invited him to assist with cleaning and sorting tablets,
which he did in the evenings. While engaged in this task he identified
a record of a payment by Jehu, King of Israel, to Shalmaneser III,
King of Assyria, in the ninth century BCE. Yet more importantly, he
identified records of a solar eclipse that had been independently dated
from astronomical records to 763 BCE – this proved important for
establishing the chronology of Near Eastern history – and of an invasion of Babylonia by the Elamites in 2280 BCE. These impressive
findings led the Museum trustees, encouraged by Rawlinson himself,
to appoint Smith to a post as Senior Assistant in the Assyriology
Department. That was in 1870. His discovery several years later of
the flood narrative in what came to be known as *The Epic of Gilgamesh*
made him world famous. The Museum thereafter sent him on three

expeditions to Nineveh, where he excavated the library of Ashur-banipal, discovering more fragments of *The Epic of Gilgamesh* and a valuable set of records of Babylonian dynasties. On the third of those expeditions he died of dysentery; Queen Victoria gave his wife and children a pension of £150 a year.

Researchers like these and others had begun to strip – both literally and figuratively – the layers of time from an immense story. Sumerian and the following Akkadian and Assyrian periods are now known with thoroughness, though much more remains to be discovered both under the ground and in the texts.

A bird's-eye view of Mesopotamia today shows the Tigris and Euphrates flowing south to meet at the point where the Shatt-al-Arab begins, their combined waters seeping and dispersing into the marshes of the region and eventually trickling into the Persian Gulf 120 miles further down. Along the course of the Shatt al-Arab there were, until recently, extensive marshes and forests of date palms, between them creating a unique environment. War and 'economic development' have reduced both the palm forests and the marshes to a fraction of their original size.[8] Six thousand years ago the courses of the rivers were different, and indeed have changed a number of times as a result of major floods and the erosion of the alluvial plains through which the rivers pass. In the fourth millennium BCE the cities whose *tells* are now miles from the rivers once stood on their banks: Nippur, Uruk, and Ur stood on branches of the Euphrates; Girsu on the Tigris. In those epochs the branches of the rivers encircled islands of fertility, and on them the cities of the world's first great civilization arose: Sumer.

Six thousand years before Sumer, in the north of Mesopotamia, people had begun to cultivate emmer and einkorn cereals – both ancient forms of wheat – and to create permanent settlements next to the cultivation sites. What motivated some hunter-gatherers to settle – given that the change was not beneficial to their health at first – is still an open question, though the advantage of a food supply that can be stored for weeks or months and not just a few days, as is the case with meat, is doubtless part of the reason. Settlement was a necessary corollary of dependency on cereal sources of food: sowing and harvesting happen at fixed times; storage of grain requires that it be protected from weather and pests in a built facility of some kind – a possible reason for the development of larger pottery vessels; and once harvested it is difficult to transport far in bulk. Over time the practice of settlement extended down the rivers as the population grew. Near the site of Ur, Leonard Woolley found just such an early village at al-Ubaid, which gives its name to the pre-Sumerian culture of the region.

'Early' can be a misleading term. The carved stones at Göbekli Tepe in south-eastern Anatolia date to a period between the tenth and eighth millennia BCE. Long-distance trade, evidenced by the presence of obsidian and shells found at Jarmo in the foothills of the Zagros Mountains, was taking place by the seventh millennium. The first-known irrigation system, found at Choga Mami, had appeared before 6000. The region was evidently bustling and increasingly populous, its organization and technological development already considerable by the time Sumer emerged from – or its founders entered – this activity in the fourth millennium.

Near the end of the third millennium a Sumerian King List was drawn up by a scribe at Lagash, probably at the behest of a currently reigning monarch to give his lineage a pedigree and thus to establish his credentials. It says that the first Sumerian ruler was one Etana of the city of Kish, described as 'he who stabilized all the lands'. He was an heroic figure who flew to the sky on the back of an eagle in search of a magic plant that would enable him to have a son. Kish, along with Eridu, Uruk, Ur, Larsa, Lagash, and Nippur, was one of Sumer's leading cities, sometimes allying and sometimes vying with the one or more of the others in what the scribes of Lagash – which seems to have been the principal city for historians at the end of the third millennium – described in terms of frequent disputes over boundary ditches.

Sumerians did not call themselves Sumerians; the name was bestowed by their successors as the chief power in Mesopotamia, the Akkadians. Sumerians called their land 'the country of the noble lords', Kiengir, and they described themselves as 'the black-headed people'. Sumerians were a non-Semitic people; the Akkadians were

Semitic. When the Akkadian Empire arose in the mid third millennium both Akkadian and Sumerian were spoken in it, and the Akkadians adopted the cuneiform system of writing that Sumer had invented. Gradually Sumerian was less and less spoken, though even in the Babylonian period in the first half of the second millennium it was one of the official languages along with Akkadian and Aramaic, reserved mainly for religious ritual, as with Latin in Roman Catholic ritual up to the twentieth century.

The origins of the Sumerians are obscure. One theory is that they were continuous with the Ubaidans, who came from northern Mesopotamia to settle in its southern reaches, draining its marshes to plant crops and trading the textiles, leatherwork, pottery, and metalwork they manufactured. This suggests an already advanced basis from which Sumerian civilization could emerge. Other theories are that the Sumerians came from India and are related to its Dravidians, or from North Africa. There is a suggestion that they came from the western shores of the Persian Gulf, and were driven from there at the end of the Ice Age, when floods drowned the littorals they occupied. Yet another theory locates their source in 'west Asia', a standard because sufficiently obscure general place to site the unknown origins of this or that group.

Perhaps most plausible is the idea of several groups combining: Ubaidan farmers, nomadic pastoralists with flocks of sheep and goats, and fishermen from the marshy mazes of the Euphrates and Tigris Delta coming together and, over a number of centuries, melding into the creative and highly organized civilization that was Sumer. The city of Eridu on the Persian Gulf coast has been suggested as a focal point for this merging of peoples, traditions, and skills; their descendants spread over southern Mesopotamia and founded the city-states of Larsa, Sippar, Shuruppak, Uruk, Kish, Nippur, Lagash, Girsu, Umma, and others.[9] The city-states of Sumer were sometimes independent, sometimes mutually hostile, sometimes united, but not until the Akkadian Empire in the late third millennium (2300–2000), especially under the great King Sargon (2270–2215), was the region a single empire.

The period between 4000 and approximately 3000 BCE is known as the Uruk Period. Some estimates say that at its height the city of Uruk had a population approaching 80,000, and its area of 6 square kilometres was surrounded by mighty walls said to have been built by its most famous king, or *lugal* ('big man'), the hero Gilgamesh. Uruk traded with Sumer's other cities up and down the rivers, which were the main means of transport, and its architectural style, pottery, and tools are found in distant urban centres, as far away as today's Syria and Anatolia, suggesting colonies or at least major influence. In the early Uruk Period the city and its neighbours were unwalled, and probably governed by a chief priest, *ensi* or *en*, advised by a council of elders consisting of both women and men. The later change to more secular rule by a *lugal* and the appearance of defensive walls suggest increasing competition and strife among the Sumerian cities, and perhaps the first military threats from Elam, situated to the east of the Tigris in present-day south-eastern Iran, which had begun rising to prominence in the early third millennium.

Cuneiform writing developed from a pictographic form of record-keeping in the late fourth millennium, during the Uruk and Jemdet Nasr periods of Sumerian history. Within centuries of the third millennium beginning, some of the city-states had come to dominate others; the *lugal* Eannatum of Lagash conquered the long-time rival city of Umma, recording his emphatic victory in the graphic Stele of the Vultures.[10] Umma had its revenge: after Eannatum's death it overthrew Lagash, and conquered Uruk also, Umma's ruler Lugalzagesi making Uruk his capital and extending his influence further up the rivers into central Mesopotamia and perhaps beyond.

This unification of the Sumerian cities presented an opportunity to their Semitic neighbours in nearby central Mesopotamia, the Akkadians. The first man known to have ruled an empire in the true sense was the Akkadian King Sargon the Great, who appears to have begun his conquests by capturing the city of Kish and then attacking Uruk, where he overthrew Lugalzagesi and 'led him in a collar' from the battlefield. Inscriptions say that he destroyed the cities of Umma and Lagash and laid waste their territories, and that his god Enlil thereafter gave him all the land between the Upper Sea and the Lower Sea (the Mediterranean and the Persian Gulf respectively), thus making the whole of Mesopotamia his own. He made the city of Akkad in central Mesopotamia his empire's capital, and from there, in a reign that might have lasted over fifty years, prosecuted successful war on Elam to the east and the Hurrians of Anatolia to the north. References in inscriptions to his annexing the Cedar Forests and the Silver Mountains imply that his territories stretched from the Levant on the Mediterranean to the Aladagh Mountains near the Caspian Sea. If so, his boast that he was 'King of the World' has a measure of justification, given the date.

Sargon's city of Akkad itself has still not been found; locating and excavating it will be one of the great jewels of Near Eastern archaeology when it happens.

As was inevitable, legends accumulated around so striking a figure as Sargon. He was said to have had humble origins, the son of a gardener at the palace of the King of Kish. One legend says this king made Sargon his cup-bearer; another legend says he was the illegitimate son of a high priestess, an *entu*, who put him in a basket of reeds in which he floated down the Euphrates before being found by a gardener, who brought him up.[11] But, because he lived in the age of writing and was so significant a figure, part of what was written about him and his family belongs to record rather than legend. The names and careers of his wife and a number of his offspring are known, including his sons Rimush and Manishtushu and grandson Naramsin, all in turn kings after him. Indeed all the kings of Assyria and Babylon who came after him, for fifteen hundred years, saw themselves as in some sense his heirs.

Sargon maintained an elite corps of five thousand men who, the records say, 'ate bread daily before him', thereby cementing their bond of loyalty. He consulted them before military expeditions, some of which he undertook to safeguard the interests of Akkadian merchants in neighbouring lands. But by the end of his long reign there was trouble in the empire: a period of famine resulted in insurrections that he was obliged to quell by force, though until his death murmurings and rebellions persisted across the empire. Nevertheless in the hands of his heirs the empire lasted for several centuries more, reaching its height between 2400 and 2200 BCE. During this time Akkadian became the region's dominant language, though written in Sumerian script. Sumerian itself was retained both as the patois of the empire's 'black-headed' subjects and as the language of temple ritual. After the fall of the Akkadian Empire its territories split into two principal parts: Assyria in the north and Babylonia in central Mesopotamia, from which southern Mesopotamia, the original Sumerian heartland, was ruled – by then a depopulating region because of climate change.

History – the period of written records, lasting from then until now – thus began in Sumer, and with it literature. *The Epic of Gilgamesh* was to its own and successive ages what Homer was to the Greeks and what the biblical texts were to the millennium between the fall of Rome in the West and the Renaissance that heralded the beginning of modern times. That is to say, it was the story, the text, *the book*, that was central to the self-conception of a civilization. In Sumerian the name 'Gilgamesh' appears as 'Bilgamesh', the hero of a set of poems dating from around 2100 BCE about a king of Uruk. These poems were combined and embellished over succeeding centuries in Akkadian into the versions of *The Epic of Gilgamesh* that have been recovered and pieced together from many clay tablets and fragments from the *tells* of Mesopotamia. The best copy of the *Epic* so far found was discovered in the library of the Assyrian King Ashurbanipal at Nineveh, dating from the seventh century BCE. I recount the *Epic* in Appendix II.

So many themes of later mythology and story are embedded in the tale of Gilgamesh that one could devote an entire book to them, from Deucalion and Noah, through every hero and giant-slayer of legend, to feral protagonists such as Tarzan and Mowgli, glancing on the way at fire-breathing monsters, zombies, visits to the underworld or the past for knowledge, allusions to the psychology of dreams and the power of sexuality, connecting the themes of friendship between Gilgamesh and Enkidu to Achilles and Patroclus, Jonathan and David, Nisus and Euryalus – and so richly and multiply on.[12] *The Epic of Gilgamesh* is the fountain of literature, and even its techniques are discernible in later work. It is of intense interest to speculate on the point that this epic existed from very early as a written text, unlike the epics of Homer, whose style and structure embody their oral origin in the use of stock epithets and repetitions required for memorization and extemporization. In the case of *The Epic of Gilgamesh* the repetitions appear to serve a different purpose; for example, in the journey to Humbaba's forest the length of each day's march and the breaks for rest and food are repeated before each of Gilgamesh's dreams, and in the same terms, but this seems to be to provide the reader (and, more numerously no doubt, the hearers of the text when read aloud) with a sense of how long and arduous the journey was. Elsewhere in the text the narrative and descriptions do not have the Homeric character of oral recitation.

So much is by way of speculation; what is undeniable is the seminal nature of the themes in *The Epic of Gilgamesh*, and the proof it represents of the degree of elaboration and maturity of the civilization from which it sprang. Combine this with the exquisite art and architectural sophistication of the Sumerian and Akkadian worlds, their advanced economies, and the technology and agriculture that underlay them, and it is a wonder that acquaintance with these several thousands of years of high history was lost until so recently.

Another striking text from these recovered millennia is *The Code of Hammurabi*. If there is a name that has most entered general consciousness of the deep past, it is that of this king of Babylon who reigned in the eighteenth century BCE. He was the sixth and greatest king of the dynasty that succeeded the Akkadian Empire. He brought under Babylon's control the territory of Elam and the cities of Larsa, Mari, and Eshunna, and made Assyria in the north of Mesopotamia a tributary state. His great claim to fame is his Law Code. It was not the first code to be promulgated – the Sumerians had a register of restitutions and compensations for losses caused by crime – but Hammurabi's innovation was to decree punishments for wrongdoers as well as restitution. He had the laws incised on a stone stele and mounted in a public square where everyone could read it or have it read to them. Elamite conquerors of Babylon in later years took the stele as a trophy to their capital at Susa, where it was found by French archaeologists in 1901. It is now in the Louvre in Paris. A sketch of some of its principal provisions is given in Appendix III.

The *Code* is educative reading for a number of reasons, not least for the window it opens on to life in Babylonia in the first half of the second millennium BCE. When the text was combined with the material culture unearthed by archaeologists, it became apparent that the Babylonian Dynasty was Amorite, the Amorites being a nomadic people who had entered the Akkadian Empire probably from the Levant in the course of the third millennium and settled in the mid-Mesopotamian region. Their town of Babylon was insignificant at first, and its ascendancy under the dynasties that produced Hammurabi was short-lived, for by the middle of the second millennium it had been conquered by the Hittites from Anatolia, and then by Kassites from the Zagros Mountains in today's Iran. The Babylon of the bible is a much later place, indeed a thousand years later; by then it was the capital of the Neo-Babylonian (sometimes called Chaldean) Empire, dating from the late seventh century BCE and having as its best-known king Nebuchadnezzar II, who destroyed Jerusalem in 587 BCE and took the leaders of the Israelites into captivity.

By the sixth century BCE we are in historical terrain already familiar before the nineteenth-century discoveries were made. A complete account of the Fertile Crescent would include Dynastic Egypt from the end of the third millennium, when tradition has Menes become the first king of a unified Egypt, having brought the Upper and Lower kingdoms together. Egypt's history from this point is a brilliant one in terms of its material culture and organization; the difference in comparison to Mesopotamia is that its occupancy of a narrow band along the Nile, and its separation from Mesopotamia and Anatolia until much later in history, made it less of a cultural crossroads and therefore less susceptible to the changes that saw the rise of Sumerian, Akkadian, and subsequent centres of civilization. The Egyptians had one direction of expansion – south into Nubia, and Nubia was its southern limit. Nubia itself had been the home of the Kerma civilization before conquest by Egypt's Thutmose I in the sixteenth century BCE. Much later, in the eighth century BCE, a Kushite power arose in Nubia and defeated Egypt, ruling it for a century – one complete dynastic cycle, the twenty-fifth – before being replaced by a native Egyptian dynasty. The Kingdom of Kush lasted for a thousand years in its own right before being conquered by Ethiopia.

All this too is recent knowledge: the searchlight of history has not shone on Mesopotamia alone in the course of the last two centuries.

The rediscovered history of the Near East contains a major mystery: the collapse of its civilizations around the year 1200 BCE, bringing the Bronze Age to an end, introducing a Dark Age of several centuries in length, from which the Iron Age emerged.

Trade, alliances, and conflicts, and later the rise and fall of kingdoms and empires, had covered the Near Eastern world in a network of relationships at least since the early fourth millennium, if not before. By the second millennium, up to approximately the year 1200, the Near East and eastern Mediterranean was an intensely interconnected region with a complex pattern of trade both in raw materials and finished goods. Key to the economies of the time was the trade in copper and tin, the ingredients of bronze. As one historian has described it, copper and tin were to that age what oil was to the world economy of the twentieth century. But they were not the only commodities traded. As the cargo found in a shipwreck off Uluburan shows, the commercial and industrial life of the time was richly varied.

In 1982 a Turkish sponge diver discovered the wreck of a ship off the Grand Cape, Uluburan, on the Anatolian coast near Kaş. Ten years of marine archaeology followed, opening a remarkable window on to the Late Bronze Age. When the ship sank, it was carrying ten tons of raw copper in the form of oxhide ingots, rectangles with protruding handles to make carrying them easier. About a tenth of the ingots were especially shaped to be carried on pack animals for long-distance overland transport. The ship also carried a ton of tin – one part tin to ten of copper is close to the ratio for bronze. Of a hundred and fifty Canaanite jars, a type widely used in the Aegean and Near East, most were found to contain terebinth, a resinous substance used as a kind of turpentine, while others contained olives and glass beads. There were close to two hundred ingots of glass, coloured turquoise, lavender, and cobalt blue. There were logs of African ebony, ivory from elephant tusks and hippopotamus teeth, tortoise shells and ostrich eggs, pottery and oil lamps from Cyprus, bronze and copper bowls, gold, silver, and semi-precious stones both in unworked and worked states as ornaments, a collection of swords, daggers, and spear-heads, a collection of tools including axes, sickles, and adzes, and foodstuffs including pine nuts, almonds, grapes, figs, olives, pome-granates, and spices. The ship carried ten stone anchors, and the cargo was packed together with dunnage – miscellaneous stuff used both for ballast and to keep the cargo from sliding about.

From this bill of lading one can picture the sophisticated world it represents. It explains, for an especially illustrative example, the wealth and elaborate architecture of a city that stood at a focal point of the trade routes by land and sea connecting Mesopotamia and the Levant with Egypt and Mycenaean Greece: the port city of Ugarit.

Ugarit was an ancient city, its oldest walls dating to 6000 BCE, but the height of its flourishing occurred in the middle of the second mil-lennium, by which time it had become the perfect entrepôt between the cities of the Tigris and Euphrates, Egypt, the Hittite Empire on the Anatolian Plateau with its capital at Hattusa, and the sea lanes of the Mediterranean to Crete, Mycenae, and elsewhere. It was the leading international port of the age, and its grand palace was built in stone, its earlier avatars having been made of clay.

The site of Ugarit is on a headland in the outskirts of today's Lata-kia in northern Syria. It was rediscovered in the 1920s by a farmer whose plough opened a tomb on the ancient city's necropolis. Exca-vations uncovered an enormous palace complex, mansions of wealthy citizens, and temples to the gods Baal and Dagon. Best of all was the discovery of fifteen hundred mainly cuneiform texts dating to the thir-teenth and twelfth centuries, among them nearly fifty epic poems, the most famous being *The Legend of Keret*, *The Tale of Aqhat*, and *The Baal Cycle*. The first tells of King Keret of Hubur, something of a Job figure whose many misfortunes are compounded by the failure to keep a promise to a god. *The Tale of Aqhat* relates how the eponymous hero spurns the advances of the goddess Anat even after she offers him immortality and a chance to sleep with her, and is killed at her vengeful instigation; Aqhat's father, Danel, a righteous man, who had gone through much to have a son in the first place, recovers Aqhat's 'bones and fat' from the stomach of the queen of the vultures so that he can bury him properly. *The Baal Cycle* recounts Baal's victory over Yam, who was ambitious to be king of the gods; there follows a series of conflicts in which Baal and his chief rival Mot are respectively killed and ground into fragments, but resurrect, Baal eventually being triumphant. As with *The Epic of Gilgamesh*, the themes are strongly reminiscent of stories that recur in myth and scripture in later centuries, including the bible.

The significance of the Ugaritic texts, apart from the literature they contain, is that they are written in the Sumerian, Hurrian, and Akkadian languages in cuneiform, with some in Egyptian hieroglyphs and Cypro-Minoan. Just as Sumerian had survived as the language of sacred ritual, Akkadian survived as the language of law. In the texts are found cuneiform alphabet lists and early signs of the Phoenician alphabet.

Ugarit was under Egyptian control for part of its history – the Egyptian Empire by the second millennium extended up the Mediterranean coast of today's Israel, Lebanon, and Syria beyond the site of Ugarit, a region it contested with the Hittite Empire ruled from Hattusa. As a result of war between the two empires Ugarit changed hands, becoming a vassal of the Hittites, and was so at the time of the great collapse around 1200 BCE. Among the texts recovered from the site of Ugarit is a letter from Ugarit's last king, Ammurapi, to the king of Cyprus, reporting that he was under threat from invaders and begging for help; shortly afterwards the city is reported as sacked and its food stores and vineyards destroyed.[13]

This was one event in a widespread devastation, engulfing the cities of Greece, the Levant, Anatolia, and Egypt, and disrupting kingdoms further east also. It happened with suddenness in the decades either side of the year 1200. A reference to a mysterious invader called the 'Sea Peoples' in an Egyptian text led historians to wonder who these locust forces could be, and where they came from. Egypt was the only centre in the western part of the Near East that partially survived the attacks, in a much weakened and diminished condition;

Assyria and Elam further east were less affected and survived, though also diminished, one reason being that trade with the Western centres of civilization ceased.

The collapse was dramatic. In Anatolia the Hittite capital Hattusa was burned to the ground and permanently abandoned. So was the city of Karaoğlan; its dead were left unburied. Troy was destroyed and lay derelict for a thousand years. The towns of Enkomi, Sinda, and Kition on Cyprus were sacked, burned, and abandoned. Ugarit was just one of a number of Levantine centres to fall; Kadesh – site of a major battle between Egyptians and Hittites in the thirteenth century – Qatna, Aleppo, and Emar were among the others. Further south the towns of Gaza, Ashdod, Acre, Jaffa, and Ashkelon on the coast, and the inland towns of Bethel, Eglon, Debir, and Hazor, were destroyed. Mycenae in Greece disappeared, and along with it the settlements of the Peloponnese in its vicinity. Thebes, Tiryns, and Pylos were destroyed likewise.

In 1855 CE a French Egyptologist at the Louvre found an inscription from Medinet Hebu describing an invasion by 'Sea Peoples' during the reign of Rameses III. That text and others named the 'Sea Peoples' as a combination of 'Denyen, Ekwesh, Peleset, Shekelesh, Sherden' and others, whom scholars much latter associated respectively with Danaans, Achaeans, Philistines, Sicilians, and Sardinians – 'Peleset' suggests Palestine, hence Philistines; 'Denyen' and 'Ekwesh' are reminiscent of Homer's 'Danaans' and 'Achaeans', the near-homophony of 'Sherden' and 'Sardinia' is taken as suggestive – and so on. Some of the identifications are plausible, but the inference that a federation of peoples from the Mediterranean and western Eurasia combined to mount an invasion of the Near East is unsupported by any other evidence. Instead a consensus has grown that a combination of factors resulted in the collapse of the Bronze Age: earthquakes, food shortages resulting from climate change with resulting migrations of refugees, unrest, and uprisings, would have so weakened the polities of the region that attackers, even in relatively smaller numbers than a putative massive 'Sea Peoples' invasion, could have triggered a collapse – because the degree of interdependence of the region through trade was such that disruption to it would rapidly bring down all of the cities and states involved. One need only imagine if, today in our highly interdependent world, energy- and food-supply chains were cut, resulting in empty supermarket shelves, cars and trucks unable to move, and no electric power, to see how thin is the veneer of civilization without the basics that keep it going. Inhabitants of the Late Bronze Age were probably more resilient than we are today, from the point of view of feeding themselves and making do; but the Dark Age that followed the collapse indicates how serious it was.

For one thing, writing vanished from the Mycenaean world. Linear B, the only Bronze Age Mycenaean script to have been deciphered so far, ceased to be used, the decorations on pottery became simpler, stone building all but ceased, and the largest dwellings in villages, assumed to be chiefs' houses, were no longer huge palace complexes of stone but thatched mud huts like the others round them. Differences between localities increased, evidence that travel and regional inter-connectivity were absent or much rarer. The very first indications of recovery are some of the burial practices of the tenth century BCE on Greek islands, but it is chiefly in the eighth century, when alphabetic systems of writing originating in Phoenician script spread across the eastern Mediterranean, that the shoots of returning civilization are clearest. Burial practices involving ornaments, horses and weapons placed in a grave alongside a corpse, showing that the dead person had been a significant individual, speak of a society where there is hierarchy and enough surplus wealth to allow some of it to be permanently lost in a burial. Such a grave was found at Lefkandi on the island of Euboea, dating from about 950 BCE, and graves excavated from the subsequent centuries show a continuation of the practice.

The loss of writing is a significant indicator of decline. Where there is trade, there is a need for records. Where there is centralized government and a palace culture, there is need not just for record-keeping but for laws and diplomatic and commercial communication. This requires literacy, which requires schools. Where there is education to produce scribes, literature is possible. The evidence from the Bronze Age is that the increasing sophistication of social, political, and economic organization was accompanied by literacy and literature, communication, and the exchange of ideas and knowledge, in a feedback loop that prompted yet greater sophistication.[14] The Bronze Age Collapse put an end to that for centuries afterwards.

Whatever the original cause or combination of causes, the end of Bronze Age civilization was a 'systems collapse', as the anthropologist Joseph Tainter aptly called it, in which disruption to the high degree of mutual interdependencies that sustained it caused the entire edifice to fail.[15] Eric Cline and others have pointed to the troubling similarities between the Late Bronze Age world and our own: too complexly administered centralized societies, over-specialization in almost every aspect of life and economy, fragile dependencies on essential resources such as food and energy – all these characterize our world now, and disruption to this tissue of connections, whether in the form of cyber-attack, major climate disaster, large-scale conflict, or pandemic disease, could readily knock the whole structure to pieces.

As noted, what emerged from the post-collapse Dark Age was – to continue using these over-simplifying but serviceable labels – the Iron Age. And, as the label implies, a significant feature of the newly emerging dispensation was the utilization of iron for weapons and tools. It was the agent of a number of important changes.

Iron had already begun to be worked, principally in the region of south-eastern Europe abutting the Black Sea. Although requiring higher temperatures to smelt, the advantages of iron over bronze – other than aesthetic – were many: it was easier to work and much less expensive, which meant that more people could acquire sharp-edged weapons, and more such weapons could be made. Perhaps one aspect of the Bronze Age Collapse was an inward migration of people, driven by hunger and armed with iron swords and spears – perhaps Dorians, or Scythians, and others; people from the north and west of the Bronze Age world. The warriors among them would have been more numerous, because the supply of their cheaper, better, and more plentiful weapons was greater than the supply of bronze weapons to those opposing them, whom they defeated.

Such an advantage would sooner rather than later be cancelled by the adoption more generally of iron for weapons. Of greater eventual significance therefore was the use of iron for tools, especially in agriculture, a resulting increase in productivity helping to repopulate regions depleted in the collapse's aftermath. But that took centuries.

In the discussion in Part I, Section 1 above of the evolution of technology, the domestication of the horse and invention of the wheel were given birthplaces in the Pontic–Caspian Steppe in the fourth millennium BCE. Given the significance of greatly increased mobility made possible by horses and their later combination with wheels, not least in relation to questions of the spread of technology, culture, and languages, the *where* and *how* of such spread are both important questions. This has been the locus of considerable debate not just among historians and archaeologists but, increasingly in the decade before these words were written, geneticists also.

A sketch of the matter goes as follows. Archaeological and linguistic evidence – the latter on the relationships and spread of Indo-European languages from a source in an original Proto-Indo-European (PIE) language – suggested competing hypotheses about the source and manner of technological and cultural spread. One hypothesis, formed using linguistic evidence, was that PIE-derived languages spread by migration of their speakers following the turn to sedentary farming in the Neolithic revolution, a process in which pre-PIE speakers began to spread both in north-westerly and south-easterly trajectories from Anatolia, into central and western Europe and India respectively (the regions in both directions until then occupied by hunter-gatherers). This is hypothesized to have happened from about 7000 BCE onwards. The chief proponent of this view is the archaeologist and palaeolinguist Colin Renfrew.[16]

The competitor hypothesis, more widely accepted than Renfrew's view, was put forward by the Lithuanian-American archaeologist Marija Gimbutas, who, following earlier suggestions to the same effect by Gordon Childe, argued that the homeland of PIE speakers was the Pontic–Caspian Steppe among the people of the Kurgan culture (*kurgans* are burial mounds).[17] Later, David Anthony developed the theory by focusing on a particular kurgan-burial group, the Yamnaya, locating them on the northern shores of the Black Sea, where they evolved culturally from hunting and gathering to herd-ing, developing horse-management skills in the process. Then, when in the late fourth millennium the climate of the steppes became drier and cooler, they employed their equestrian skills and the wheel to became mobile and to move, first out on to the steppes and then towards Europe in one direction and Central and South Asia in the other, taking the root of Indo-European languages with them.[18]

The linguistic evidence of PIE-derived languages indicates spread from a single original source. Before genetics came along with some answers, the question was: how did it spread? Archaeologists looking at the *material* remains of Neolithic cultures had been exercised by the question whether their spread occurred as the result of migrations of people or by cultural diffusion. Did Anatolian farmers move their families and tools into Europe, or did Balkan neighbours of the Ana-tolians acquire the idea of farming from them, their own further-off neighbours following suit later, and so on by a relay of imitation? This latter view was summarized as 'pots, not people' – that is, pot-tery styles, not their makers, doing the moving via imitation. In the period after the Second World War there was a desire – thanks to the promptings of psychology at the end of a dreadful war – to think that if cultures had indeed spread by physical movements of people, then it was peaceful 'demic diffusion', rather than violent conquest. The ground for this view was that, whereas pottery styles and tool types might spread by imitation, it seems less plausible to think of lan-guages doing so; a certain tension therefore arises between what can be inferred from linguistic evidence and material evidence. One way some archaeologists chose to avoid the difficulty was to eschew big-picture hypothesizing about competing models of diffusion, and to focus on what is unearthed at specific sites: what you might call a 'local history' version of archaeology.

But, when sequencing of ancient DNA became possible, not only did big-picture hypothesizing re-emerge but with a surprising result. It not only showed that people had indeed moved, but that they significantly – in some cases almost wholly – replaced the popula-tions that had been there before their arrival. In central and northern Europe the appearance of Corded Ware Culture (so called because pots were decorated by pressing cord or string on to clay to leave patterns) correlates with Yamnaya migration into the region. The genome of 75 per cent of the subsequent population is derived from Yamnaya immigrants. In Britain the people who built Stonehenge had been almost completely replaced by incomers within a few centuries of completing the last iteration of the great monument, that is, by 2000 BCE: 'high levels of steppe-related ancestry ... was associated with the replacement of approximately 90 per cent of Britain's gene pool within a few hundred years [of the completion of Stonehenge], continuing the east-to-west expansion that had brought steppe-related ancestry into central and northern Europe over the previous centuries'.[19] In Spain the genetic evidence shows that, following the migration, incoming males had exclusive access to the indigenous females, so that what remained of pre-migration genes in the mixed gene pool was transmitted by females alone.

The genetic data speaks of two waves of migration into Europe, strongly supporting a steppe origin for Proto-Indo-European.[20] From about the seventh millennium BCE onwards farmers moved into Europe from the east, occupying Hungary, Germany, and Spain, having reached Anatolia earlier. Pioneering farmers from the latter moved to Greece, some proceeding along the Mediterranean coast to Iberia and some up the Danube into Germany. Their descendants retained 90 per cent of their DNA, showing that they did not intermingle with the indigenous hunter-gatherer populations they met. But over time – after about two thousand years – hunter-gatherer genes began to make a comeback in the population, attaining an additional 20 per cent of the genome.[21] One factor in this was that the spread of farming stopped short of a region some several hundred kilometres deep along the Baltic coast, whose climate and soil were less suitable for agriculture. The area was occupied by megalith-building hunter-gatherers – called the Funnel Beaker Culture because of their pottery style – and their adoption of the practices and technology of the incomers was slow, taking over a thousand years.

The predominantly farmer-genome balance of Europe's population remained stable for two thousand years, until 3000 BCE. Hunter-gatherer groups with no admixture of farmer genes were increasingly rare and living in remote and mutually isolated areas. Farming populations, especially in southern Europe, developed sophisticated social structures that Marija Gimbutas described, on the evidence of elegant female figurines and other archaeological evidence, as matrilocal.[22] Then a dramatic change occurred. 'In remote Britain, the megalith builders were hard at work on what developed into the greatest man-made monument the world had seen: the standing stones of Stonehenge,' writes the geneticist David Reich. 'People like those at Stonehenge were building great temples to their gods, and tombs for their dead, and could not have known that within a few hundred years their descendants would be gone and their lands overrun. The extraordinary fact that emerges from ancient DNA is that, just over five thousand years ago, the people who are now the primary ancestors of all extant northern Europeans had not yet arrived.'[23] These 'primary ancestors' are the Yamnaya, whose massive migration into Europe replaced the population they encountered there and came to dominate 75 per cent of the successor genome.

On the way to these conclusions the gene evidence threw up some ancillary surprises. One is that today's Europeans not only have genes from the Yamnaya and a mixture of their hunter-gatherer and farmer predecessors but also a third, 'ghost' gene linked – astonishingly – to Native Americans. Because it was exceedingly unlikely that ancestors of the Apaches and Sioux had made their way across the Atlantic and mingled with Bronze Age inhabitants of Europe, the idea was proposed that there had been an ancestral Siberian population, some of whom had gone east and crossed the Bering land bridge into the Americas, and some of whom had gone west and mingled with hunter-gatherers in Europe. The ghosts were given the name 'Ancient North Eurasians', and researchers waited to see if archaeology could provide confirmation. It came in the form of DNA from the 'Mal'ta Boy', found in Siberia, who lived twenty-four thousand years ago.[24]

Another surprise offered by genetics concerned the ancestry of modern horses. The first indication of a human–horse relationship that went beyond humans merely hunting and eating horses is found in the fourth millennium Botai Culture situated in ancient Kazakhstan. The Botai, who are genetically more closely related to Ancient

North Eurasians than to the Yamnaya, corralled, milked, and butchered horses, and there is evidence that they harnessed and rode them, a likelihood given that this would be the best way of herding them. But genetic analysis of Botai horse remains reveals that they are not the direct ancestors of modern horses, contributing only about 3 per cent of the modern horse genome. Instead they are related to Przewalski's Horse, otherwise known as the *takhi*, or Mongolian wild horse. This suggests that Yamnaya horse culture is at most the result of learning from neighbours; they captured and bred horses independently of the Botai.[25]

Another oddity concerns the Bell Beaker Culture of the third millennium, named for the bell shape of the pottery, which superseded Corded Ware Culture, not smoothly across Europe but in random pockets. It appears to have begun in Iberia and spread in saltatory fashion to various parts of central and north Europe and the western (i.e., British) islands, in the beginning existing alongside Corded Ware but then replacing it.

In the later stages of the Bell Beaker phenomenon it produced a regionally diverse richness of craftwork, not just in pottery but in gold and copper ornaments, house-building styles, and funerary practices (in some places burying, in others cremating the dead), nevertheless suggesting shared social outlooks too.

The puzzle is that the sporadic nature of Bell Beaker appearances in the archaeological landscape revives the question of whether it was spread culturally or by movements of people. One suggestion is that it was spread by a *few* people – craftsmen, travellers, highly skilled and therefore welcomed and imitated in the places they visited. This combines the cultural and demic diffusion models in a plausible way.[26]

The two sure handholds in the puzzles of this period are genetics and linguistics. What the former reveals is sketched in the foregoing; the language evidence merits more of a mention, as much for its interest as its importance. It relates to the fact that the languages spoken in Europe, Iran, and the Indian subcontinent are members of a family of languages, the Indo-European languages, all descended from PIE.

The relationship between the languages of Europe and India was first noted by a British judge in India, Sir William Jones. Like all men of his time and class, he had been brought up in the classical tongues of Greek and Latin, and after he arrived in India he taught himself Sanskrit. It did not take long for him to note the striking similarities between it and ancient Greek and Latin. In his third presidential address to the Asiatic Society of Bengal in 1786 he made the much quoted remark that

> The Sanskrit language, whatever be its antiquity, is of a wonderful structure; more perfect than the Greek, more copious than the Latin, and more exquisitely refined than either, yet bearing to both of them a stronger affinity, both in the roots of verbs and in the forms of grammar, than could possibly have been produced by accident; so strong indeed, that no philologer could examine them all three, without believing them to have sprung from some common source, which, perhaps, no longer exists.[27]

Even the inexpert eye can see affinities. The word 'father' in Sanskrit is *pitar*, in Greek *pater*, in Latin *pater*. 'Two' in Sanskrit is *dva*, in Latin *duo*, in Greek *duo*. (Try saying the English word 'two' by pronouncing the *w*, and making the *t* heavy, like the American *t* in 'butter' – *budder*.) 'Seven' is *sapta* in Sanskrit, *septem* in Latin, *hepta* in Greek. 'Foot' is *pad* in Sanskrit, *ped* in Latin, *pod* in Greek. The brothers Grimm, famous for their fairy-tales, were linguists who noted how certain consonant sounds change between languages, in particular how *p* becomes *f*, *t* becomes *th*, and *k* becomes *ch*.[28] One can intuitively see how the transformation works. Say *t* and note the position of the tongue: it presses against the back of the teeth, then releases to make a little explosion of air. Now do it lazily, very softly approximating the tongue to the back of the teeth. The result is *th*. Try this with *p* or *b*: the lips are pressed, then released for the air to carry away a plosive sound. Now do it lazily, not pressing the lips together but leaving them slightly apart. The result is *f*. Apply these changes to the Sanskrit word for 'father', *pitar*, and you get 'father'. Not all consonants change at the same time: Sanskrit *brhata* is 'brother', courtesy of the *t* to *th* change, leaving *b* alone, whereas in Latin and Greek it is the *t* that is left alone and the *b* becomes *f*: respectively *frater*, *phrater*. Across contemporary languages one sees similarities between *p*, *b*, *f*, and *v*; think of how an English speaker must adjust to pronunciation of certain Spanish words where *d* and *th* are practically the same: 'Madrid' sounds like *Mathrith* in a Spaniard's mouth. Variations on *k* and *ch* are common between Germanic and Romance descendants of Indo-European: the latter either pronounced as if gently clearing the throat, as in Greek *xi*, or more emphatically as in 'church' – the sounds vary according to whether the tongue is pressed on the palate for *k* or kept from it for *ch*; Old English *kirk* (German *kirche*) thus becomes 'church'.

The difference in the pronunciation of certain consonants helps with constructing a family tree for the Indo-European languages. Western branches of the family are called 'Centum Languages', *centum* being Latin for 'hundred', while Eastern branches are called 'Satem Languages', *satem* being the Avestan word for 'hundred'. (A *c/s* difference appears in English with how we say, for example, 'candle' and 'centre'.) Centum Languages include Greek, Latin, Celtic, and the Germanic languages (Dutch, Danish, Norwegian, English, Swedish, Yiddish), together with some interesting extinct languages such as Hittite and Tocharian. The Satem Languages include the languages of Iran and India – Farsi, Bengali, Gujarati, Hindi, Urdu, Marathi, Sindhi, Punjabi, and so on – and the Slavic languages, including Latvian, Lithuanian, Polish, Czech, Russian, Slovenian, Bulgarian, Serbo-Croat, Sorbian; and on a branch of their own, Albanian and Armenian also.

Only when written forms of language appear, preserved on materials that can be dated along with other artefacts, can their chronologies be ascertained. An important point is that a language spoken or written in a given area is not by itself proof that its speakers moved into that area in large numbers, still less that they replaced a previously indigenous population there. Look at the example of French as the language of law, government, and the aristocracy in England after the Norman conquest. The Normans themselves were not French but Scandinavians by descent, who had adopted French, and French accordingly became the smart language of England for a couple of centuries after 1066, with English then reasserting itself because it was the everyday language of the majority. The same happened with

English in Ireland and Scotland, and in India and all other regions of the British Empire. From the seventeenth to the nineteenth centuries in Europe, French was the language of diplomacy, and in Poland and Russia it was the preferred language of the upper classes and the nobility. Many people will have forgotten that there was a debate in the newly formed United States about whether the country's official language should be that of the just-defeated enemy, English, or German – given the number of settlers of German origin then in Pennsylvania and New York (the 'German Belt' now stretches numerously across to Oregon via the northern states.)[29] As all this shows, language and population do not necessarily go hand in hand.

At the same time questions about the *origins* of this large family of languages and its first spread are analogous to questions about the Big Bang and the universe's origins: extrapolating backwards, so to speak, the family shrinks and shrinks to a parental point. The coupling of the genetic data with the linguistic data is very strong evidence that the subsequent history of Indo-European languages is largely a matter of movements of its speakers and their descendants in the way described above. PIE – Proto-Indo-European – is thus given an *Urheimat* on the Pontic–Caspian Steppe among the people of the Yamnaya culture, and was carried from there by them and their descendants eastward and westward – and then by their westward descendants, very much later, in the time of colonialism and European empires in recent centuries, all over the world. It is quite something to have a series of successively world-transforming developments traced back to a spot on the map at a date in time.[30]

2.   The Coming of Humanity

The foregoing addresses the recently rediscovered past of humankind since about twelve thousand years ago, the epoch known as the Holocene. This is prehistory and – with the advent of writing – history. Another aspect of the past that has even more recently come into view is the evolution of humankind itself, in the tens of thousands, hundreds of thousands, millions of years leading to the present – a story inscribed in bones and stones, and not very many of the former in sum: a fascinating, ambiguous, puzzling, and surprising new story, even more recently discovered than the story of history before classical times.[1]

For all that the more that is known about humanity's evolution the more puzzling it becomes – the paradox of knowledge at work – the broad outline of the story is familiar. The hominid lineage that eventuated in *Homo sapiens* diverged from its common ancestor with the chimpanzee about 6 to 7 million years ago. Anatomically modern humans, *Homo sapiens*, began what proved to be a far-reaching migration out of Africa before fifty thousand years ago. By about fifteen thousand years ago *Homo sapiens* existed in every corner and clime of the planet, and was the only species of *Homo* left out of a number.[2] In the early phase of investigation into human evolution it was assumed to be a linear story, illustrated by the familiar queue of ape to ape-man to modern man ('man', *in sensu* male of the species, is the icon in these depictions), successively standing straighter and taller. But the detailed and impressive sciences of palaeoanthropology and genetics tell a much more complicated story.

The *immediate* prehistory of humanity might be regarded as the period from the migration of anatomically modern humans, *Homo sapiens*, out of Africa from about 60 to 50 kya. This was not the first time anatomically modern humans had moved out of Africa into and through the Near East, reaching as far as the Balkans; indeed there might have been more than one such previous endeavour. The remains of anatomically modern humans were found at sites in today's Israel, Mugharet-es-Skhūl and Jebel Qafzeh, dating between 90 and 120 kya – a period of warm climate – sharing tool technologies similar to Neanderthals. These forerunners were extinct in that region by 75 kya. Other forms of archaic *Homo*, notably but not only *Homo erectus*, had long since spread throughout Eurasia, starting at least 1.8 mya.

These remarks relate to one of the key questions about human evolution: the difficulty of ascertaining which, among a number of hominin species, are to be identified as standing in direct line of ancestry to *sapiens*. As palaeoanthropological fossil hunting has increased in intensity and scientific exactness, more and more hominins – hominins or hominin species – have been found, in at least one case with ambiguous mixtures of more modern and less modern features while yet dating from relatively recent times, this being *Homo naledi* from the Rising Star Cave System in South Africa. This variety, though, should not be a surprise; today there are at least 78 species (some say 132, some 148) of Old World monkeys in two subfamilies, the *Cercopithecinae* and the *Colobinae*, spread across sub-Saharan Africa and South and East Asia, with an outlier group on either side of the Strait of Gibraltar. Their number and variety bespeak specialization and separation among ancestral groups. To understand specialization, think of the beaks of Darwin's Galapagos finches, one of the thirteen or more species having powerful stubby beaks for nut cracking, another with longer thinner beaks for reaching into flowers, another adapted to catching insects, and so on.[3] Variation allows different food niches to be exploited in the same area. Evolutionary divergence owing to geographical separation is illustrated by chimpanzees, the *panins*. There are two kinds, the larger *Pan troglodytes*, known as the 'common' or 'robust' chimpanzee, and the more gracile bonobo *Pan paniscus*. The difference between them is in part accounted for by the fact that their territories lie north and south of the Congo River respectively, and that panins are not good swimmers.

It would accordingly be a surprise if *sapiens* had a single ancestral line; nature does not appear to work that way. Today among the great and lesser apes (the *Hominidae* and *Hylobatidae* respectively) there are separate populations of gorillas, chimpanzees, orangutans, and gibbons that differ more than any *sapiens* individuals differ from each other. The lesser apes consist of four genera and sixteen species of gibbon; the great apes also consist of four genera, comprising three species of orangutans (one very recently discovered and nearly extinct), two of gorillas with either four or five subspecies, two of chimpanzees (*Pan troglodytes* has four or five subspecies also) and one of humans. These are the extant species of each genus; each has a number of extinct branches in its family tree. In comparison with the 7 billion members of *Homo sapiens* on the planet today, the other genera of great apes are tiny in number.

Many challenges face palaeoanthropology. Although thousands of hominin fossils have been found in the last century and a half, at increasing rates both of finds and of developments in the techniques of analysing them, and, although the finds reach back over 6 million years to the point where hominins began to diverge from other hominids, most are from more recent stretches of this period. Moreover all of them are from sites most conducive to being found – in caves or in places where geological events have exposed strata and sedimental layers, as in the east African Rift Valley. Given that less than 3 per cent of the land area of Africa has been searched for fossil hominins, the story of human evolution is of necessity a very patchy one. The African Rift Valley is formed (is continuing to be formed) by the separation of three tectonic plates, exposing layers of geology stretching back over millions of years; it is the pool of light under the lamp-post where the lost car keys of the evolutionary past are most sought, because most other places are just too dark.

The focus on Africa is, however, not exclusive; South and East Asia, the Near East and western Eurasia all figure in the story. Africa was identified as the probable home of humankind by Darwin in the *Descent of Man* on the basis of T. H. Huxley's anatomical comparisons of human beings with gorillas, chimpanzees, and orangutans, showing humans to be closer to the African great apes, gorillas and chimpanzees, than to the East Indies' orangutans.

Palaeoanthropology and in particular genetics provide strong evidence that this is correct. The dispersion of early *Homo erectus* out of Africa 1.8 mya was followed considerably later by *Homo heidelbergensis* – the likely ancestors of Neanderthals, Denisovans, and *sapiens* – about 0.7 mya. Other hominins might also have dispersed into Eurasia, as suggestive but ambiguous genetic traces in the named hominins seem to indicate.

Location is not the only challenge. Postcranial remains – the skeleton below the neck – are much rarer than skulls (and especially jaws) and teeth, for two main reasons. One is that skulls and teeth are heavier, denser, and more durable than bone from vertebrae, ribs, and limbs. Skulls and teeth are less likely to be crunched up by predators and scavengers, and, unlike what else is left by scavengers, they are less likely to be washed away by rain and scattered. Larger hominins are more likely to leave something of themselves behind than smaller ones; and the nature of the environment in which they died will affect how well they are preserved – some environments promote fossilization, others promote dissolution because of soil acidity. In the former type of environment there might be significant quantities of hominin remains, in the latter few or none – at some sites numerous hominin-made stone tools are found but no bones – and therefore one cannot infer much from the bones' absence. The dating of remains, although now rather a precise science, cannot exclude the possibility that they may have been displaced from where death occurred, and penetrated into earlier or later geological strata because of earthquakes, floods, or animal activity.

All these factors put question marks in the margins of any assertion about hominin evolution. Nevertheless the emerging picture is not completely blurry. At about the time that hominin ancestors began to diverge from chimpanzee ancestors, some 6 mya or more, the planet as a whole was slowly starting to grow cooler and drier. In Africa the effect was a replacement of forest and woodland with grassy savannah. Correlation of hominin fossils with evidence of their then-contemporary habitats suggests that to begin with they lived on the boundaries of the wooded and open terrains, exploiting the resources of both. They might have resorted to the trees to gather fruit and to nest at night, out of the way of predators; bipedal movement would have better suited their forays on to the grasslands. At the period of the ancestral hominin–panin split there might have been other pre-hominin and pre-panin ancestors about – remember how many species of monkeys there are in the Old and New Worlds today. On the best evidence so far available, the evolutionary picture looks as follows.

Visualize a capital letter 'Y', with a shorter right-hand fork than left-hand fork. Starting at the bottom of the stem of the Y, and dating this at 7 mya, write *Sahelanthropus tchadensis*. (Each of these will be explained below.) A little further up the stem, between 6 and 5 mya, write *Orrorin tugenensis* and, a fraction higher, *Ardipithecus kadabba*. At 4.5 mya on the stem write *Ardipithecus ramidus*.

Above this is the point at which the stem splits into two, at 4 mya and after. In the region of the split there is something of a scrum. Draw a circle round the split-point and in it make a ladder from 4 to 2.5 mya, writing in succession *Australopithecus anamensis, Australopithecus bahrelghazali, Australopithecus deyiremeda*, and at about 3.5 to 3 mya *Australopithecus afarensis*. Around 3 to 2.7 mya the world was cooling from a long warm period in which sea levels had been up to 6 metres higher than today. Extensive glaciation was occurring in the northern hemisphere, sea levels were dropping, and the interior of Africa was drying. The beginning of developments that led to *sapiens* may well lie here: hence the forking of the Y.

We can now climb the respective branches of the Y. Up the shorter right fork, at 2.5 mya, write *Australopithecus africanus*, and, alongside this, off to the side, *Paranthropus aethiopicus*. At 1.8 mya write *Australopithecus sediba*, and alongside write *Paranthropus boisei*. And between 2 and 1 mya put *Paranthropus robustus*.

As the sequence of names suggests from *Australopithecus* to *Paranthropus*, the right fork is diverging away from the part of the Y that leads towards *sapiens* at the top of the left fork.

But up the left fork there is as much of a scrum as at the point where the stem splits. For here we see: between 2.3 and 1.65 mya *Homo habilis*, about 1.9 mya the Dmanisi fossils, between 1.9 and 1.5 mya *Homo ergaster*, at 1.8 mya the beginning of the very long and successful run of *Homo erectus* – some have suggested that *erectus* survived in Asia until just 30 kya, which means that between 40 and 30 kya there could have been four *Homo* species extant: *erectus*, *neanderthalensis*, *floresiensis*, and *sapiens*. That would be remarkable. With its long run *erectus* was still around when *Homo antecessor* and *Homo heidelbergensis* appeared at 1.2 mya and 800 kya respectively, followed by *Homo neanderthalensis* about 700 to 500 kya, *Homo naledi* about 3.5 kya, and *sapiens* coming fully on scene by about 300 kya. Denisovans are contemporary with Late Neanderthals and anatomically modern humans.

This is a crowded picture, but, depending upon whether one is a *lumper* or a *splitter*, one can make it less or more crowded respectively. The problem concerns individual variation: have different-looking skulls and teeth come from members of different species of *Homo*, or are they the same species but just different from one another in the way that people today are different in being short or tall, broad or narrow in face, chinless or with a big jaw, a high or low forehead and a round or long head? Click on the link in the footnote here[4] to see an array of five skulls recovered at Dmanisi; the inexpert would almost certainly think they are different species of *Homo*. Detailed analysis of skulls – craniometry – and bone morphology is a strong indication of how to classify them, but the lumping and splitting dilemma remains, even though detailed studies of cranial anatomy, and how different aspects of it evolve differently, place constraints on how much variation there can be within a given population.[5] Teeth and feet can be even more significant indicators in this respect, because there is little scope for individual variation in the basic arrangement of dentition and jaw structure within a species, and likewise with the basic structure of a bipedal foot.

Sahelanthropus tchadensis, as the name suggests, was found in Chad, far from the eastern and southern African sites of most fossil finds relevant to human ancestry. A Franco-Tchadienne group recovered remains from nine individuals, including one near-complete skull. Dating is uncertain, but the fossil environment in which the remains were found – the bones of other animal species – suggest a date of 7 to 6 mya. This is where genetics places the hominin–panin divergence.

Sahelanthropus tchadensis' teeth and relatively flat, short face show hominin-like hints, though the skull overall does not.

Between 6 and 5 mya a candidate for the first bipedal hominin was found at Cheboit in Kenya. This is *Orrorin tugenensis*, accorded its significant position courtesy of its teeth and the robustness of its femur – strong thigh bones are required for upright locomotion. Close in time is *Ardipithecus kadabba*, pieces of which were found in Ethiopia's Awash River Valley and at first thought to be an australopithecine until examination of its dentition and fossil environment suggested a date too early for that.

At 4.5 to 4.3 mya *Ardipithecus ramidus* offers the most complete picture of an early hominin to this point; more than seventeen individuals including a complete skeleton – the oldest hominin skeleton found to date – were recovered in the middle Awash River Valley. For approximate comparisons with an anatomically modern human standing 1.8 metres tall and weighing 70 kilos, 'Ardi' stood 1.2 metres tall and weighed 50 kilos. A modern human brain is about 1,400 cc in volume; Ardi's brain was 300–375 cc. The evidence of teeth, hands, feet, and the fossil environment suggests that Ardi lived partly in trees and partly on the ground in woodland. His feet had an opposable big toe, good for grasping branches, but also a bipedally apt strong mid-foot and a pelvis suited to standing. His hands were not adapted for knuckle-walking as in gorillas and chimps.

After Ardi the forks of the 'Y' start their respective rise from the stem. This is when the australopithecines appear; at 4.2 mya *Australopithecus anamensis*; at 3.6 to 3.0 mya *Australopithecus bahrelghazali*; at 3.5 to 3.3 mya *Au. deyiremeda*. The first of these is somewhat insecurely based on fragments found at Lake Turkana and Kanapoi in Kenya. *Australopithecus bahrelghazali* is also sparsely attested, but is important because of where it was found: far to the west of the African Rift Valley in South Sudan. *Australopithecus deyiremeda* is the most recent of this group to be named, after discovery of its remains at Woranso-Mille in Ethiopia's Afar region in 2015. It is important because it confirms what palaeoanthropologists had begun to suspect two decades before: that a number of hominin species coexisted, and the lineage to *Homo sapiens* was less direct than had been assumed. A major indicator of this was the discovery, at the beginning of this century, of *Kenyanthropus platyops* at Lomekwi near Lake Turkana in Kenya, dating to 3.5 mya.

But the australopithecine thought most likely to be in direct line of ancestry to modern humans is *Australopithecus afarensis*, living 3.5 to 3 mya. Found in abundance in Ethiopia's 'Afar Triangle' and also at Laetoli in Tanzania and Lake Turkana in Kenya, *afarensis* is known from several hundred individuals of both sexes and all ages, including the famous 40 per cent-complete skeleton 'Lucy', more recently joined by 25 per cent-complete 'Big Man', also called 'Kadanuumuu', from Korsi Dora in Ethiopia. *Afarensis* males stood about 1.5 metres tall and weighed over 40 kilos, with a brain volume of 400–500 cc. The Laetoli footprints show that *afarensis* was fully bipedal. The footprints are a remarkable find: three individuals made their way across a field of volcanic ash deposited by a recent eruption, leaving their marks clearly behind them; a rain shower, followed by hot sunshine, solidified the marks. Then further layers of ash were spread over them by more eruptions, burying the prints until 3 million years of erosion exposed them again. Other prints – one set from a three-toed horse, the *Hipparion* – are visible in the layer too. Comparison of 3-D images of the prints' contours was made with prints by modern humans using different gaits, one with bent knees and one with extended legs. The toe-to-heel depth of the Laetoli prints indicates extended (straight) leg walking. A curiosity in this connection is that, until the early twentieth century at least, many Chinese employed a bent-knee gait; in Lu Xun's *Ah Q* we read of the eponymous antihero seeing a modern-minded contemporary, who had been educated in Japan, mimicking the straight-legged stride of Westerners, and aspersing him for it. (He gets a beating for his pains.)[6] If the Laetoli walkers' prints had indicated a bent-knee stride, it might mistakenly have been taken as a primitive feature. This is an indication of how assumptions can lie too deep to be noticed.

The first australopithecine named was *Australopithecus africanus*, discovered by Raymond Dart in South Africa in 1924. He found the skull of a child – the 'Taung Child' – with all its milk teeth in place. At first Dart was ridiculed for claiming to have identified a creature 'intermediate between living anthropoids and man', but he was vindicated by other *africanus* finds, comparison with East African australopithecines discovered later, and fossil environments dating from 3.5 to 3.0 mya associated with them. The particular significance of his find was the weight it gave to Darwin's view that Africa was the likely homeland of humanity. Until then the origin of humanity was assumed to be Europe, as suggested by Neanderthal finds, or East Asia, as suggested by 'Java Man' (*Homo erectus*).

From the gaggle of these and other named australopithecines the right fork of the Y leads away to their more robust relatives and descendants designated *Paranthropus*, to suggest divergence from the hominin line. The left fork thereafter takes on added ancestral significance, beginning with *Homo habilis*, 'Handy Man' because of his association with tools, living 2.3 to 1.65 mya. He was first found at Olduvai Gorge in Tanzania, supplemented by other fossils from Koobi Fora in Kenya and the Omo River in Ethiopia. Smaller than australopithecines but with a markedly larger brain, a human-like foot and a thumb capable of a precision grip, *habilis* occasions debate over whether he was a hunter or a scavenger, hunting requiring a higher degree of social coordination, planning, and communication. In light of his being the immediate forerunner, at least in time, of *Homo* groups, who almost certainly possessed such capacities, it is plausible to attribute them to him too. For the next in the temporal sequence of hominins found is the enigmatic collection from Dmanisi in Georgia, 1.9 mya, representing the earliest securely dated hominin fossils outside Africa.

Called *Homo georgicus* by some, the Dmanisi fossils represent a number of individuals of both sexes and various ages. Their limbs are proportioned like those of modern humans, but their skulls are no bigger than those of *habilis*. An intriguing feature is that one of the skulls belonged to an old and toothless individual, who would have needed care and support by others to survive as long as he did.

The first hominin most closely to resemble an anatomically modern human is *Homo ergaster*, this name meaning the 'workman' because of his advanced tool technology, dating from 1.9 to 1.5 mya. His fossils are best known from the Lake Turkana region, a famous specimen,

'Turkana Boy', being almost complete save for his extremities and left clavicle. Other *ergaster* fossils show a human foot, and in some specimens a brain volume of up to 900 cc, though others are smaller at 500 cc.

To treat *ergaster* as his own species is to say that he is not an African version of *Homo erectus*, as some scholars believe him to be. *Erectus* was first discovered in Java, but his remains occur throughout Eurasia, from Spain to China and the East Indies. He is probably the first of the *Homo* group to leave Africa, and his long survival – some say until 30 kya; but even if it were to 110 kya, having appeared before 2 mya – is remarkable. Citing 110 kya is not arbitrary; it is the date for some of the *erectus*-assigned fossils recovered in the Solo River Valley in Java.

On one mainstream view *erectus* is the direct ancestor of both *Homo heidelbergensis* and *Homo antecessor*, with the former in turn being the direct ancestor of Neanderthals, Denisovans, and modern humans, emerging in Europe, Asia, and Africa respectively. In appearance, with a brain volume of up to 1,200 cc, sophisticated Acheulean tools, the ability to undertake sea travel, social organization, possibly speech, and possibly the making of ornaments and even art, *erectus* is either a direct ancestor of *sapiens* or closely related to *sapiens'* direct ancestor. If it was for decorative purposes that *erectus* made the patterns of scratches on shells found in Java and dating to 500 kya, then *erectus* made art; and the presence of symbolic representation is a significant indicator of much else. In fact European *erectus* is known to have collected ochre, the only use for which is as a pigment.

The discoverer of *erectus* was Eugène Dubois, a medical officer with the Dutch Army in the East Indies. His speciality was anatomy and his consuming ambition was to find the ancestors of humankind. His reward was to find the first *erectus* fossil, in 1891. Subsequent substantial finds in China ('Peking Man'; forty specimens were found at Zhoukoudian) were connected with Dubois's Java Man to encourage the idea of East Asia as the origin of humankind. *Erectus* individuals show considerable variety in space and time, but type specimens have a pronounced brow-ridge, a low but thick keel of bone along the skull's midline, and teeth larger than those of modern humans. At 1.6

to 1.8 metres tall and weighing between 40 and 65 kilos, *erectus* was nearly as big as a modern human, and some specimens had a brain volume of 1,300 cc, comparable to that of modern humans. They are regarded as the inventors of the Acheulean tool industry, as the first *Homo* to control fire, and as perhaps the first to build huts.

Homo heidelbergensis is hypothesized by some to be, as noted, the last common ancestor of Neanderthals and modern humans. He – in the form of his jaw – was first found in a cave at Mauer near Heidelberg in 1907, and thought to be a Neanderthal. By the 1970s, despite much uncertainty and controversy, he had come to serve for some palaeoanthropologists as representing all *Homo* fossils in Africa and Europe dating between 700 and 300 kya. Other scholars think that these fossils represent different subspecies of *Homo*, one of either *erectus* – indeed for a long time *heidelbergensis* was classified as *Homo erectus heidelbergensis* – or *Homo rhodesiensis* ('Broken Hill Man', from Kabwe in Zambia), this latter regarded by some as *the* direct ancestor of *sapiens*. Genetic evidence from remains of over thirty individuals found in the Sima de los Huesos ('Pit of Bones') Cave in northern Spain's Atapuerca Mountains suggests that *heidelbergensis* should be classified as a pre-Neanderthal or an archaic Neanderthal, and that the divergence between Neanderthals and modern humans should be regarded as occurring about 800 kya. On one view, *heidelbergensis* is a 'chronospecies', an interim and developing form indicating a lineage changing over time from a now-extinct earlier form to recognizably descended later forms, serving as the possible link between *antecessor* and *ergaster* before it, and Neanderthals, Denisovans, and *sapiens* after it.

The relationships between *Homo* types in the period from 1 mya onwards is exacerbated by the relative paucity of fossil evidence from 400 to 250 kya, confounded by the surprising discovery of a *Homo* species that mixes primitive and modern forms – this being the remarkable *Homo naledi*, found in 2013 in the Rising Star Cave System in South Africa. Over seven hundred skeletal parts representing nearly twenty individuals were recovered in an extraordinary feat of speleology (the art, for so it is, of cave exploration), which required small, very slender, and non-claustrophobic palaeoanthropologists to squeeze themselves through a series of tiny fissures into a deep cavern where the bones lay.[7] *Naledi* has a small brain, 450–600 cc, a height of 1.4 metres and weight under 40 kilos, similar to *Australopithecus afarensis* males, and although bipedal they had the high-shouldered adaptation typical for tree-dwelling and branch-hanging. At first they looked australopithecine to their discoverers. No tools were found with them, but their hands were nimble and suitable for making and using tools. The presence of their remains so deep in an inaccessible cave, requiring artificial light to navigate, suggests that the dead were purposely secreted there, a conscious funerary practice implying a degree of social development even if it was just to protect the remains from scavengers.

From appearance alone the discoverers of *naledi* surmised that they might have lived between 1 and 2 mya. A series of tests, including uranium–thorium (U–Th) and electron-spin resonance (ESR) dating on teeth and the sediments they were recovered from, showed that in fact they lived between 335 and 236 kya. This was a major surprise. It meant that small-brained *Homo* with primitive anatomical features were living alongside larger-brained, more anatomically modern relatives in Africa, further confusing the question of *Homo* lineages and relationships. The possibility that early *Homo* and australopithecines might have interbred, the likelihood that individuals from different *Homo* lineages interbred, and genetic proof that, much later, Neanderthals and modern humans interbred suggest that mingling across *Homo* strains was frequent enough to explain the ambiguity of lineages displayed by the fossil record. With the progress of genetic investigation evidence has emerged that introgression from one *Homo* genome to another is both more common and more complex than thought; modern Africans, for example, have some Neanderthal genes, showing that Neanderthals migrated back into Africa after 150 kya; and Neanderthals show gene input from multiple lines of African ancestors. Likewise there is strong evidence of a backflow of Eurasian *sapiens* into Africa after 50 kya, and of Eurasian inflow into southern Africa from 5 kya.

Diminutive *Homo floresiensis*, for which the Tolkien Estate forbids use of the name 'the Hobbit', was extant until 50 kya (when modern humans reached Flores) and possibly later. It is another indicator of the complexity of the *Homo* picture. If this little creature is a genuine dwarf species of *Homo*, it is an example of the miniaturization that can occur in island habitats – Flores also had miniature elephants. (Unluckily, insularity also seems to promote giantism in some creatures: Flores had giant rats.) *Floresiensis* certainly lived long enough for 'insular dwarfism' to occur; craniometric analysis shows that it is more similar to *Homo* living nearly 2.0 mya than to modern humans. Although its brain volume of around 400 cc is like that of an australopithecine, *floresiensis'* body morphology is reminiscent of *habilis*, but, as it were, *bonsaied* down to a single metre in height and a weight of 25 kilos. Stone tools dating to 190 kya are evidence of the earliest occupation of Flores by *floresiensis'* hypothesized early *erectus* ancestors, who must, remarkably, have sailed to the island, because there has never been a land or glacier bridge to the Indonesian mainland in the period of *Homo* evolution.

As with much else in the palaeoanthropological sphere, there is controversy about *floresiensis*. Remains from nine individuals have been found, the most complete specimen being LB1 (named for the Liang Bua 'cool cave' where this female skeleton was found, buried in volcanic ash). Is it possible that they were not a miniature form of human at all, but small for other reasons – for example: could endemic disease be the reason for their size? Microcephaly and cretinism have been suggested; the former is a rare condition, occurring in just one in several thousand births, but cretinism can occur in as much as 10 per cent of isolated populations, often as a result of congenital hypothyroidism. Another possibility is that the recovered remains were of individuals suffering from Laron syndrome, which is insensitivity to growth hormone; it is heritable, and might conceivably have spread through the Flores population to become endemic. Those with this syndrome are less susceptible to cancer and type 2 diabetes than the general population. Impressionistically, reconstructions of *floresiensis* faces are suggestive of the protruding forehead and saddle nose characteristic of Laron-affected individuals.

The most famous of our *Homo* relatives are the Neanderthals. They were also the first to be discovered, in 1856 in the Neander Valley in

Germany, hence the name. There had been an earlier Neanderthal skull found at Forbes Quarry in Gibraltar in 1848, but not immediately recognized for its significance. Note the date of the Neander find: it was three years before the publication of Darwin's *Origin of Species*, and fifteen years before his *Descent of Man*. When the latter was published, it occasioned little controversy, because by then the idea that everything, including humans, had evolved was on the way to being familiar, even to those who did not wish to accept it. Darwin was shown the Gibraltar skull by Charles Lyell in 1864, just before it was presented at a meeting of the British Association for the Advancement of Science; he mentions the Neanderthal finds briefly in *Descent*.

The earliest date given to Neanderthal fossils is 430 kya (their divergence from the *sapiens* lineage is thought to have occurred about 800 kya), but there is some uncertainty about the attribution. From 130 kya onwards there are many Neanderthal remains, and, as noted, the mainstream surmise is that they descended from *heidelbergensis*, placing the Sima de los Huesos fossils at 430 kya as archaic Neanderthals and a slightly later *heidelbergensis* fossil from Aroeira in Portugal as a transitional example. This accords with the emerging picture of various *Homo* strains coexisting and sometimes intermingling for a long period. Neanderthal–*sapiens* divergence is thought to have begun 800 kya, but, as noted, interbreeding was still occurring after modern humans left Africa before 50 kya. By 30 kya Neanderthals were extinct except for a remnant of them – about 3 per cent – in the genome of *sapiens*, a little more in Denisovans.

The Neanderthals' familiar large brow-ridge and powerful, stocky build made them apt for caricature as the grunting stupid cavemen of popular imagination, a reputation it took a long time for them to lose. But their advanced Mousterian stone-tool industry, possible body painting, ornaments, large brains (1,300 cc in females and 1,600 cc – larger than *sapiens* – in males), red hair, fetching or trading resources up to 300 kilometres from their settlements, zoning of their settlements for different purposes, use of fire to cook and smoke meats, some evidence of music (suggested by the Divje Babe Flute, dating perhaps from 140 kya), and possibly art (apparent engraving of stone objects, pigment shapes, and hand-stencils in caves dating to 66 kya, which is 20 kya before *sapiens* reached western Europe) give a different picture, the cultural indications sharply contradicting their brutish reputation.

Did they speak? They had the hyoid bone in the throat, the ear morphology, and the FOXP2 gene, all associated with language-using capacity in *sapiens*; and the zoning of their settlements – cooking here, stone-working there – suggests levels of social organization of a kind arguably predicated on language-use. On one reconstruction of the Neanderthal upper-respiratory tract it seems that they might not have been good at producing nasal sounds, but it is questionable whether this is a barrier to articulate language. On the contrary, the evidence points more rather than less to language-use, an added consideration being that their interbreeding with *sapiens* might have been less likely if they were mute.

There is, however, some evidence of conflict and cannibalism among Neanderthals too, so the picture of a being who, like the Beast in *Beauty and the Beast*, is belied by his appearance must not be painted too rosily. The real poignancy of the Neanderthal story is that there seems to have been rather few of them overall; their small numbers made them a fragile population, and whatever caused their extinction following the arrival of *sapiens* in their terrains – whether violence by *sapiens*, lack of immunity to *sapiens*-carried diseases, outperformance by *sapiens* in the commandeering of resources, or just the fact that in population and adaptation terms they had reached the end of the road anyway – extinction is the norm for all species of everything – it would appear that they were not sufficiently numerous to resist, compete, or survive. Theirs was a hard lot in any case: 80 per cent of Neanderthal remains are of individuals who did not survive to the age of forty, and their skeletons show much injury and hard wear.

To the question: what is *the* story of the evolution of *Homo sapiens*? there is no single or clear answer, but instead an intriguing, complicated, and very under-evidenced picture of multiple species of hominins and *Homo*, probably separating, then intermingling, then separating again but with a dramatic denouement that itself awaits much clarification – namely, the appearance not merely of anatomically modern *Homo* but of *behaviourally* modern *Homo*, contemporaneously with the final disappearance of all other species related to it, and with a rapid dispersion over the entire planet and a steady, then increasingly rapid expansion in numbers. At first blush this is not just a remarkable story but, from a certain perspective, one with rather sinister aspects, portraying *sapiens* as a predatory and exploitative species that had come a long way indeed from the small bands of hominin ancestors of various kinds that scavenged the carcasses of other predators' kills and scrabbled in the dust for something to eat.

Twenty-first-century genetics, and in particular ingenious developments in techniques of recovering and sequencing ancient genetic material pioneered by the likes of Svante Pääbo at the Max Planck Institute in Leipzig, throw intriguing and sometimes conceptually revolutionary light on the story of human origins. But it was much earlier work, in the 1980s, that succeeded in identifying the point at which the most recent common female ancestor of all today's human beings lived: about 150 kya, and in Africa. This is 'Mitochondrial Eve', so called because mitochondrial DNA is passed only through the female line, and the chain of ancestry through our mothers, grandmothers, great-grandmothers, and so on back converges on a single woman living at around that time. What in almost all cases distinguishes male from female humans is their possession of a Y chromosome (humans usually have 46 chromosomes in 23 pairs, one of which is XX in females and XY in males[8]). Tracing humanity's patrilineal line back to 'Y-chromosomal Adam' has proved more difficult, yielding an unhelpfully wide range of 180 to 580 kya, with best estimates at 120 to 156 kya. This latter range puts Adam more in line with Eve.

There are at least two interesting implications of these findings. One is that they support the single-origin 'Out of Africa Theory', in opposition to the multiregional hypothesis, which has humanity developing from *Homo erectus* lineages separately in Asia, Africa, Europe, and Australia. They also support the view that the emergence of behaviourally (not just anatomically) modern humans might be linked to a population bottleneck, a small subset of anatomically modern humans from whom behavioural modernity arose, as a result – some argue – of a development in genetic features distinctive of that subset. This idea is complicated by the fact that, as noted, modern human genomes include genes inherited from *Homo* populations more ancient than Eve and Adam and already living outside Africa, for example Neanderthal genes and, in the case of Melanesians and Australians, Denisovan genes. But these facts would not controvert the idea that genetic modification in a given subpopulation was the prompt for the development of behaviourally modern traits.

The question of behavioural modernity in *sapiens* evolution – the question of the very source of *sapience* itself, indeed – is hotly debated; I discuss it in Section 3 below. 'Behaviouraly modern' humans are those with language, art, symbolic thought, practices of transferring knowledge capital from one generation to the next so that it accumulates, capacity for flexibility and innovation that develops and varies this capital according to new challenges and circumstances, and use of analogical thinking to notice similarities and parallels and to transfer practices from one sphere to another accordingly. Between the evolutionary process resulting in anatomically modern humans about 300 kya and (using this as a shorthand, with caveats) Gordon Childe's 'Neolithic Revolution' around 12 kya – and most particularly the period of *sapiens'* emergence (again, but this time permanently) from Africa around 60 to 50 kya to spread over the entire planet by about 15 kya – there is a story to be told in its own right. This spread was accompanied by the disappearance of all other *Homo* species and many animal species – the effect on other species of *sapiens'* worldwide dispersion looking uncomfortably similar to what would be achieved by an undiscriminating pandemic disease.

Palaeoanthropology applies scrupulous archaeological methods and advanced forensic science to making sense of several million years of mystery. It is a detective story on the largest scale. Perhaps it is inevitable that confusion should grow even as more evidence is gathered, until the jigsaw pieces of bone and artefacts start to come together to form a clearer picture. Contemplating the skulls of our various forefathers, who look back at us with such enigmatic silence from across aeons, is a dizzying experience. Human history has been a rush in the last twenty thousand years or so, but the route to that point was a long one.

The evolutionary story resulting in *sapiens* almost exactly coincides with the geological epoch known as the Pleistocene, lasting from 2.8 mya to 12 kya. In hominin-to-human evolutionary terms this period is called the Palaeolithic or Old Stone Age, with later subdivisions into Upper Palaeolithic (Late Old Stone Age, from *c.* 50 kya) and Mesolithic (from *c.* 20 kya, the transition to the Neolithic or New Stone Age beginning 12 kya). These labels refer to the stone-tool technologies that developed over this period, showing slowly increasing sophistication as time passed, and at each stage suggesting much else about the beings who made and used them. But, as we see, by the Upper Palaeolithic the story has ceased to be more about biological than cultural evolution, and has become the reverse; now interest focuses upon the emergence of behaviouraly modern humans. Although this is a contested term, it is a suggestive and useful one, and it is the last chapter to be written of the past that was unknown until the nineteenth century.

3.   The Problem of the Past

The foregoing is a survey of what has been learned about the past since the nineteenth century, much of it in just the last half century, even in the last few decades. Humanity has gone from knowing a mere fraction of its history to reaching back millions of years beyond that. From a psychological point of view, the leap is immense: from a biblical framework in which the beginning of the world occurred six thousand years ago to one in which the universe began 13.72 billion years ago, planet Earth 4.5 billion years ago, life 4 billion years ago, hominins 6 million years ago, *Homo* 2 million years ago, anatomically modern humans 300 kya, behaviourally modern humans (say) 100 kya, settled agriculture 10 kya, cities 5 kya – none of this formerly known or even guessed. From that perspective, the advance is – in the literal sense of this term – stupendous. At the same time, and indeed in part explanation of the success of the endeavour itself, we have become greatly more critical and scrupulous in our thinking about how we know all this. Some of the points that, as a result, arise for consideration are as follows.

If one wished to begin with a dramatic flourish, following upon the remarks at the beginning of Section 1 of this Part, one could say that history began in classical Greece in the fifth century BCE – that is, history *as historical enquiry*, as the activity of enquiring into and describing what happened in the past. To claim this is to claim that history as a self-conscious endeavour began at that time in that place in the writings of Herodotus. But this is too much of a flourish; if we count as history what we now know about such records as king lists, anecdotes about non-legendary great figures, and records of battles and conquests carved on monuments, then history is at least as old as writing itself, for we find such things inscribed in cuneiform on Mesopotamian steles and clay tablets. And no doubt orally preserved and transmitted history predates these forms of record by thousands of years.

History as the *period of past time* accessible through both archaeology and written records began in the fourth millennium BCE, as revealed through the recent discovery of Sumerian Mesopotamia. Before that, history is 'prehistory', addressing what humans did, where they did it, and how they did it, reaching back into the Neolithic Period before writing, 10,000–3000 BCE. In relation to everything before that in the ambiguous and fragmentary tale of the evolution of humankind, going far back to what are perhaps the first identifiable stone tools 3.3 mya, we are no longer in history in the same sense as the term is applied to what happened in the past from Sumer onwards.

We immediately see from this that the word 'history' is ambiguous. It can either denote *the study of past events*, or it can denote *the past events and circumstances themselves*. In this second sense 'history' comprehensively refers to everything that has happened in the millennia that have elapsed before our own time, on to which we can cast light by all means of research, including archaeology – that is: by means of 'history' in the first sense. The further back into the past we enquire, the greater the difference in research technique there has to be. 'Recorded history' takes us back only to the late fourth millennium BCE, when writing and therefore documentary records first become available. Here texts and archaeology supplement each other. 'Prehistory', as the indeterminate period of time before then, concerns the emergence of settled agricultural life and the growth of urbanization taking place in that region of the world where evidence of the first agriculture-related settlements is found.[1] Archaeology is on its own here. That takes us back to about 10 to 12 kya. Before then the story of humankind and its evolutionary antecedents are the province of palaeoanthropology, in which further sets of archaeological techniques take part.

As we shall see, every statement that one makes of this kind is subject to qualification and indeed contest: a prime example is the claim that human beings were engaging in forms of agriculture and settled life more than ten thousand years earlier than 12 kya, and that it is incorrect to see the move from hunter-gatherer lifestyles to settled agriculture as a linear, one-directional movement, still less as 'progress'. In what follows this caveat must be taken as read, and some of the chief reasons for it will be discussed below.

A key general question about history is this: is history in the second sense (namely, *what happened in the past*) the creation of history in the first sense (namely, *the activity of historical enquiry*), or can we be sure that we are capable of discovering what happened in the past – a question that can alternatively be put as 'Can we discover the objective truth about what happened in the past?' The reason for raising this question is that the way the past lingers into the present – as documents, monuments, ruins above and below ground, traditions, memories – is incomplete, unclear, equivocal, often cryptic. We find these remains surviving in the present, and we build interpretations around them. Both the remains and our interpretations are contemporary with ourselves. Because history in the first sense – enquiry into the past – familiarly consists in narratives and explanations forged in the present, it is no wonder that historians often disagree among themselves about what happened in the past. Think about witnesses to a traffic accident and the different stories they can tell, the contrasting interpretations they offer, the way that shock, or interest, or sympathy, or antipathy can affect memory and point of view. If eyewitnesses to an event that has just happened can differ among themselves, indeed can even contradict each other, what chance have we of getting 'the correct' account of something that happened long ago?

It is seductive to think of history – the realm of past events – as a place spread out 'behind' us that we could visit and explore if only we had a time-machine. This metaphor dominates our imaginations because it is associated with a pre-reflective belief in history's objectivity: we take it that there are facts of the matter about what took place in the past, and history (as enquiry) has the job of discovering them. Our sense of realism is offended by the thought that the past is a creation of the present, particularly when we meet with certain kinds of revisionism about history, and most especially when it is obviously tendentious, for example as when it is denied that the Nazi

Holocaust of Europe's Jewish population happened. As this example shows, the question of whether history is an art that creates or a science that discovers is not an idle one. It comes down to asking whether there is such a thing as historical truth, and, if there is, to what extent we can know it.

The significance of this last point cannot be overstated. History is very much alive in the present, as international tensions, ethnic rivalries, conflicting traditions and world-views, national self-perceptions and lingering grudges testify. What happened in the Atlantic slave trade in the period between the sixteenth and nineteenth centuries; what happened to the Armenians in the final decades of the Ottoman Empire, most especially in the genocide beginning in 1915 that the Armenians call the *Medz Yeghern* ('Great Crime'); what happened to the Jews of Europe in the 1930s and the 'Final Solution' of the 1940s; what happened in and following Pol Pot's 'Year Zero' of 1975 in the Cambodian killing fields – these things matter. History groans under a burden of tragedy and suffering, and not only are these things not forgotten but arguably should not be forgotten, because they make many kinds of difference to our present and future.

Yet this intensifies the historiographical dilemma, the problem of ascertaining the truth about the past. It has been well said that the study of history tells us more about ourselves than about the past. Each generation and society looks at the past through its own lenses – indeed through a multiplicity of competing lenses, whose refractive indices depend on a variety of political and social positionings – which amplify some things and screen out others; and the prescriptions of these lenses change with time and circumstance.

The problematic ambiguity in the word 'history' is a long-standing one. The word derives from ancient Greek *istoria*, which means 'enquiry'. But by the fourth century BCE the word *historikos*, meaning 'reciter of stories', was in use alongside *historeon*, 'enquirer'. This raises the question of which category the first great named historians belong to, these being Herodotus, Thucydides, Polybius, Livy, Sallust, and Tacitus, the first three writing in Greek and the last three in Latin. And indeed this remains a question still, though today there is an acknowledged division between popular narrative historians (such as

Arthur Bryant and John Julius Norwich), each of them an *historikos*, and coal-face archival research historians (almost all academic historians), who would each merit the label *historeon*.

As it happens, the historians of antiquity understood the problem very well. Thucydides (who determinedly saw himself as an *historeon*) roundly criticized Herodotus (on Thucydides's view, an *historikos*) for his anecdotal history, which jumbles together tales, facts, legends, and speculations about the great struggle between Persia and Greece and its origins. Thucydides says at the beginning of his account of the Peloponnesian War that historical enquiry should restrict itself to contemporary events that can be verified by direct observation. To a considerable extent he practised what he preached: he served in the Athenian army, and wrote about what he had experienced or could check with those who had experienced what he wrote about. Yet he had an axe to grind, and he made up – or perhaps more accurately we should say, 'creatively reconstructed' – whole speeches to put into the mouths of leading figures, the famous 'Funeral Oration of Pericles' being a chief case in point; so that even he, avowedly an *historeon*, made use of the arts of the *historikos*.

History was, until the Renaissance, far more the province of the *historikos* than the *historeon*. But from the seventeenth century CE, inspired by developments in science and philosophy and the intellectual spirit that lay behind both, the idea of a more scientific form of historical enquiry emerged. It was motivated largely by work on documentary sources, not least because principles had been established for authenticating manuscripts. By the first half of the nineteenth century, in a period of flowering scholarship in Germany, it was possible to believe that a completely objective knowledge of the past was attainable – the past 'as it actually happened' in the words of Leopold von Ranke, the doyen of nineteenth-century historians.

Von Ranke is described as a 'Positivist' about history, not only because he believed in its objectivity but because he thought that it is governed by discernible laws. John Stuart Mill agreed with this view, adding the thought that psychological laws figure among the laws of history and provide access to an understanding of the actions and choices of past people. This view of history treats it as a science, its

laws being comparable to natural laws and its truths being discover-
able by empirical research.

The Positivist view of history was vigorously opposed by a group
who came to be known as 'Idealists', notable among them Wilhelm
Dilthey. These thinkers were influenced by the philosophers Kant
and Hegel, whose views led the Idealists to argue that history is not
like natural science in studying phenomena from an external perspective but rather is a social science that studies its phenomena from
the internal perspective of human thought, desire, intention, and
experience. Whereas the Positivists thought that historical enquiry is
an empirical study of objective facts, the Idealists viewed it as an
exercise in 'intellectual empathy', designed to achieve an understanding of what earlier people felt and thought so that we can understand
why they did what they did. They further thought that this is a route
to knowledge (as opposed to opinion), because, as Vico argued, intentions and social constructions are transparent to us as agents ourselves;
human nature is sufficiently a constant.

This therefore means that not all the Idealists regarded history as
merely subjective; Dilthey argued that it remains objectively constrained by the books, letters, art, buildings, and other products of
human experience that exist in the public domain and serve to make
'intellectual sympathy' possible. However, some of the most notable
of his fellow Idealists disagreed; Benedetto Croce argued that history
is always more subjective than not, because there is no way for the
historian himself to be absent from its construction.

The views just sketched constitute *philosophy of history*. It is closely
allied to *historiography*, which is discussion of historical techniques and
methods. But it must be distinguished from *philosophical history*, a label
for grand theories of history's metaphysical significance as advanced
by Hegel, Marx, Spengler, Toynbee, and theologians. Like von
Ranke, these thinkers regard history as manifesting laws, but they
add the very different claim that these laws are moving human affairs
towards a climax or goal, that history is unfolding teleologically,
purposively, or at any rate towards some kind of destination or
culmination – for example, in Hegel the end of history is the complete self-realization of *Geist*, or the world-spirit; for Marx it is the

withering away of the state in a condition of peaceful and mutual communalism. Religious versions of the view anticipate an end of the world as it currently is – an 'end time', a day of judgement, an eschatological *dénouement* to the universe, followed by a new dispensation. For the Jews this is the coming of the Messiah; for the Christians it is the return of the Messiah; for worshippers of the Scandinavian mythological deities – if any such remain – it is the resurrection of Balder the Beautiful; and in all cases a wonderful existence for the faithful is to unfold thereafter. Such views are part neither of history nor the philosophy of history.

The inspiration for von Ranke's scientific approach to history came from a group of scholars at Göttingen University in the eighteenth century, then a newly founded institution that embodied the spirit of Enlightenment rationalism to a high degree. In their turn the Göttingen scholars had been prompted by two earlier influences. One was the idea of critical exactness in documentary studies demanded by the French Benedictine monk Jean Mabillon (1632–1707), who had single-handedly invented the science of palaeography. The other was the universalist sweep in the historical writings of Gibbon and Voltaire. The Göttingen scholars combined these influences, and the gate to historical Positivism was open.[2]

Without doubt much good for historical research resulted from this. But a baleful aspect of the Göttingen school's influence remains in the racial theory it invented in the course of bringing an anthropological dimension to historical research. Two of its professors, Johann Friedrich Blumenbach and Christoph Meiners, postulated a colour-coded set of five human 'races' and invented names for them: *Caucasian* for 'white' people, *Mongolian* for 'yellow' people, *Malayan* for 'brown' people, *Ethiopian* for 'black' people, and *American* for 'red' people. The Table of Nations in the bible's tenth book of Genesis suggested to their older colleagues Johann Christoph Gatterer and August von Schlözer a threefold classification of the descendants of Noah's sons Ham, Shem, and Japheth – thus: the Hamitic or black race, the Semitic or Jewish/Arab race, and the Japhetic or white race.

Racial theory is an obvious underpinning for racism, and it is no accident that formalization of the concept of race should occur at this point. What we now think of as globalization had begun in the fifteenth century CE, with exploration of ways to find sea-routes for trade in the exotic spices of the East. As a result of sailing west to reach the East, conquest and the enslavement of native populations began in the 'New World' under its Spanish and Portuguese invaders not long afterwards; colonization of parts of the North American eastern seaboard followed. A new and extensive slave trade across the Atlantic, dwarfing Arab slaving in East Africa, laid the basis for immense British and later American wealth but also for the first stirrings of conscience: Quaker opposition to slavery began in the eighteenth century, ending the slave trade by the early nineteenth century and slavery itself in the second half of the same century.[3]

But slavers, and later on colonizers, required justification for their activities, and they found it in doctrines of racial difference, especially in the idea of 'superior' and 'inferior' races. A significant part of this tendentious anthropology was the supposed historical basis claimed for it. The biblical authority cited above decreed that the sons of Ham were forever to be 'hewers of wood and drawers of water' — that is, servants and slaves; this meshed with Aristotle's views about the inferiority of 'barbarians' and the existence of people who are 'natural slaves', and it permitted the unthinking elision of concepts of ethnicity and race, for along the way allied ideas had joined the fray: in the Spain of the *Reconquista* the notion of 'pure blood', *limpieza de sangre*, was invoked to exclude Moors and Jews from a place in the land – the Jews having already, in medieval times, become an 'Other' for Christian Europe.

These considerations are a major part of the long roots of debates about historical meanings and culpabilities. Take for example the 'History Wars' that racked Australia over the question of its indigenous peoples. In the late 1960s the question of whether Australia had been *discovered and settled* by the British in the eighteenth century or *invaded* by them was raised by historian Henry Reynolds, sparking a furious quarrel that has never since died away. Reynolds was a faculty member of James Cook University in Queensland,

and he became friendly with one of the university's gardeners, Eddie Koiki Mabo, a Torres Strait Islander. When Mabo learned from Reynolds and his colleague Noel Loos that the plot of land he thought he owned on his native Mer (Murray Island) was legally Crown Land, and that he did not own it despite thinking he did, he took the government to court over indigenous land rights, and won – posthumously; he died before the Australian High Court decided in 1992 in favour of the proposition that Australia should never have been regarded as *terra nullius* – 'no one's land' – and that the country's indigenous peoples both had and retained rights to the land.

This decision led to the passing of the Native Title Act in 1993, recognizing the right of Aboriginals and Torres Strait Islanders to access land for living, hunting, fishing, teaching their customs, and observing their traditions – though it is important to note that this is access to *some* land, amounting to about 15 per cent of Australia's total land area, in the form of registered Land Use Agreements voluntarily entered into by land title groups.

The Native Title Act was passed by the Labour Party government of Prime Minister Paul Keating. When the conservative Liberal Party won the 1996 election, his successor, John Howard, said that he was tired of the 'black armband' version of the country's history and wished to see a reassertion of 'Judeo-Christian ethics, the progressive spirit of the enlightenment, and the institutions and values of British culture'.[4] The flames of controversy were fanned by these words, because, in light of Keating's emphasis on respecting indigenous history and rights, these words directly assumed the superiority of Eurocentric values, and an underlying racist implication could not be obscured by arguments to the effect that reopening wounds would impede the formation of a unified national identity, and that current Australians were in any case not to blame for what happened in the past.

To inflame matters further, the controversy occurred just at the time the report on the 'Stolen Generation' was published – referring to the children of Aboriginal and Torres Island descent, most of them categorized, in a term that is now unacceptable, as 'half-caste' – meaning

mixed-race – who were removed from their families and transferred
to state and Church institutions to be brought up. The policy, which
lasted from 1905 until the 1960s, was predicated on the belief that the
Aboriginal peoples were dying out, and that children, especially
girls, were at risk of abuse in their communities. As many as one in
three children was removed; the report stated that they numbered at
least one hundred thousand.

Henry Reynolds thought, as most in his profession would, that the
questions raised by these matters had to be properly placed in their
historical context, and therefore pursued his researches, distinguishing the question of whether Australia was 'invaded or settled' by the
British in the late eighteenth century from the question of whether
violent conflict occurred between settlers and Aboriginals both then
and for more than a century afterwards. For, even if the invasion had
not involved violence, it would, he argued, still have been an invasion; on that point turns the issue of rights and land rights. But even
if Australia had been a *terra nullius*, so that what occurred was settlement and not invasion, it remains that there was a century of conflict
as a result. Reynolds's argument is that what happened was both an
invasion and a violent one.[5]

Conservative reaction to Reynolds's argument was predictable,
and took one or both of two main forms: either that it impeached the
view, long orthodox, of a heroic settlement of a new world involving
the taming of a wild land and the exploration of difficult and often
dangerous frontiers; or (once again) that it was divisive and destructive, reviving animosities, turning Aboriginals and white Australians
against each other and thwarting the effort to go forward in unity.
For anyone sympathetic to Reynolds's endeavour, the issues also turn
on two main points, this time in combination: the truth about what
happened, and restoration of Aboriginal rights.

Deciding the question between Reynolds and his critics is a perfect illustration of how differently things can be made to look when
viewed down different ends of the historical telescope. The historian
described as the 'leader of the conservative history warriors', Keith
Windschuttle, argued that what Reynolds described as Tasmanian
Aboriginal resistance to settler expropriation of their land was

merely criminal activity. Consider therefore the incident with which Reynolds opens his book *Forgotten War* about the settler–Aboriginal conflicts.

In September 1831 a leading figure among Tasmanian settlers, Captain Bartholomew Boyle Thomas, and the manager of his estate, James Parker, were killed by three Aboriginals. One of the local newspapers, the *Launceston Advertiser*, angrily announced that Boyle Thomas and Parker had been 'barbarously murdered by the inhuman savages . . . Thus two more respectable and highly respected individuals have been added to the list of those who have fallen victims to the barbarity of a race which no kindness can soften, and which nothing short of utter annihilation can subdue.'[6] A year before this an attempt had been made to move bands of hostile Aboriginals away from settled territory; a cordon of over two thousands soldiers and settlers was formed to drive the bands before them, an event known as the 'Black Line' and one of the culminating events of the Black War, 1803–32, between settlers and Aboriginals in Tasmania, in which over two hundred settlers and possibly as many as eight thousand Aboriginals had died directly or indirectly.[7]

Settler anger at the killing of Boyle Thomas and Parker was exacerbated by the fact that the former had been something of a hero in the Napoleonic Wars and the independence movements in South America, and after settling in Tasmania had been among those who wished to reach an amicable arrangement with the Aboriginals. Unmoved by his attitude, another local newspaper called for his death to be revenged, not just on the perpetrators 'but upon the whole race'.[8]

The Aboriginals who had killed Boyle Thomas and Parker were caught and taken to Launceston. The guilt of all three was clearly established by a coroner's court. The next question was what to do with them. Out of this debate came a singular and striking document, a letter to the local press by an anonymous correspondent. It merits quotation.[9] After acknowledging that his or her first reaction was to desire the 'extermination of the Blacks', the correspondent wrote of having serious second thoughts that challenged this sentiment, asking,

Are these unhappy people the *subjects* of our king, in a state of rebel-
lion? Or are they an injured people, whom we have invaded and with
whom we are at war? Are they within the reach of our laws; or are
they to be judged by the law of nations? Are they to be viewed in the
light of murderers, or as prisoners of war? Have they been guilty of
any crime under the law of nations which is punishable by death, or
have they only been carrying on a war *in their own way*? . . . We are at
war with them: they look upon us as enemies – as invaders – as their
oppressors and persecutors – they resist our invasion . . . What we
call their crime is what in a white man we should call patriotism.[10]

Reynolds argues that the frontier war was constant and ubiquitous,
that every year for a hundred and forty years, from the beginning of
settlement until the 1920s, people were dying violently in remote parts
of Australia – 'remote' changing its meaning over time – most of whom
were Aboriginals, though a significant number of settlers died too. His
original estimate had been that twenty thousand Aboriginals were
killed in these conflicts, but later said that this was too low 'even for
Queensland alone'. Newspapers, letters, and other documents show
that there was widespread awareness of the conflict right into the early
twentieth century, but, by the time Reynolds was a young history lec-
turer in Queensland in the 1960s, the textbook he taught from contained
nothing at all about Aboriginals – the word did not even occur in the
index. Yet in the Queensland of that time tense race relations and conflict between whites and Aboriginals were commonplace.

A key point for Reynolds is that the conflict between settlers and
Aboriginals was a war in the full sense of this term, a war fought *in*
Australia that was a war *about Australia*. It was a war about the ownership and control of the best land in the country; it was a war about
sovereignty. He writes, 'This was our Great War. What could be
more important than the ownership, the control, of a whole continent? And it was a war of global importance, because it was a war over
ownership of one of the world's continents.'[11]

Reynolds's argument about Australia could, *mutatis mutandis*, be
applied to the United States but multiplied many times over. The

longest war ever fought by the United States was the war against the Native American nations as western expansion, fuelled by repeated gold rushes and the coming of the railroads, invaded the home territories of Shoshone, Cheyenne, Sioux, and other peoples on the Great Plains and in the Rocky Mountains. In the south-west, in New Mexico, Texas, and Nevada, the Comanche and Apache tribes resisted the settlers; when the United States acquired Florida from Spain the result was a series of bitter struggles with the Seminole people.

The period between 1860 and 1890 was said by Dee Brown, a leading historian of the Native Americans, to be when

> the culture and civilization of the American Indian was destroyed and out of that time came virtually all the great myths of the American West – tales of fur traders, mountain men, steamboat pilots, goldseekers, gamblers, gunmen, cavalrymen, cowboys, harlots, missionaries, schoolmarms, and homesteaders. Only occasionally was the voice of an Indian heard, and then more often than not it was recorded by the pen of a white man. The Indian was the dark menace of the myths, and even if he had known how to write in English, where would he have found a printer or a publisher?[12]

But the conflicts of the nineteenth century were not new. Native Americans first resisted colonists in the early seventeenth century in Virginia, in the Powhatan Wars (1610–14, 1622–32, and 1644–6), and in New England in the Pequot War (1636–8). Throughout the seventeenth and eighteenth centuries repeated conflicts – at least a dozen – from New York to North and South Carolina, from Nova Scotia to Kentucky and West Virginia, took place as Native American tribes fought the invaders of their land. The US Army was still fighting Native Americans in the first quarter of the twentieth century, even after treaties, forced deportations, and sequestration of almost all Native American nations to 'reservations' had been completed; the Apache Wars officially ended only in 1924.

This is a remarkable story of resistance and dispossession, and atrocities were committed on both sides. One example will suffice as an indication of the three centuries of strife involved. In 1851 a treaty was signed at Fort Laramie between the US government and seven 'Indian nations', including the Arapaho and the Cheyenne, recognizing their right to a vast tract of land between the North Platte River at the northern limit and the Arkansas River at the southern limit, and between western Kansas to the east and the Rocky Mountains to the west. The territory overlay parts of what are now the states of Wyoming, Nebraska, Colorado, and Kansas. In 1858 gold was discovered in the Pikes Peak area of the Rocky Mountains in Colorado, triggering a flood of prospectors and settlers into the territory. Pressure was applied to the federal government to review the Fort Laramie Treaty in order to redefine the extent of Native American territory. In the Treaty of Fort Wise in 1861 four Arapaho and six Cheyenne chiefs signed away over eleven twelfths of the land given to them in the Fort Laramie Treaty. Their own people, furious with them for this, and regarding them as having been bribed and tricked into signing, refused to recognize the treaty, and continued to live and hunt over the 1851 treaty lands. Some of the militant warrior bands of the Cheyenne, the Dog Soldiers, were especially hostile towards white settlers, and tensions rose around the gold-fields trail in the Smoky Hill River region of Kansas.

A regiment of Colorado volunteers had been raised to aid the Union side in the Civil War that began in 1861, under the command of a Methodist preacher turned US Army colonel, John Chivington. After it won a victory over a Texan force in the Battle of Glorieta Pass in New Mexico in March 1862, the regiment returned to Colorado, where Chivington, in collaboration with the Colorado territorial governor John Evans, decided to make use of it to deal with the Cheyenne.

Chivington's animus towards Native Americans was heated. 'Damn any man who sympathizes with Indians!' he is reported to have said, when his plan for the massacre of Cheyenne at Sand Creek was opposed by one of his officers. 'I have come to kill Indians, and believe it is right and honourable to use any means under God's heaven to kill Indians.' He urged his troops, 'Kill and scalp all, big and little; nits make lice.'[13]

In the spring and early summer of 1864 US soldiers attacked, without warning, several large settlements of Cheyenne, and, on one occasion, when a group of Cheyenne chiefs approached the regiment to open discussions with them, they were shot to death by the soldiers. The militant bands aside, most of the Cheyenne were eager to avoid conflict, and when offered peace and the protection of the US military if they would relocate to Fort Wise (known also as Fort Lyon) in south-eastern Colorado, they agreed. When they arrived there, the group of Cheyenne and Arapaho under the leadership of Chief Black Kettle were told to make camp some 40 miles from the fort, in the bend of a river called Sand Creek.

It was here that Chivington and his force of almost seven hundred troopers attacked them on 29 November 1864. Believing that they were safe under the protection of the authorities, the Cheyenne had not mounted guards. Many of the men were away hunting, and about two thirds of the six hundred or so inhabitants of the camp were women and children. Chivington launched his assault at dawn; the first that the camp's inhabitants knew of them was the approaching drumbeat of hooves as the troops charged. Black Kettle had raised the American flag on a pole because he had been told by a US Army officer, one Colonel Greenwood, that as long as the US flag was flying above him no soldier would fire on him.[14] But the flag proved worthless; the troops proceeded to enact an indiscriminate slaughter.

One of those who testified at the subsequent enquiry, a trader and mediator called Robert Bent, who was married to a Cheyenne woman, and who had been unwillingly taken along with Chivington's force, described what he saw:

> When the troops fired, the Indians ran, some of the men into their lodges, probably to get their arms . . . I saw five squaws under a bank for shelter. When the troops came up to them they ran out and showed their persons to let the soldiers know they were squaws and begged for mercy, but the soldiers shot them all. I saw one squaw lying on the bank whose leg had been broken by a shell; a soldier came up to her with a drawn saber; she raised her arm to protect herself, when he struck, breaking her arm; she rolled over and raised her other arm, when he struck, breaking it, and then left without killing her. There

seemed to be indiscriminate slaughter of men, women and children. There were some thirty or forty squaws collected in a hole for protection; they sent out a little girl about six years old with a white flag on a stick; she had not proceeded but a few steps when she was shot and killed. All the squaws in that hole were afterwards killed, and four or five bucks outside. The squaws offered no resistance. Every one I saw dead was scalped. I saw one squaw cut open with an unborn child, as I thought, lying by her side. Captain Soule afterwards told me that such was the fact. I saw the body of White Antelope [one of the chiefs] with the privates cut off, and I heard a soldier say he was going to make a tobacco pouch out of them. I saw one squaw whose privates had been cut out . . . I saw a little girl about five years of age who had been hid in the sand; two soldiers discovered her, drew their pistols and shot her, and then pulled her out of the sand by the arm. I saw quite a number of infants in arms killed with their mothers.[15]

Bent's account was corroborated by other witnesses, one of whom was Lieutenant James Connor: 'In going over the battleground the next day I did not see a body of man, woman or child but was scalped, and in many instances their bodies mutilated in the most horrible manner – men, women and children's privates cut out, &c . . . I heard one man say that he had cut the fingers off an Indian to get the rings on the hand.' After describing other even more graphic and disgusting desecrations of bodies, Connor said that 'to the best of my knowledge and belief these atrocities that were committed were with the knowledge of J. M. Chivington, and I do not know of his taking any measures to prevent them.'[16]

Chivington's troops were a militia, ill-disciplined and poorly trained; they had been drinking whiskey on their night ride to Sand Creek, and some of the casualties they sustained in the brawl were said to have been the result of their own inaccurate shooting. Some estimates say that about 130 of the Cheyenne and Arapaho died, all but a few of them women and children, and the rest escaped; Chivington claimed that he had killed 500–600. He would have been able to make a more accurate estimate, for he and his men returned to the site of massacre the next day to take more scalps and body parts, including male and female genitalia and foetuses, with which they decorated their saddles and hats and which they exhibited in saloon bars and even in Denver's Apollo Theatre.

Relatives of the massacre's victims, together with many Arapaho and Cheyenne from other areas, joined the Dog Soldiers in several years of retaliatory attacks in Colorado and Nebraska. As with almost all the resistance offered by Native Americans, the superior numbers, firepower, and organization of the conquering white men was too much for them. Exceptions like the victory of Sioux (Lakota), Cheyenne, and Arapaho warriors at the Battle of the Little Bighorn, 'Custer's Last Stand', in June 1876, were just that: exceptions.

But the immediate consequence of the Sand Creek Massacre was different. Hailed at first as a victory over a numerous and dangerous enemy, the true nature of the events at Sand Creek quickly became known, because of eyewitness accounts from some of the troops involved – officers of two of the squadrons in Chivington's force had refused to take part, and there were civilians like Robert Bent with the force. Two military enquiries and one Congressional enquiry were held, the latter concluding that Chivington had 'deliberately planned and executed a foul and dastardly massacre which would have disgraced the veriest savage among those who were the victims of his cruelty. Having full knowledge of their friendly character, having himself been instrumental to some extent in placing them in their position of fancied security, he took advantage of their inapprehension and defenceless condition to gratify the worst passions that ever cursed the heart of man.'[17] Amazingly, that was all the punishment Chivington received: criticism.

The Federal government made a new agreement with the Cheyenne and Arapaho, the Treaty of the Little Arkansas of 1865, pledging reparations to the survivors of the Sand Creek Massacre and giving the tribes open access to lands south of the Arkansas River (but excluding them from lands north of the river). Two years later the federal government backtracked, replacing these undertakings with a new arrangement, the Medicine Lodge Treaty of 1867, reducing the land area of the reservation by 90 per cent. It was not the last of the area reductions imposed.

★

When one views the full array of conflicts between native inhabitants of North, Central, and South America against the Europeans who invaded and took possession of their lands, one sees how readily they are construed in the same way as Reynolds describes events in Australia. The imperialistic endeavours of European powers in India, the East Indies, and Africa were characterized just as much by oppression and exploitation, though their assumption of power took the *explicit* form of armed expropriation only when resistance to it was analogous to Aboriginal or Native American endeavours – as, for example, with the Zulus of South Africa, the Afghans of the North West Frontier, and the Boxers of China.

These aspects of the history of empire and colonialism touch only the question of their beginning. What followed, in the horrors of the Atlantic slave trade, in centuries of slavery itself, and in the subjugation of entire populations in almost every corner of the world, is a tragic tale indeed. But reconfiguring the beginning of this tale not as a glorious achievement of civilization over savage or primitive peoples and places but as invasions, and invasions that were bitterly resisted, is a major step to refocusing how the past is to be understood.

The history of America, the history of the British Empire, the history of Australia, as these were taught and learned when the author of these words was at school, included nothing about these dire struggles but concentrated on what were offered as positive aspects, in line with a given set of value judgements: the heroism of settlers, the 'civilizing' effects of education, health care, religion, civil order, and government.[18] Cinema portrayed the white man in North America as an heroic figure, the Native Americans as savages; nothing in the howling, whooping, painted warriors descending on the wagon train or the cavalry troop told of their desperation, of their struggle to protect their independence against the organized theft of their land and livelihood, their traditions and security.[19] Added to the violent expropriation of these things was the repeated betrayal of broken treaties and broken promises, the perfidy that land hunger and gold hunger made honourable for the putatively civilized person over the putative savage.

History as the study and recovery of the past is thus again shown to matter for the present and future, and the degree of honesty in it is again shown to be crucial. Much history is polemical – has a point to make – and partial in the sense of taking sides. But there is a degree of both polemic and partiality designed not merely to persuade and illustrate but to change the past, to make it look very different. In the case of histories written from the victor's perspective, this is achieved by leaving things out: leaving out the other side's point of view, leaving out inconvenient facts and culpabilities. Or it can be done with a still more dangerous intent: by the deliberate introduction of falsehoods and distortions. Nothing demonstrates this more dramatically than when revision of history takes the form of – for the most troubling examples – *denial* and attempted reversal of victims' memory and the convergent findings of responsible researchers. Denial of the Holocaust of European Jewry in the Second World War and denial of the Ottoman genocide of Armenians in the First World War are prime examples.

Holocaust deniers either claim that the 'Final Solution' was not a plan to kill Jews but merely to deport them and that there were no extermination camps and gas chambers, or that, if there were, the number of Jews killed in them was a tenth or less of the 6 million recorded in mainstream and official histories. In trying to make their case, the deniers have to deal with a mountain of evidence to the contrary.

In the manner typical of how history enters popular consciousness in simplified form, the Holocaust is regarded as having been triggered on 20 January 1942 at the Wannsee Conference chaired by Reinhard Heydrich, head of the Reichssicherheitshauptamt ('Reich Main Security Office', 'RSHA'). In fact the Holocaust was already under way; the point of the Wannsee Conference was to intensify it, instituting the mass transportation of Jews to specially built death camps at Treblinka, Chelmno, Bełżec, Sobibor, and Auschwitz-Birkenau. This point matters because the Holocaust has a larger and more various context than Wannsee and its outcome, which makes it more difficult for Holocaust deniers to deny.

Since coming to power in 1933 the Nazis (Nationalsozialistische

Deutsche Arbeiterpartei, 'NSDAP') pursued a policy of intimidation, expropriation, and 'encouragement to emigrate' against Jews, who with Gypsies were officially classified as 'aliens', not just in Germany but in Europe as a whole. In 1937 a 'final solution to the Gypsy question' was implemented, involving arrests, deportation, and internment of Roma in concentration camps at Ravensbrück, Mauthausen, Buchenwald, and Dachau. Following the annexation of Austria by Germany, offices were set up in Vienna and Berlin to 'facilitate' emigration by Jews; in 1940 after France fell into German hands Adolf Eichmann proposed that Jews be deported to the French colony of Madagascar. It was, however, chiefly the invasions of Poland and two years later Russia that prompted the Nazi leadership to think in more ambitious terms, for these events brought several million more Jews into the picture, increasing the problem from the Nazis' point of view. Jewish communities were forced into ghettos in occupied Poland, but, as fighting on the eastern front intensified, so did the need to deal more thoroughly with the Jewish populations in those areas. The policy of a 'Final Solution to the Jewish Question' was accordingly adopted to a design by Heinrich Himmler, Reichsführer of the Schutzstaffel ('SS'). On 31 July 1941 Hermann Goering instructed Heydrich to make 'necessary preparations for a total solution of the Jewish question'.[20] This was six months before the Wannsee Conference.

Impetus for an intensification of efforts came from recognition that the means then being employed for effecting a 'solution' were insufficient. When Hitler invaded the Soviet Union (Operation Barbarossa began in June 1941) death squads of SS Einsatzgruppen ('Deployment Groups') and Ordnungspolizei ('Order Police') were sent into occupied territories behind the front lines to kill Jews, whom Himmler had classified 'as a matter of principle . . . as partisans' and therefore as enemy combatants. Jews were rounded up and shot to death; one of the most emblematic occurrences was the massacre at Babi Yar on 29–30 September 1941, when approximately 33,700 Jewish men, women, and children were murdered by firing squads of SS, Order Police, and Ukrainian Auxiliary Police in a ravine outside Kiev. In Odessa in October 1941 over 35,000 Jews were killed by German and Romanian troops. These were the largest-scale massacres; everywhere in the occupied zone similar atrocities occurred.

As an added justification for the mayhem thus unleashed, the Nazis stated that their attack on the Soviet Union was aimed at extirpating Bolshevism, and that Jews were all Bolsheviks; so 'all Jews and Communists' were the targets of the activity. Between them the SS and Order Police had fifteen thousand men at work on this endeavour in the east, and their numbers were supplemented by willing helpers in Poland and the Ukraine, either among existing police forces or in specially recruited squads. Even as Operation Barbarossa was launched, the squads began work; 5,500 Jews were massacred in Białystok in Poland in June 1941, among them hundreds who had been locked inside the Great Synagogue, which was then set on fire. But Heydrich criticized the squads for their low rate of work; they were not killing enough Jews, so he ordered them to kill all the women and children too.[21]

By the end of 1941 about 440,000 Jews had been killed in the eastern occupied zones. Deaths by shooting had doubled by the end of the following year, to 800,000.[22] Yet, in light of the fact that there were millions of Jews in the Nazi-controlled territories, this rate was insufficient. Attempts to speed up the extermination endeavour with lorries fitted with exhaust pipes that fed fumes into an enclosed interior, or leaving locked freight cars on railway sidings for those inside to freeze, dehydrate, or starve to death, were too slow and cumbersome.[23] Moreover to control the large numbers destined for death, and in a strange awareness of the possibility of a reckoning with history, methods were sought that would make the slaughter both manageable and invisible to posterity. The desiderata were four: that those destined for death should be unaware that this was about to happen; that their killers ideally should not have to see, touch, or even hear them; that what killed them should be instantaneous or at least quick; and finally that no visible mark of harm should appear on their corpses.[24]

The Nazi authorities therefore began to build specially designed extermination centres. The first was at Bełżec in Poland; its foundations were laid in October 1941 and it became operational in March 1942. Sobibor and Treblinka became operational in May and July 1942 respectively. 'Became operational' meant that transports were arriving and mass murder was in progress. This was Operation Reinhard, the programme for eliminating all Jews from Poland. The victims were gassed in purpose-built bunkers by fumes from tank engines, and buried in mechanically excavated mass graves. At Auschwitz-Birkenau the methods used were gassing by cyanic poison, Zyklon B, and cremation in purpose-built ovens. Sobibor, Bełżec, and Treblinka were closed down, bulldozed and grassed over by the end of 1943, having taken the lives of 2.7 million Jews; Auschwitz-Birkenau was thereafter the principal destination of trains from all over Europe.

Auschwitz-Birkenau had been operating since June of 1942, in which month 16,000 French Jews, over 10,000 Silesian Jews, and 7,700 Slovakian Jews were gassed in the newly completed Bunker I. This was the beginning of a greater level of efficiency in industrial murder. In the course of the following two years various 'improvements' were made at the camp, with the addition of new railway sidings and extended ramps for disembarkation and marshalling. The further increase in efficiency thus made possible is illustrated by the fact that in fewer than eight weeks, in June and July 1944, a total of 320,000 Hungarian Jews were gassed to death in its bunkers and turned to ashes in its ovens. In just six months, from the spring to the autumn of 1944, Auschwitz-Birkenau saw the murder of some 585,000 Jews from all over Europe.[25]

The facts of the Holocaust reside in official documents kept by the Nazi authorities themselves – they were meticulous record-keepers – and in the recorded speeches, letters, diaries, and memoranda of the principal actors in the Nazi regime, including Hitler, Himmler, and Goering; and even more so in the memories and testimonials of survivors. The overwhelming evidence from so recent a set of such excoriating events makes 'Holocaust denial' seem incomprehensible. Its challengers see it as motived by anti-Semitism, by racist attitudes, by pro-Nazi and far-right sentiment – and indeed most Holocaust denial, though not all, has those roots. It is accordingly an emotionally fraught matter.

The first deniers were the Nazis themselves. As defeat started to become a possibility, Himmler ordered that evidence be destroyed. In addition to the death camps of Sobibor, Treblinka, and Bełżec, the camps at Majdanek, Poniatowa, and Trawniki were also closed in late 1943. To get rid of the inmates at these three last named camps, squads of SS, Order Police and Ukrainian Sonderdienst ('Special Services') and auxiliary police killed 42,000 Jews in the bitterly named Aktion Erntefest ('Operation Harvest Festival'). In late 1944, with the advancing Soviet forces not far away, the crematoria at Auschwitz-Birkenau were demolished. At Rumbula, near Riga in Latvia, the corpses of 25,000 Jews who had been shot to death in the autumn of 1941 were exhumed and burned, and the same happened at other mass graves, such as those at Bełżec and Treblinka. At some of the death camps bone-crushing machines had been installed to deal with the skeletons of the dead – photographs survive of Sonderkommando ('Special Unit') operatives posing beside them.[26] These efforts at covering the traces of the Holocaust were patchy, because of the rapidly collapsing situation; only those camps that had been closed and demolished earlier were thought to be safe from discovery. They were not.

The next earliest deniers were active in the years immediately after the war, when a generalized scepticism about atrocity stories – a hangover from the First World War, in which propaganda about 'enemy evil' turned out to be false – and the fact that many of the documents of the Holocaust had yet to be published, offered encouragement to some to challenge what was being said by military witnesses of liberated camps, and by camp survivors.

The person who well merits the title he is accorded of 'Father of Holocaust denial', the French politician and writer Paul Rassinier, was in a position to lend credibility to those who followed in the denial tradition: he had been a member of the Resistance, and had survived incarceration in the camps at Buchenwald and Mittelbau-Dora (the V2-rocket factory). He said that reading Jean Norton Cru's study of witness statements by French soldiers in the First World War had been an inspiration to him, showing him how unreliable, distorted, exaggerated, and contradictory the testimony of witnesses can be.[27] Although the mainstream of Holocaust denial has been associated with racist, neofascist, and generally far-right political orientations, Rassinier had been a communist and an anarchist and a pacifist. After the war he sat as a deputy in the French parliament. He was an inspiration for a French left-wing group of Holocaust deniers centred on the Parisian bookstore and Holocaust-denying publishing house of La Vieille Taupe, run by Pierre Guillaume.[28] This group later distanced themselves from him when he moved to the right; but he had bequeathed a tradition of Holocaust denial to some sections of the French left.

Rassinier's arguments ranged from saying that German concentration camps were no different from prisons in France or camps in Russia; they were all expressions 'more or less severe, according to circumstances, of the essence of the state as such not just of the Nazi SS state . . . the underlying logic of the essence of the state is the logic of war and enslavement,' so he saw his argument as a warning against the 'Manichaeism that places all the blame on one side, thus provoking war . . . It is war itself that is the absolute evil, not one warmongering party or another.'[29] He accepted that there were gas chambers at some of the camps, but said they might have had purposes other than killing people – perhaps, disinfecting them; they were, he said, next to the sanitation facilities and not next to the crematoria. However, 'that the gas chambers were used for extermination cannot be completely denied' but could well have been 'the work of one or two mad SS men or camp bureaucrats'.[30] His principal contention was that there was no systematic mass murder of Jews under the Nazis.

But, as noted, most Holocaust denial came from the political right. One of the earliest was Maurice Bardèche, a self-avowed anti-Semite and brother-in-law of a leading figure among Vichy-supporting French fascists, Robert Brasillach, who had been executed as a *collabo* in the post-war reckoning in France. It was Maurice Bardèche who brought Rassinier to prominence, despite their being on opposite political sides at first, by publicizing his work and making use of his reputation as a former concentration camp inmate.

Another follower and promoter of Rassinier was the American

historian Harry Elmer Barnes, who had worked with an institute founded in 1921 and funded by the German government, the Zentralstelle zur Erforschung der Kriegsursachen ('Centre for the Study of the Causes of the War' – i.e., the First World War), committed to proving that Germany had been the target of British and French aggression in 1914, and that articles 231–48 of the Treaty of Versailles forcing Germany to accept blame for the war, to give up territory, and to pay onerous reparations, were morally invalid.

After the Second World War, Barnes was determined to argue that stories about the Holocaust were false and that Germany was the victim, not the perpetrator, of the collapse into war in 1939, just as in 1914.[31] His claim was that Holocaust propaganda had been fabricated as a justification for American entry to the war and that it 'defamed the German national character and conduct'.[32] In a 1964 article entitled 'Zionist Fraud', Barnes – citing Rassinier as a 'courageous' authority on the matter – claimed that the Holocaust was a fiction created by 'those whom we must call the swindlers of the crematoria, the Israeli politicians who derive billions of marks [dollars] from non-existent, mythical and imaginary cadavers, whose numbers have been reckoned in an unusually distorted and dishonest manner'.[33] It might in passing be said that one technique for detecting tendentiousness in such claims is to note that if cadavers are non-existent then the numbers of them cannot be 'distorted'; the claim to that effect is a tacit admission that there were indeed cadavers to be counted.

Rassinier, Bardèche, and Barnes are examples of outright deniers; there are many more. Later deniers were able to profit from them and from near-deniers and those whose observations could be invoked in support of the denial cause, such as the German historian Ernst Nolte and the American historian A. J. Mayer. Nolte expressed sympathy for those whose motivation for being deniers was their hostility to Israel over the Palestinian question. He held that there had been no Wannsee Conference and that its minutes were post-war forgeries by Jewish historians, whom he blamed for promoting a 'negative myth' of the Third Reich. Without denying the existence of death camps and gas chambers, he nevertheless argued that some of the deniers' claims were 'not without foundation'.[34] This was less committal than the controversial claim by Arno Mayer that most of the fatalities at Auschwitz were the results of disease, not killings, and that, because the SS had destroyed all documentary evidence relating to gas chambers, the basis for claims about them have to be regarded as 'rare and unreliable'. [35]

One sees how such a statement can be meat to enthusiastic deniers. One of the more publicly visible of them was David Irving. An historian of the Holocaust who challenged Irving by name, Deborah Lipstadt, was sued by him for saying in her book *Denying the Holocaust*, published in 1993, that he was 'one of the most dangerous spokespersons for Holocaust denial', charging him with distorting and even falsifying evidence to make it conform to 'ideological leanings and political agenda' that she described as bigoted. Lipstadt's lawyers invited two authorities to examine Irving's publications and recorded *viva voce* utterances; these were the historian of the Third Reich Professor Richard Evans of Cambridge University, and Professor Christopher Browning of the University of North Carolina at Chapel Hill, historian of the Holocaust. They also invited the architectural historian Robert Jan van Pelt to report on evidence for the existence of gas chambers at Auschwitz-Birkenau. Irving represented himself in court, cross-examining the defence witnesses. The upshot was a judgment of three hundred and fifty pages by the judge, Mr Justice Gray, exhaustively detailing the evidence presented in court and, on the basis of it, concluding 'that no objective, fair-minded historian would have serious cause to doubt that there were gas chambers at Auschwitz and that they were operated on a substantial scale to kill hundreds of thousands of Jews'.[36] Irving lost his case.

The Holocaust denial controversy raises the important question of distinguishing between *denial* and *revision*. The examples of bringing to light the perspective of Australian Aboriginals and Native Americans given above are examples of revision, an essential corrective to the one-sided and triumphalist accounts of 'victor's history' that have distorted understanding of the past, very often deliberately. Another example is the 'New Conquest History' of Central and South

America, based on scholarship that challenges the exclusivity of the narrative about conquistadores' military successes, the 'Spiritual Conquest' of conversion, and the colonization process. By fresh archival work and palaeographic resurrection of Mesoamerican voices and perspectives drawing on 'New Philology' as applied to American ethnohistory, the previously omitted side of the story of indigenous and black men and women is being told, casting the conquest period in a new light.[37]

The same is true of colonialism in general. A major spur to reveal suppressed truths about the past everywhere was the outrage that followed the killing by police of George Floyd in Minneapolis in May 2020. This tragic incident was nothing new in a United States where racism and racial injustice are endemic and very many have suffered as George Floyd did; but the incident occurred in the full light of virally rapid social media transmission of news and images, a relatively new phenomenon, and the shock wave that spread across the world inspired a determination among many to challenge the silence or distortion that contributes to perpetuation of problems for historically disadvantaged communities – African-Americans being a major example. In the United Kingdom one reaction was to challenge the complacent view of the past in which statues of slave-traders stood in the public spaces of cities such as Bristol, which had benefited enormously from slavery on West Indian plantations and the slave trade itself. It would have come as a salutary shock for many to learn of those whose not-too-distant ancestors were paid the equivalent of millions of dollars to manumit their slaves, while the freed slaves themselves received nothing. Yet more significant is the educative effect of understanding the legacy of racism and disadvantage left by the history that victors' narratives had hitherto so successfully blanked (a term with more than one appropriate resonance).

Deniers describe themselves as revisionists, and repudiate the label 'deniers'. One way of marking the difference between the two categories was offered by members of the history faculty at Duke University, responding to efforts by a Holocaust denier who repeatedly placed advertisements in campus newspapers calling for what he described as an 'Open Debate on the Holocaust'. The Duke historians wrote that it is true that historians engage in revision, but that revision 'is not concerned with the actuality of those events; rather, it concerns their historical interpretation – their causes and consequences generally'.[38] A fuller account would include recovering lost voices and perspectives, evaluation of interpretations offered, and assessment of choices of emphasis and lacunae. Deniers, on the other hand, might not merely suppress material but actively distort it, falsify it, incorporate falsehoods and outright denials that certain events even occurred. In their study of denial Michael Shermer and Alex Grobman described revision as refinement of knowledge about an event as a result of new evidence or re-evaluation of existing evidence; the event itself, where there is a convergence of well-attested data about its occurrence, is not called into question.[39] Denial is the claim that there was no such occurrence, or that it was of a very different character and one that is conformable to the denier's personal, and often political, agenda.

The complexity of debate about the nature of history and historical knowledge goes further than the question of denial *versus* revision, given that revision is an element in the fact, noted above, that history matters to the present and future, and therefore often has a polemical angle targeting a concern contemporary to the historian. What is the distinction between the *polemical* purposes of the denier and the revisionist? The essence of the answer is whether the claim is 'X did not happen' (the denier's claim) *versus* the revisionist's claim that 'Although X happened, there is more to it; there are other sides to the story; notice these aspects of it which are important to our understanding of it; this is what it can tell us about what is happening now'; and the like. Denial is what has sometimes been called 'negationism'; revision is the effort to persuade to a reinterpretation.

A classic example is Christopher Hill's work on what had come to be described as 'the Puritan Revolution' of the mid seventeenth century in England. In a set of lectures delivered in 1962 and subsequently published as *The Intellectual Origins of the English Revolution* (1965), Hill argued that this revolution was not primarily a religious event, as had been thought, but the first great political and social revolution of modern times, setting the pattern for the American, French, and

Russian revolutions that followed in the next three centuries. Thirty-five years later Hill issued a much expanded version of the *Intellectual Origins* in response to the vigorous debate his claims had prompted.[40] Further study had confirmed him in the view that the English Revolution wrought a significant change not just in England but – because of what it made England become – the world; not just in giving later revolutionaries their model but in creating the conditions for England's imperial expansion in the eighteenth and nineteenth centuries, during which it exported its institutions, its economic ideas and practices, and its language across the globe.

The revolution's significance lay in the combination of changes involved: regicide, radical shifts in the ownership of land, mass democratic movements, and the bringing of taxation more firmly under parliamentary control, to name just a few. Together they altered England's constitutional and social character. Executing the king for treason was a practical repudiation of two doctrines: the divine right of kings, and the idea that sovereignty lay in the crown alone. Even at the Restoration there was no going back to earlier views of monarchy, as James II found to his cost. The basis for constitutional government was laid by Charles I's decapitation, which proved to be as symbolic as it was literal. When the Englishmen of the American colonies defied George III, and the French guillotined Louis XVI, they were not just conscious of this precedent but cited it.

Understanding consequences requires understanding what triggered them, Hill argued; and therefore the subsequent history of England and the global effects it generated compel us to see their trigger as itself the effect of much more than a dispute over confessional practices. For example: repudiation of feudal land tenure meant that landowners could consolidate their position and plan longer-term investment in agriculture. Some of the seeds were thereby sown for the accumulation of capital that later financed the Industrial Revolution. As importantly, the tax revenues controlled by Parliament were used to build a formidable navy, which gave control of the seas and therefore had a significant influence on international trade. This in turn motivated further imperial expansion, contributing yet more to the wealth that fuelled the Industrial Revolution, which in turn fuelled yet further imperial expansion: a snowball effect.

In arguing that the events of the mid seventeenth century were a turning point in world history, Hill did not claim that the men of the time either intended such consequences or guessed that they would follow. They did not even have a name for what they were doing: Oliver Cromwell was the first to use the world 'revolution' in the modern sense – and only after it had happened. There were no plotters or conspirators; Hill argued that revolutions do not need them, because they occur when a people has had enough and a sentiment for radical change arises among them.

The chief focus of Hill's study is the *ideas* that led to the English Revolution. Developments in philosophy, science, and medicine, and in theories of economics and history, combined with diverse literary influences – among which the bible in English was significant – effected a sea-change in outlook. Sixteenth-century arrangements had become impossible for the people of the seventeenth century. They had witnessed the Dutch throwing off the yoke of a foreign oppressor – Spain – although this was not exactly applicable to their own case. They felt they had to attempt something bold and novel, without yet knowing what it was or where it would lead. But some of them – Thomas Hobbes is an example – suspected that whatever changes came would be far-reaching; and they were right. This was the point Hill sought to make.

Hill's work therefore exemplifies revisionist history. The facts and dates of the reign of Charles I and the civil war that led to his execution are fully documented and agreed upon. Hill did not dispute them; what he did was to argue for a way of seeing the causes and consequences of them in a new light. Establishing the facts, and how to understand their causes, meanings, and consequences, is the meat of historical scholarship; revising our understanding of an historical topic by fresh evidence or new argument is the stuff of historical debate. A denier, quite differently, is one who would say that Charles I was not executed, or did not attempt to avoid summoning Parliament during eleven years of 'personal rule', was never defeated by the Scots in the 'Bishops' Wars', was the innocent victim of a

Puritan-inspired *putsch* because he had married a Catholic princess, and so on.

Political and diplomatic history, military history, social history, the history of ideas – the variety of domains of historical enquiry is broad, and the resources and techniques for their pursuit various. Nevertheless in each the contrast between denial and revision remains the same, turning on the effort respectively either to reject, or to start from, a convergence of views about a topic. To overturn or significantly to change a convergence – an orthodoxy – on some topic is legitimate if convincing evidence, responsibly gathered and marshalled and presented with careful arguments, can be made for doing so. Denial is not *ipso facto* wrong. Neither is the fact of service to an agenda. But when the methods are those of distortion and falsification, and the agenda is one of questionable moral partiality, the eyebrow of suspicion rises.

In the arguments advanced by the Centre for the Study of the Causes of the War, championed by Harry Elmer Barnes, the claim was that Germany was the victim of aggression in 1914, and that the Treaty of Versailles was an historical injustice therefore. Given that the First World War's causes are complex and contested, ranging from railway timetables to over-accumulations of armaments to the confusions of Great Powers diplomacy, the Centre's arguments might be regarded as one side in a polygonous debate, to be evaluated on their merits. But when Barnes much later claimed that the Holocaust was a fiction made up by post-Second World War Zionist conspirators, something more than revision is envisaged; this is where a line is crossed.

Bearing the foregoing in mind, the link between questions about the objectivity of history and questions about the uses of history comes into troubling focus. To illustrate why, one need do no more than consider arguments and political views about the teaching of history in schools in China, Japan, the United Kingdom, the United States, France, Canada – indeed almost anywhere. What is to be included, what left out, what emphasized, what should the overall tenor be? Should history celebrate a nation's achievements, should it be frank about its less savoury activities and choices, should it emphasize social history or political history, should it treat the past as sequences of 'causes, courses, and consequences' or as a complicated, ambiguous, and turbulent sea of events stirred more by accident than the intentions of agents?

In the last decades of the twentieth century in the UK, debates about the content of history lessons were complicated by debates about how history should be taught. From the 1970s onwards, history was taught as a method, not as a chronological narrative; for example, pupils were asked to compare census returns for a given town with the returns for the same place a decade later, and to identify significant trends or changes. Earlier in their school careers pupils would have learned about the Tudors and Stuarts, safely remote, portrayable as possessing a measure of glory if the litany of betrayals, murders, and the gruelling lives of the majority were dressed up or played down. More recently the Second World War has been such a focal topic that some pupils find themselves studying it several times over at different stages of their school careers.

Conservative politicians in the United Kingdom have repeatedly pressed for history to be a chronology of Britain's heroic rise to empire and tenure of it. Having become a naval power late in the day (Spanish Armada apart, it is really only with Samuel Pepys in the late seventeenth century that Britain's two centuries of naval supremacy began), the 'island story' of a sea-faring race was invented. National heroes from King Alfred, to Lord Nelson, to Winston Churchill exemplify the virtues that eventually made the Sun never set on the British Empire. Romans and Normans invaded, but are treated as having done so beneficially; Boudicca and King Harold, who lost their respective resistances to these invasions, are relegated to the also-ran category. Triumphalist disdain attends thoughts of those who contemplated invasion but failed – Napoleon, Hitler.

But what about Britain's treatment of indigenous peoples in different parts of the empire – say, in Kenya during the Mau Mau uprising, and 'insurgents' in Malaya – or the way immigrants from India, Pakistan, and the Caribbean have been treated? What about the slave trade that made Britain rich, and the cities of Bristol and Liverpool flourish? What about the chronic unpreparedness of the UK for war in 1914 and 1939, its habit of government by amateurs, the perennial problem of the deeply class-divided nature of its society?

In the United States what one might call the official, normative, identity-creating story of 1776 and what followed has been an important aspect of nation-building. Immigrants were encouraged until the early 1920s, and their assimilation turned on accepting the melting-pot idea that *e pluribus unum*, 'out of many, one', is possible. The history of independence and the expansion to the west has drama, heroism, large spaces, and challenging conditions as the background to the rise of great cities, great wealth, and great power in the world. In this story the dispossession and genocide of Native Americans, slavery, post-slavery segregation and injustice, the terrible slaughter of the Civil War, and the unflinching *realpolitik* of US foreign policy, especially in the Cold War – in which the CIA served as one arm of that policy's decision-makers about which foreign governments were acceptable to Washington and which were not – do not figure as prominently. Veneration of the flag and celebration of the Fourth of July are about the positives, not the negatives.

As with arguments about teaching biological evolution in US schools, so the content of history lessons has been a bone of much contention. Among the states of the Union that are more socially and politically conservative, it was required by law that history teaching should aim at encouraging patriotism. In the early 1990s the problem of history course content came to a head. President George H. W. Bush established a working group to devise a set of 'National Education Goals', with history nominated as a core subject alongside science, maths, geography, and English. The National Council for History Standards – the very title of which suggests all the difficulties it would face – produced an advisory recommendation about US and world history that gave increased prominence to multiculturalism, black history, and women's history. The predictable result was a backlash from Republican politicians and media, which saw it as 'Political Correctness gone mad', and claimed that it portrayed the US as an 'inherently evil' state. A candidate for the presidency, Senator Robert Dole, went so far as to claim that the proposed curriculum was 'treasonous' and would 'do more damage to the US than its external enemies'.

To the credit of historians, history teachers, and more sober minds, the recommendations were eventually adopted – in the main, though with the (welcome) addition of a section encouraging reflection on the nature and uses of history itself.

But, even as the controversy was intensifying, it embroiled the august Smithsonian Institution in difficulties over its plan to mount an exhibition for the fiftieth anniversary of the end of the Second World War. It had in its collection the B-29 Superfortress airplane that on 6 August 1945 had dropped on Hiroshima the world's first atom bomb. Named *Enola Gay*, after the pilot's mother, it became the target of a ferocious debate, not because it was the instrument of *the* ultimate 'area bombing' attack on a civilian population (a form of warfare already outlawed by the 1977 First Protocol to the 1949 Fourth Geneva Convention), but because the curators proposed to invite exhibition visitors to ponder the morality of the weapon's use. Some artefacts from Hiroshima and Nagasaki and photographs of victims were to be included. Heated objections came from the media and politicians, with the US Senate intervening to describe the proposed wordings of exhibition material as 'revisionist and offensive to many World War II veterans'. The Senators were here contradicting the US Air Force Association, which had already acknowledged that the proposed exhibition treated bomber force veterans with respect. When the Smithsonian suggested watering down the exhibition and removing evidence of suffering of Japanese victims, the government of Japan objected strongly. Caught between the outrage, the diplomatic tensions, and a sense of responsibility to history, the Smithsonian backed down on all fronts, and its director resigned.

Much the same happened with efforts in Canada to revisit the question of Second World War aerial bombing of Germany. Even during the war there had been criticism of indiscriminate 'area bombing' attacks on civilian populations – the Bishop of Chichester asked in the House of Lords in 1943, 'We are fighting barbarians; why are we behaving like them?' – but veterans of bomber forces, most vocally the members of No. 6 Group Royal Canadian Air Force, which had suffered high casualties operating heavy bombers over German-held Europe, were in no mood to be redescribed as perpetrators of a war crime. When Ottawa's War Museum was opened in 2005, the interpretative material relating to Canada's bombing involvement was entitled 'An Enduring Controversy'. In fact these words intentionally captured not only the fact that argument about the campaign had started during the war itself, but that it had revived with unabated heat every time occasion prompted. There had been an uproar in 1992 when Canada's Broadcasting Corporation ran a television series about the war in which the moral acceptability of area bombing was questioned, provoking veterans' ire.[41]

National debates about such matters are made more complex by almost always being pieces in a larger jigsaw. In Canada there was controversy over the renaming of Dominion Day, marking the anniversary of Canada's becoming a dominion in the British Empire in 1867; the new name chosen was Canada Day, to mark the final severing of Canada's judicial ties with Britain's Privy Council in 1982. In the US Columbus Day attracted the attention of those concerned to point out that Columbus's arrival in the Americas was not a matter of celebration for the descendants of peoples already living there or subsequently brought there as slaves. The battle lines were familiarly drawn between conservatives and liberals, and only somewhat defused by a change of designation – from *celebrating* Columbus's 'discovery' of what Europeans had called 'the New World' to *marking* it. For those on the liberal side of the debate, invasion, genocide, dispossession and slavery certainly required remembering.

Germany's acceptance of its twentieth-century past and its determined efforts to come to terms with it make a contrast with France's uneasy handling of its Second World War experience. Deep wounds from the terrible loss of life in the First World War played their part in France's reluctance to repeat the experience, but the worst aspect was the collaborationism of the Vichy Regime, and the nervous postwar efforts of many not just to disassociate themselves from it but to claim that they were in the Resistance. There were doubtless a number – among those who tarred and feathered women who had had German boyfriends – who were no less 'guilty' than they, in any of a number of ways. For the French, with their tumultuous history, there is as much to ponder – the French Revolution's Reign of Terror, 1793–4, Napoleon (hero or marauder?), the Algerian War – as the British with slavery and imperialism, Russians with Stalin's terror, Spain and the legacy of Franco, Austria contemplating its post-*Anschluss* part in the Third Reich, Europe as a whole with colonialism and racism.

Europe is not alone in the burden of its historical culpabilities. What of Turkey and its Armenian massacres, or Japan dealing with the atrocities committed against Korean and Chinese people during the 1930s and 1940s? Diplomatic tensions increase every time a new Japanese history textbook fails to meet Chinese demands that events such as the Rape of Nanjing in December 1937 be included, or when a Japanese prime minister pays respects at the Yasukuni Shrine honouring Japanese war dead. The Chinese government itself has firm views about what can be taught in history classes: anything that fails to teach 'correct history' and impugns 'socialism and the leadership of the Party' is proscribed as politically and ideologically 'heretical'. Such events as the Chinese invasions of Tibet and Vietnam, incursions into India, irredentist threats against Taiwan, and land-grabbing of the Spratly Islands are represented as China's noble effort to liberate those in need of liberation, as acts of self-defence, or as reclaiming territory that had wrongfully been taken from China in former times. Such events as the devastating famine of the 'Great Leap Forward', 1958–62, in which 30 million starved to death, and the Cultural Revolution, 1968–76, in which yet more tens of millions died, many suffering bullying and persecution in the process, are not mentioned.[42] As one Chinese historian put it, 'Going very deeply into the history of Mao Zedong, Deng Xiaoping and some features of the Liberation is forbidden.'[43]

There are places where a reckoning with history has been made, or an honest attempt to make one has been made. Germany, as noted, has addressed its Nazi past. The Republic of Ireland has largely ceased to frame its history solely in terms of the course and outcomes of eight hundred years of oppression and periodic atrocities by the English (or more generally the Protestant British from the neighbouring

island of Great Britain), even though this is, alas, a major part of its
story. In South Africa the grim history of apartheid and the subsequent endeavour to create a 'rainbow nation' – at time of writing the
balance in favour of white wealth and living standards is still hugely
disproportionate – have taken the form of a 'truth and reconciliation'
project that seeks to be frank about what happened and to include all
the communities of South Africa – a complex society comprised of
Xhosa tribes, San tribes, Zulus, Indians, Afrikaner whites of Dutch
descent, whites of English descent, and the mixed-race ethnic group
once officially designated as 'Cape Coloureds' who form nearly half
the population of the Western Cape. These examples show that the
questions so profusely littering history's landscapes like landmines
can be given answers. On the other hand, whether old wounds will
tear open and haemorrhage, as happened in the Balkans in the 1990s,
or whether seemingly intractable problems such as the Israel–Palestine
conflict can ever be resolved, remain hard questions. For some
peoples, the greater tragedy is not only that history has hurt them but
hidden them: Timorese, Kurds, Naga, Rohingya, Uighurs, Tibetans,
Artsakhtsi, might all explain how that has happened to them – and
atlases obscure the fact that national boundaries are artificial, most
drawn in the blood of wars, collecting and sometimes unwillingly
corralling diverse peoples and traditions: the 'minority peoples' of
China's west and south-west, the Dalits of India and Nepal, are salient examples, but there is no difference in principle (though much in
material fact) between their situation and that of Catalans and Basque
in Spain, Scots and Welsh in the British Isles, Flemings and Walloons
in Belgium. Nevertheless, historical examples of resolution and reconciliation are not just our best hope but a good hope, even here.

A different question concerns judgements about progress and regress
in history. Consider the question of the 'Dark Ages' following the
collapse of the Roman Empire in the west of Europe. Across whole
swathes of life the quality of social and civic affairs – literacy and the
associated matters of schooling and book publication, engineering
skills, the functioning of aqueducts, the keeping of records, personal
health and security, even population – declined. From Petrarch in the

early Renaissance to Edward Gibbon in the eighteenth century, an orthodoxy arose to the effect that the post-classical period was one of superstition, ignorance, and civilizational decay. Petrarch thought that he was still living in the Dark Ages in the fourteenth century; later the Dark Ages were restricted to the period between the fifth and tenth centuries CE. Much blame was laid upon the spread of Christianity after an orthodox form of it had been achieved in the fourth and fifth centuries, because of its imposition of a hegemony over thought and the destruction of literature and material culture associated with pre-Christian 'pagan' thought and beliefs.[44] The banishing of philosophers from Athens in the year 529 by the Christian Emperor Justinian is an example.

If there were indeed Dark Ages after the fall of Rome in the West, this would be an example of regression. The historical consensus has turned against the Dark Ages trope: scholars who study the period defend it against the negative imputation of the label by pointing to the rise of monastic education and preservation of texts, and to the Carolingian Renaissance, in which the deterioration of literacy and education was to some degree stemmed by Charlemagne's reforms in his domain. For just one example: look at the intricacy and beauty of Anglo-Saxon art and crafts, in metalwork, ivory carving, textiles and manuscript illumination; this is a corrective in itself.

It is hard to deny, though, that general literacy declined and most of the literature, history, and philosophy of the ancient world was lost altogether, a fraction surviving though largely unread (because 'pagan') in the Byzantine world. Some of it had to await the Arab conquest to be brought back to light.[45] It is a speaking fact that, between the construction of the Basilica of Maxentius in the Roman Forum and Brunelleschi's dome for Florence's cathedral more than a thousand years later, the engineering skills that made the former possible were lost.

Although the period in question was neither so dark nor so homogeneous as implied by the pejorative label attached to it, at the same time it is obvious that, in relation to the high standards of life, art, literature, and organization of the classical and Roman periods of the thousand years that preceded it, the same regions of Europe in the centuries following the Western Empire's collapse scarcely compare. The charge of regress stands, and has the same descriptive utility as when the same phrase, 'Dark Age', is applied to the centuries following the Bronze Age Collapse around 1200 BCE. By the same token it is arguable that, without going to the full extent of the so-called 'Whig interpretation of history' in which everything that has happened in a given period represents inexorable progress towards sunlit uplands, there are times and respects that merit being described as instantiating progress by a measurable standard. For example: any ordinary resident of Europe or North America in the early twenty-first century enjoyed the freedoms, rights, and almost all the opportunities that four centuries beforehand were available only to aristocrats, gentry, and senior clerics – a tiny minority of the population. Our forebears, four centuries ago, unless they were themselves members of such groups, would more likely than not have been illiterate peasants who never or rarely went far from their places of birth, and who lived lives of gruelling restriction in almost all dimensions. Their lives might have been contented ones despite that, if the quality of their personal relationships, health, and security of food supply permitted; but in the objective measures of their possibilities as individuals there is no comparison to today. In one good meaning of the term 'progress', the difference between then and now might well be regarded as exemplifying it in spades.[46]

Another and associated instance might be the philosophical and scientific revolutions of the sixteenth and seventeenth centuries, which can be said without hyperbole to have led to the modern world – an alloy, to be sure, as by no means without extremely disagreeable aspects, but arguably in many more respects an improvement; only think of communications, computing, medical advances, the applications via technology of science in general – the list is very long. The flowering of thought of which these revolutions consist was made possible by the Reformation, not because Protestantism was friendly to innovations in science, but because an unintended by-product of its occurrence was that large parts of Europe were liberated from the hegemonic control of a Church that was hostile to ideas that impeached doctrinal orthodoxy. The cases of Giordano Bruno,

Cesare Vanini, and Galileo Galilei in the years 1600, 1619 and 1632 respectively illustrate the Catholic Church's attitudes to Copernican ideas; the relative weakness of the new Lutheran (less so the Calvinist) churches in controlling what people could think and publish inadvertently freed the mind of Europe.[47] This too will be thought of as progress by many, though doubtless not by all.

Making a case that a given period or set of events embodies regress or progress is to use history polemically but at the level of interpretation. This is more than merely legitimate, it is important; for, in the constant reassessment and negotiation of how we understand the past and its relation to the present, debates of this kind are essential.

4.   'Reading-in' to History

One of the cautions required in historical enquiry is prompted, obviously enough, by the Map Problem. A geographical map at a scale of 1:10,000 is regarded as 'large scale'; this might roughly equate to one page of a history book per year of time (one page per 10,000 hours, say). The limitations imposed by an indiscriminate record of every event, every second, everywhere, from the accidental and trivial to what changes the course of a civilization, would be worse than useless: it would be blinding; all trees and no wood. History as enquiry is selection and organization, an attempt to make sense. But accepting that history at best bears the same relation to what happened in the past as a map does to a country or continent in turn generates another problem: the Reading-in Problem. For the question now is: on what basis do we make our interpretations?

To 'read-in' is to interpret data according to assumptions and interests local to the investigators; it is to see things coloured and shaped by the conceptual and experiential spectacles worn by the investigators. As such, it is a major source of potential distortion. Consider the debate between those who espoused forms of *Verstehen* theory in the social sciences to distinguish them from the presumed objectivity of the natural sciences.[1] The fundamental idea was that in the latter the aim is description and explanation, while in the former it is understanding and interpretation. The tools available for description (measurement and repeatable experiment) are different from those available for understanding; here our principal resource is the investigator's insight, sympathy, and experience. And to say this is to accept from the outset that reading-in is inescapable. Is this right?

The answer to whether this is right or wrong is far from clear-cut. For example, in response to the relativist claim that we cannot understand the past – or for that matter people of different cultures, speakers of different languages, than our own – because we are locked inside our own culture's conceptual framework is readily met with such examples as our ability to sympathize with the agony of Achilles's grief for Patroclus, as recounted in Book 18 of the *Iliad* : 'a black cloud of grief enfolded Achilles, and with both hands he took the dark dust and poured it over his head and defiled his fair face . . . and he himself in the dust lay outstretched . . . and with his own hands he tore and marred his hair . . . then terribly did Achilles groan aloud', and his comrades held on to his hands in case he harmed himself.[2] He could not sleep; at night he paced the beach where the ships of the Achaeans were drawn up, lamenting his beloved friend. One could cite any number of examples of love and grief, anger and resentment, hunger and pain, comfort and fear, from literature and history across the ages and across cultures, by which we are moved or for which we feel sympathy and understanding. Human commonalities are great, and profound. There appear to be genetically encoded abilities to recognize, and to respond to, smiles and laughter, weeping, expressions of pain, terror, and anger. We are all descendants of the population from among whom came behaviourally modern humans.

This is not to deny that there are barriers between cultures that can make aspects of each invisible to members of other cultures. There can be just such barriers between (say) men and women, and older and younger generations, within the same culture. In the latter case it would be supremely pessimistic to believe that the barriers are intrinsically insurmountable. By the same token it is standardly hoped that the human commonalities can be a bridge for intercultural understanding also. On that supposition *Verstehen*-type theory is based.

The question is how critical it can or must be. The following examples illustrate how pressing a question this is for history in all its forms, but especially as regards its most remote regions.

First suppose that an archaeologist of the distant future is excavating our present world, perhaps after some mighty disaster has destroyed all libraries and computer hard drives so that little or nothing remains of written records, and she has only the physical remains of devastated urban centres as evidence of our time. She will find buildings of all sizes, the smaller ones far more numerous than the larger ones, so she will attach significance to these latter, speculating on their use and, by extension, what they say about the nature of our society. Let us equip her with membership of various social and cultural milieus. Suppose her time is one in which people spend eight hours of every day exercising in gyms; she and her contemporaries are super-fit beings, their gyms are enormous and richly endowed. Or, suppose her time is a highly militarized one, in which the entire population between adolescence and senescence is engaged for the largest part of each day in a wide variety of military preparations and training in specially designed barracks and armouries. Or, suppose her time is a hyper-religious one, in which the bulk of each day is spent in ritual and devotional observances, with whole streets of churches as in some American towns today. How will she interpret the large buildings she uncovers in her digs? To what uses will she assign them? Will they be gyms, barracks, or churches?

In archaeology of the pre-classical age large buildings are interpreted as temples or palaces. This is because large buildings raised in the period between classical times and the beginning of modern times – say, between the sixth and sixteenth centuries CE – are one or the other. Large buildings are not that way now; they are libraries, theatres, concert halls, schools, universities, art galleries, hospitals, government offices, blocks of flats, barracks, factories, and department stores as well as cathedrals and palaces. From the rise within three centuries of each other of Christianity and Islam in their hegemonic forms – Christianity was made the official religion of the Roman Empire by the Edict of Thessalonica in 380 CE; Islam began to spread around 650 CE – the largest buildings in the territories under their sway were almost all cathedrals and mosques, the latter modelled on Byzantine examples of the former. The amount of wealth and labour absorbed from communities for the raising and maintenance of these embodiments of cultural dominance was prodigious, but, even in the thousand years or so during which this was so, the daily lives of people were not exclusively focused upon them.

Yet standardly the large building at the centre of a settlement dating to, say, 6000 BCE is assumed by archaeologists to be either the chief's house or a religious building. It is not standardly assumed that such buildings were schools, or central storehouses for grain or weapons, or dormitories for adolescent boys preparing for manhood, or places reserved for menstruating or postpartum women, or guest-houses, or assemblies of elders discussing matters of government, or homes for widows, or refuges for the sick, or places set apart for manufacturing clothing, ornaments, weapons, or agricultural equipment. Likewise art, from cave paintings to stone carvings, is almost universally assigned a religious significance. It is not hard to understand why this is so – namely, that the only evidence available for interpreting the purpose of the largest building is what the largest buildings were typically for in the periods of history for which there are also other resources for interpreting them. Even in our own day there is a kind of reflex reluctance to allow that people can make and contemplate artworks for the sheer pleasure of doing so, *and* that both are very important in their own right (think of the controversies that erupt over public spending on support for the arts). The assumption made is that a community would not make the effort to have a larger than normal building in its midst unless there were the sort of reason for it familiar from – well: what is its benchmark? – apparently, the age of kings and popes and the buildings they raised to announce who and what they were.

Come to think of it, this benchmark assumption is not even true for the classical period, or (for example) for pre-classical Egypt, whose largest buildings were tombs. In Minoan and Mycenaean sites the largest structures seem to have been palaces – which means something more than the residence of a ruler, for they were centres of government and justice, thus serving a multiplicity of purposes. In Greece the largest structures were theatres; in the Rome of the Late Republican and Imperial periods they were civic fora and arenas such as the Colisseum. Roman temples – even important ones like the Temple of Vesta in the original Forum – were small by comparison.

Is the evidence for interpreting a large building in a 6000 BCE settlement as a temple better therefore than the evidence available for interpreting it as (say) a wrestling arena? Only *reading-in* from what we know or think about later times, and our own preconceptions about why a community would devote resources to raising a larger than normal structure in its midst, directs us to the standard interpretations.

It is indeed hard to break away from these assumptions. Why would people quarry, shape, and then drag huge stones over hundreds of kilometres to a particular site that, from accompanying evidence, had obvious significance for centuries? Such was the case at Stonehenge; the people who did this shared some significant conceptual commitments with others from all over Europe and the Near East, from the tip of Scotland and the west of Ireland – Europe's furthest reaches – to Scandinavia and the islands of the Mediterranean; and they did so for thousands of years – from Göbekli Tepe in Anatolia in the tenth millennium to Stonehenge in the third millennium BCE. The ready conclusion is that they did it for reasons similar to those that inspired the building of great cathedrals. The motivation was something that really mattered to them; they invested a vast effort in it; there must have been an expectation of great and meaningful reward.

The discoveries at Göbekli Tepe and Çatalhöyük are a prime instance of this assumption at work. On the website of the Smithsonian Institution's magazine one will see dramatic claims to the effect that the remarkable Göbekli Tepe site is 'the world's first temple', which 'upends the conventional view of the rise of civilization', and that it is 'early evidence of prehistoric worship'.[3] This is a view that has become commonplace; articles and documentaries (the latter typically beginning with mysterious music and long atmospheric nocturnal shots of the site lit from within) promote the idea that the site is home to the world's first religion, or at least the world's first sacred site.

The tallest stones on the site stand 6 metres high, weigh 20 tons, and were shaped and carved without iron tools. They are T-shaped, and stand in sockets dug out of the bedrock. Excavations and geophysical surveys reveal about two hundred pillars in twenty circles, three different enclosures of them connected by the sides of an equilateral triangle. There also totem-pole-like steles sculpted into humanoid figures, and the pillars have figures of animals skilfully carved in bas-relief upon them, featuring snakes, lions, bulls, gazelles, foxes, donkeys, spiders, and birds, especially vultures. These last might be significant in that some peoples bury their dead as skeletons after the flesh has been removed – a practice known as *excarnation* – as in the 'sky burials' of Tibetan tradition and the Zoroastrians' 'Towers of Silence' (*dakhma*) atop which corpses are exposed for vultures to pick clean.

The earliest layers of Göbekli Tepe predate agriculture, pottery, and metallurgy, and by thousands of years they predate the invention of writing and the wheel. The great stones on the site were quarried nearby, up to half a kilometre away, but the effort required to prepare them and erect them on the site, and even more so the shaping and carving of them, bespeak a high level of social organization and tradition. Klaus Schmidt of the German Archaeological Institute, the leader of excavation on the site, hypothesizes that it was a sanctuary, a 'pilgrimage destination', and described it as a 'cathedral on a hill'.[4]

This view was contested by the Canadian archaeologist Edward Banning, following discovery of evidence of flint-knapping and preparation of foodstuffs at the site, which had hitherto been said to have no marks of continuous or domestic habitation. His principal argument is that prehistoric people did not distinguish sharply between the sacred and profane, and that what would now be considered sacred, religious, or superstitious was integrated into people's general world-view and activities. 'The presupposition that "art", or even "monumental" art, should be exclusively associated with specialized shrines or other non-domestic spaces also fails to withstand scrutiny,' he argued. 'There is abundant ethnographic evidence for considerable investment in the decoration of domestic structures and spaces, whether to commemorate the feats of ancestors, advertise a lineage's history or a chief's generosity; . . . or record initiations and other house-based rituals.'[5]

It is a further question, according to some, whether the idea of 'religion' makes sense in application to specialization of built structures and the meaning of artworks before about 5000 BCE.[6] The standard view to the contrary is inferred; no unequivocal evidence of belief systems is otherwise available before writing: 'The first written

records of religious practice date to *c.* 3500 BCE from Sumer. Mesopotamian religious beliefs held that human beings were co-workers
with the gods and laboured with them and for them to hold back the
forces of chaos.'[7] However, burial of the dead and cave art suggest
religious feeling and practice from long before, interpreting 'religion'
copiously to mean belief in agency beyond the natural world but
operative within it. In the case of burials, such attitudes might perhaps be apparent as long as 300 kya ago, given that Neanderthals and
Homo naledi purposely buried or secreted their dead. In the case of
cave art, if it has religious meaning or intent, then such attitudes are
apparent well before 30 kya.

All this is possible. But it might also be a function of reading-in on
a comprehensive scale, potentiated by the *desire* to find evidence of
particular kinds – in this case, religion – in the archaeological record.
Critics point to what they regard as a salient example of reading-in,
associated with the remarkable archaeological site of Çatalhöyük in
southern Anatolia and the generous funding of research there by the
John Templeton Foundation.[8] Templeton has been described as an
organization dedicated to promoting the plausibility of religious
belief by encouraging scientists, archaeologists, and others to endorse
it, with resulting controversy about the effect this has on the dispas-
sion and objectivity of the research it funds.[9] As an example of the
reading-in problem, it is instructive, especially in relation to
Çatalhöyük.

Çatalhöyük, which means 'Fork Mound', is an extensive Neolithic
site dating from 7000 to 5000 BCE, consisting of a large assemblage of
houses, among them no large structures that might have served a
public purpose of any kind. The settlement had no streets, the dwellings being contiguous and accessible through their roofs. The
inhabitants buried the skeletons of their dead under the floors of their
houses after excarnation; sometimes the skulls were removed, and
had faces painted on them in ochre. Some of the rooms in these houses
were decorated with murals.

The lead archaeologist at Çatalhöyük, Professor Ian Hodder, was
asked about Templeton funding for research at the site in an interview by journalist Suzan Mazur.

Suzan Mazur: Templeton is known for its pairing of religion and science, inserting the divine in science. I was wondering if you see any conflict of interest in serving on the Templeton board and accepting these four grants from the foundation relating to Çatalhöyük, three of which pertain to so-called 'religion' at Çatal considering that there was no religion 10,000 years ago?

Ian Hodder: Yes. Well there are a lot of issues there. I was not serving on the Board that gave out money, so I was not in any conflict of interest. I was on an Advisory Board that advised in terms of research and that sort of thing. I was not involved in any way that determined how money was spent. So I didn't see a conflict of interest. I found that Templeton was very careful to avoid anything like that. As far as the question of religion is concerned. It rather depends how you define religion. But I've now written or published three books that talk about religion in prehistory and I think it's quite acceptable to define religion in such a way that religion is something that does happen amongst all humans, even amongst non-humans. The idea of the spiritual is a very general notion.

Suzan Mazur: I'm asking this because Templeton has come under fire for putting its fingers all over science from the investigation of the origin and evolution of life to space science. It's perceived that the foundation is compromising the work of scientists and retarding science. Maurice Bloch, one of your own Çatal book authors, has said pursuing a religion angle at Çatal is 'a misleading wild goose chase' because humans only thought up religion 5,000 years ago at the earliest.[10]

The purport of Mazur's questions relates directly to the reading-in problem. Her point was that the Templeton Foundation funds projects which it considers as identifying a spiritual dimension in what is under investigation, and they will award their $1.4 million annual prize to any scientist or philosopher who makes such a significant link: beneficiaries have included astronomer Martin Rees, physicists Paul Davies (an adviser and board member of the Templeton Foundation), Marcelo Gleiser (rewarded 'for his work blending science and spirituality'), and philosopher Charles Taylor. Its funding areas include 'Theology and Science' and 'Science and the Big Questions'; its support for Ian Hodder and the Çatalhöyük investigations has produced, among other things, the series of books *Religion in the Emergence of Civilization* (2010); *Religion at Work in a Neolithic Society* (2014); and *Religion, History and Place and the Origin of Settled Life* (2018).[11] Templeton's use of its considerable wealth to promote a potentially distorting agenda in research has prompted much protest.[12] Its remit is well described in an *Inside Higher Ed* article from 2013: 'The Templeton Foundation's grants are intended to fund the study of the intersection of theological and scientific questions. Grants for medical research have looked at the power of prayer on health . . . it sponsors the annual Templeton Prize, given each year to a person who has made "an exceptional contribution to affirming life's spiritual dimension".' When 'religion' is described as Hodder describes it – as something shared 'even amongst non-humans. The idea of the spiritual is a very general notion' – then anything is permissible, and the Templeton agenda, of reading-in 'religion' as a phenomenon to be met with in any kind of research, succeeds.[13]

The point of this discussion scarcely needs emphasizing. To make deliberate reading-in a condition of funding a research project subverts the integrity of enquiry. There can be no objection to funding that supports research into religion, religious practices, religious history, and the like; problems arise when funding (especially when it is hard to resist because it is so large) is offered as an inducement not merely to look for *but to find* support for independently questionable phenomena in areas of enquiry whose findings should be dictated *only* by the evidence found, not by what someone would *like* to find. An archaeological investigation of an ancient site aims to discover what is there and what it tells us; Templeton-type enquiry sets out to look for religion in an ancient site before the first spadeful of earth is turned. In other ventures it has sought to promote revision of the concept of 'life' in biology to make it more consistent with Creationist accounts of life's origins;[14] it has paid for conferences and publications aimed at making religion more than consistent with science. This is reading-in as policy.

Göbekli Tepe appears to have been abandoned by 8000 BCE, and

Çatalhöyük by about 5000 BCE. In the latter case inferences to the site's religious and spiritual aspects are drawn from the symbolism of its mural art and funerary practices. A reviewer of Hodder's *Religion at Work in Neolithic Society* remarks, 'As might be expected from Hodder's history of risk-taking in archaeological interpretation, the volumes skirt, sometimes uneasily, around the problems of the legitimate limits of deriving knowledge of the past from its material remains' (and adds the caveat, in mentioning that Hodder's work was supported by Templeton, 'Readers should be aware that the Templeton Foundation is a philanthropic foundation that funds research which will discover "new spiritual information" ').[15] A result of the determined effort to see the past through such assumptions is that almost every reference to Neolithic culture, or Upper Palaeolithic art, or burial practices of the earlier Palaeolithic, is a reference to beliefs and attitudes too similar to those held by subscribers to the religions of today.

The reply that reading-in is a legitimate use of *Verstehen* on the grounds of shared humanity is not without merit. But, as the foregoing shows, it has to be used with great care, not least when one remembers that the subject-matter of historical and archaeological enquiry is constrained by a further three of the problems that can beset enquiry: the Lamplight, Map and Hammer problems. Archaeologists are looking only where they can see; they are generalizing from samples; their array of tools, although increasingly sophisticated, is designed to help in the examination of what they expect or hope to find. If nothing else, awareness of the combination of obstacles and distractions helps to discipline the conduct of the enquiry and the inferences drawn from what it uncovers.

Consider one of the kinds of evidence from which inferences are drawn as to the nature of Palaeolithic humanity's mind and outlook: cave art. Many of the images found in caves at, for example, Altamira in Spain and Lascaux and Chauvet in France are truly remarkable for the fineness of observation and execution they display, which is to say: their artistry. The excellence of the figures implies practice. Where did the artists practise their art, and with and on what materials? We find art inside caves: might it have been abundant outside caves too – and perhaps even more so – on exposed rocks, thus long since effaced by rain, wind, and time? Could one think of ways of trying to detect traces of pigment on exposed rocks? The idea of practice in drawing and painting suggests the use of perishable materials as canvases – pieces of bark, or skins. The ornaments found at Palaeolithic sites, such as shells and birds' talons bored to be strung, are durable items; what about feathers, leaves, skins, and furs, or durable objects, such as shells and birds' feet, that were not bored for stringing but attached to body coverings or even bodies in some other way – for example, by piercing?

An aspect of the Lamplight Problem is that much, perhaps even most, of the material culture of Palaeolithic people doubtlessly consisted of perishable materials. Ponder the implication of the column in classical architecture. It is plausible to see columns as reminiscences of, and advances upon, the wooden posts that supported roofs in earlier periods. Wood is easier to work than stone in the absence of metal tools; might the first henges have been 'woodhenges' rather than 'stonehenges'? There is indeed such a site; Woodhenge near Stonehenge in Wiltshire, England, was detected by aerial photography and found on examination to contain six concentric circles of post-holes inside a ditch and bank enclosure. The site was almost obliterated by centuries of agricultural activity; how many more such sites might there be? Might there be whole civilizations built upon the use of perishable materials who therefore have left no clue that they existed, no invitation to seek their bones there?

These thoughts prompt others. It is plausible to think that the figures so expertly carved in relief at Göbekli Tepe must have been preceded by a significant history of wood-carving. Can one be more speculative? Imagine, say, giant figures made out of vegetable matter periodically created either as self-standing items or to be superposed on, or to decorate, the stone structures at ancient sites, transforming their appearance and endowing them with meanings for the ancient people there that might or might not have had to do with relationships to agencies we now describe as 'gods' and 'spirits' – for example, the end of an initiation period, the choosing of a chief, a festival to mark the beginning or end of a hunting season, a feast, a judicial proceeding of some kind. The possibilities for speculation are many. The focus on a very small repertoire of possibilities – and mainly indeed just one, viz. 'religion' – might be to exemplify the Lamplight, Hammer, and Reading-in problems too literally. That 'cave art' is only found in caves, and sometimes in deep, dark, and inaccessible parts of caves, is a potent motivation for interpreting their purpose as the interaction with things 'sacred' or chthonic. If such works were not only – or perhaps indeed not very often – restricted to caves, our understanding of them might be very different.

There is a natural segue from the foregoing thoughts to the question of *behavioural modernity* in human evolution, postponed from discussion of the period in humanity's history between the migration of anatomically modern humans from Africa around 50 kya and the 'Neolithic Revolution' beginning around 12 kya. The key distinguishing marks of behavioural modernity are art, ornament, sophisticated tool-making, and funerary practices that suggest advanced social structures and conceptual schemes enabled by symbolic thought. Somewhat in parallel with the former orthodoxy that saw the beginning of the Neolithic Period as a revolutionary one, some think that the emergence of behaviourally modern traits was revolutionary likewise – the result of a genetic neurological modification.[16] Others see the process as gradual, of a piece with the general modernization of anatomy.[17] The revolutionary view is that the catalyst was the emergence of language. It entails that non-*sapiens* did not have language – Neanderthals, Denisovans, and their predecessors.

The only means of adjudicating between these alternatives is what can be inferred from the differences between the material leavings of *sapiens* in the period around and after the Out of Africa episode, and the period before (say) 100 to 90 kya. Those who claim that behavioural modernity emerged relatively late, say, from 40 kya, as suggested by the datings of cave art, have to explain how the different groups of *sapiens* fanning out around the world – first into the Near East and Asia, by 40 kya reaching Australia, arriving in western Europe about the same time – shared so much, and principally language, art, and symbolic thought. On the face of it, it appears more plausible to suppose that the development of behaviourally modern traits was the prompt, rather than the result, of the Out of Africa migration. On that basis a different explanation for the migration might be sought; environmental factors perhaps, population stresses, resource problems. Or it could be that the cognitive advances emerging at this point in the *anatomical* evolution of modern humans were themselves the driver, involving an increase beyond a tipping point in cognitive capacity motivating curiosity, ambition, and the confidence to venture.

All sides of the debate would agree on this: that the art, architecture, and technology of the first civilizations in Mesopotamia are unequivocally 'behaviourally modern' by our standards. This remark should alert us to the possibility of reading-in: 'by our standards'. There is a tacit assumption here about the position at which human evolution currently stands. If we feel that we contemporary humans are a point of arrival in some sense – and that is what much discussion of the past unreflectingly seems to assume – we would be mistaken. For, on the contrary, it is reasonable to suppose that if humanity can survive its near future it will continue to evolve, and generations in a more distant future might score *this* present point in their backstory – this time at which these words are written – rather low on a number of metrics of 'advancement', given that war, social and economic injustice, tribalism, racism, sexism, poverty, and ideological divisions not merely continue but flourish among us. These are arguably primitive marks; they have no relish of maturity and wisdom about them. On one view, war – as socially organized conflict that licenses killing and destruction on a significant scale – is an artefact of civilization, non-existent beforehand above the level of brief, local, and limited conflict.[18] If so, the emergence of war in human history can scarcely count as advance. Adorno's remark about humanity growing cleverer but not wiser, as shown by the development of the Palaeolithic spear into today's computer guided missile, is relevant here: technological advance is only one kind of advance.

The chief point for present purposes, however, is that what is taken to count as behaviourally modern is decided by a comparison in which one marker is the stone-tool industry associated with anatomically modern humans living about, say, 120 to 100 kya, and the other marker is the most advanced technology possessed by we humans today. The question being answered in characterizing behavioural modernity in the Upper Palaeolithic is: what looks like it could plausibly lead *to us* in the period between?

With that caveat in place, it remains that the available evidence points to art, ornamentation, long-distance trade, increasing population, broadening of the technological repertoire to include bone, antler and ivory for tools and artefacts, and zoning of settlements as significant and differentiating developments in that period. These developments are attested by the material evidence, and they underwrite inferences to capacities for planning and abstract thought, and levels of social organization that are not easily explained in the absence of language. Cognitive development – mental powers – underwriting 'symbolic behaviour' lies at the core of these phenomena, and for most scholars engaged in the debate 'symbolic behaviour' is the key notion.[19]

What remains uncertain is how long it took for cognitive development in anatomically modern humans to reach the point at which the relevant palette of capacities and their associated behaviours emerged. Was it rapid or slow? Did it happen all at once, or incrementally and piecemeal? For some investigators, recognizing the variability of developments in the Upper Palaeolithic and the way innovations appear and disappear at different places and times suggests a mixed picture but with 'consolidation' of behavioural modernity present after 40 kya.[20]

The inverse relationship between the degree of flexibility in the concept of 'behavioural modernity' and the amount of definitive material evidence keeps this debate alive. There is, however, an emerging consensus on at least these points: one cannot make direct inferences from anatomical modernity to behavioural modernity; whatever else is distinctive of behavioural modernity, the use of symbols lies at its core; behavioural modernity did not originate, as was once thought, in western Europe; and finally, Late Neanderthals display evidence of behavioural modernity too, at least to some degree.[21]

Differences of view of the kind that arose between Positivists and Idealists in the philosophy of history – between the outlooks represented respectively by Leopold von Ranke and Wilhelm Dilthey as described earlier – have reprised themselves almost exactly in archaeology. A debate arose in the decades after the Second World War about the very nature of archaeology, prompted by the increasing availability of scientific techniques as aids in archaeological research. At its most basic, the question at issue is whether archaeology belongs to science or the humanities. Can it aspire to objectivity, or must it ultimately rely on interpretation and therefore a degree of subjectivity? This is the key question about the study of history and its neighbouring fields of archaeology and palaeoanthropology in general; the debate in archaeology provides a good opportunity to revisit it in conclusion.

The science available to archaeology is impressive, and has made a great difference. The use of geophysical survey and remote-sensing technology has stripped away the surface of the ground without a single spade being used, revealing much whose presence might otherwise not even have been guessed. Environmental analysis of past climate and landscape features provides an informative framework for understanding a site, along with analysis of rocks, metals, the remains of flora and fauna, and samples of dust, pollens, and spores captured in ice or sediments. Forensic and genetic examination of human remains yields evidence of diet, health, injury, lifespans, population affinities, and migrations. Conservation techniques preserve artefacts and make them more available for analysis.

Perhaps the most important development is in dating techniques: radiocarbon dating of organic material and thermoluminescence for inorganic material; potassium–argon dating of rock associated with artefacts or fossil remains; dendrochronology, or tree-ring dating, used to help calibrate radiocarbon dating; electron-spin resonance (ESR) spectroscopy, and luminescence dating (optically stimulated luminescence, OSL) for detecting ionizing radiation in sediments and ceramics, all help in the search for more accurate and clearer indications of the timelines of the past.

Science in archaeology is called archaeometry. On the face of it the objective, quantitative methodologies involved would seem to settle the argument between the Positivists and Idealists – or, in the terminology of current debate, processual and post-processual archaeologists – in favour of the former. On the other hand, conflict between what the sciences themselves say can arise. A notable example is the claim that radiocarbon dating solved the puzzle about how developments in technology and farming spread in prehistoric Europe, the candidates being diffusion of ideas or movement of populations, the latter either peacefully or by invasion. What was called 'the second radiocarbon revolution' – the result of more precisely calibrated dating techniques that became available in the 1960s – appeared to support the idea that some of the innovations occurred locally and not as a result of population movements. This seemed to refute the idea of an invasion of Europe from the steppes – recall that in the period immediately after the devastating 1939–45 worldwide war, archaeological sentiment was against the idea of violence-involving migrations. But this in turn was refuted by the genetic data recently provided by David Reich and others concerning the Yamnaya entry into Europe, and the wholesale replacement of the population of the British Isles soon after Stonehenge reached its current form.

Dating techniques do not always settle matters of vital significance in cultural terms. The Dead Sea Scrolls, containing some of the earliest-known versions of Hebrew bible texts, were dated to a point between 400 BCE and 400 CE, eventually giving the highest probability to a date in the third century BCE. Critics pointed out that the Scrolls had been treated with oil to make the text more legible, which would interfere with the analysis by making them appear younger than they might be. For some it mattered that they should be recognized as considerably older than the dates suggested. An analogous case is the Shroud of Turin, believed by some to be an image of Jesus Christ miraculously imprinted on the linen in which his crucified body was wrapped. Three different laboratories tested samples of the linen in 1988, finding that it dated not from the first but the fourteenth century CE. It is not known how many minds this changed; faith is a typically more convincing source of certainty than science.

The prevailing orthodoxy in archaeology until the beginning of the second half of the twentieth century treated it, in the main, as a branch of history, in which cultures are identified, labelled, and recorded in something like (so its critics said) antiquarian 'stamp collecting'. In the 1960s a group of archaeologists in the United States led by Lewis Binford (maker of the 'stamp collector' remark) argued for a different approach in which explanatory models are developed through emphasis on scientific and ethnographic analysis of sites and the materials found at them.[22] This was called 'new archaeology' or 'processual archaeology', the latter because it argued that what is found in archaeological investigation should not be treated as an end-product identifying a cultural type but as something that was in use, in process, dynamically and naturally, in its own time.

The Positivistic aspect of the new approach consisted in the application not just of scientific techniques but of scientific methodology, collecting data and testing hypotheses. How ancient settlements were laid out spatially, and the evidence for economic activity of trade and manufacture, could tell much about the behaviour and social structure of the people of the time, just as their middens tell much about their diet, health, and domestic affairs. A way of characterizing the new archaeology is to see it as a turn from history to anthropology, moving from a particularist approach aimed at recording and classifying finds to a generalizing approach aimed at understanding the cultural and socio-political dimensions of the human past.

'Post-processual archaeology' is a reaction to the objectifying and science-involving nature of processual archaeology. Its proponents reassert the view that in the social sciences, to which archaeology in their view belongs, methods are – and have to be – different, because the phenomena addressed are not such as can be studied in repeated laboratory experiments but are human and social phenomena, variable, transient, and subjective. Some of the leading post-processualists are influenced by ideas in structuralism, postmodernism, and Marxist anthropology. They see archaeology as essentially involving interpretation, and they see interpretation in its turn as essentially subjective, because it brings archaeologists' biases and inclinations to bear on what they find. The Marxist element in this view emphasizes the idea that subjective perspectives are also inevitably political ones, and that therefore the uses of archaeology can bolster views of society that are oppressive, for example, by claiming that the way societies are formed is natural and objective, thus excusing and even validating social injustice.[23]

For the post-processualist, archaeologists are in the centre of the picture, responsible for the interpretations they make and not hiding behind a screen of technologically acquired measurement. This recognizes that interpretation – an active, creative endeavour – happens in the present, to which the deep past has survived in incomplete form, and can be understood only from the viewpoint of what is available to archaeologists in the present, both in their own capacities and experience and in what they have found. The key question concerns what archaeological finds mean, and what sense archaeologists can make of them. 'Meaning' and 'making sense' are essentially interpretative. Because this is so, there can be no definitive account of anything discovered in archaeology, only a succession, and often a competition, of interpretations.[24]

A difference between the post-processual view and its predecessors in *Verstehen* and other theories about methodology in the social sciences is that post-processualism explicitly repudiates the idea that archaeology is a source of *knowledge*. This follows immediately from its commitment to archaeology's being 'multivocal', that is, a set of different and even competing interpretations, none of which has a claim to be exclusively right.[25] The contrast with objectivist approaches could not be sharper: here the idea is that radiocarbon dating, ESR, geophysical survey, genome sequencing, and the like can provide hard data – facts – and that facts are authoritative. For this positivistic view, archaeology – and history more generally – is knowledge.

The sharpness of the contrast is, however, misleading. Recall that the natural sciences regard themselves as defeasible, that is, open as a matter of principle to refutation or modification by further and better evidence or argument. But defeasibility is not a barrier to cumulative progress, using well-supported theory to extend enquiry further. In seeking to draw a defensive barrier around social science by pointing to the contrast in the respective *kinds* of targets of enquiry – in the natural sciences these are geological formations, genes, spectra, galaxies, and proton collisions; in the social sciences they are institutions, families, marriage, burial practices, hierarchies, and beliefs – there is no reason for its defenders to regard defeasibility as a reason for denying that a given view can make a strong case for being right unless and until shown otherwise. To leap to irreducible relativism as a defence is to turn social science into a parlour game.

An alternative is to emphasize the question not of particular methods but of how any method is applied. T. S. Eliot remarked that there is indeed only one method in any activity, and that is 'to be intelligent', and this offers the key. Normative principles of disciplined rational enquiry, scrupulous handling and evaluation of evidence, and intellectual integrity are the super-methodological requirements for making a case, on any subject-matter, so that, if it bears scrutiny and stands up, it can serve as a brick in an edifice of theory under construction. Such edifices not infrequently require dismantling, but not all the bricks in them are tossed away when it happens.

It has to be said, however, that some of what can be brought to the process of enquiry is more helpful, and some less, to 'being intelligent'. To bring mathematical skills to physics and to apply them there is one thing; to bring a religious or political set of preconceptions to history and apply it there is another. Defenders of applying ideological commitments to the work of interpretation will say that *some* ideology will always be at work, so a consciously adopted ideology with its own credentials is better than an unconscious one. The second half of the last sentence is true, so the question has to be about the credentials of the ideology. Self-critical efforts at correcting biases, and in particular for guarding against reading-in that distorts one's view of the target of enquiry, are not merely possible but a component of the discipline of enquiry itself. Moreover enquiry in the natural and social sciences is a public matter, open to debate and criticism – and therefore to supplement and correction – and this constrains the extent to which even subjective elements are irreducible.

In intellectually responsible enquiry no resource that promises to advance understanding would be rejected. In archaeology, scientific aids to enquiry and interpretative skills on the part of practitioners are equally indispensable; it is hard to believe that an archaeologist would ignore the findings of geophysical survey or radiocarbon dating on the principle that their capacity for sympathy is enough by itself, or that their being products of modern Western science unfits them for use in interpreting prehistory.

Knowledge, as contrasted with a Babel of opinions, requires at very least the imprimatur of intersubjectivity and arrival by disciplined means at a recognized level of consensus. To get there, a claim to knowledge has to prove itself – in history, archaeology, and palaeoanthropology as anywhere else. To repeat: everything that can be summoned to assist must be welcome, and in responsible enquiry always is so.

This leaves it open whether the problem of history as *enquiry about the past* will go away one day because a final and definitively true account of history as *what happened in the past* is achieved. History *as enquiry* lives, develops, changes, fluctuates in focus and meanings, and the best hope of grasping history *as the past* is to cleave to the evidence, be scrupulous in reasoning, dispassionate in judgement, and never tempted to start from conclusions with the intention of bending facts to fit them. In that direction lies the possibility of convergence on a best-supported understanding of the past.

PART III

The Brain and the Mind

The topics of the preceding two parts of this book have respectively been enquiries outward in space and backward in time. It is hard not to resort to hyperbole in thinking about the advances they represent. To recapitulate: since the nineteenth century humankind has recovered knowledge of a past forgotten or wholly unknown beforehand. Since the beginning of the twentieth century humankind has made hitherto inconceivable discoveries about the physical universe at the smallest and largest scales we can so far reach. And now, in just the few decades prior to these words being written, humankind has been able to look inside the brain and begin mapping not only its anatomy in levels of detail impossible before but to observe it at work in real time. The technologies that make this possible, and what they reveal, deserve a literal application of that overworked term 'amazing'.

For until now very little has been known about the brain. The same is not quite true of the mind. A certain amount has long been known or anyway believed about mental phenomena – after all, almost all literature and art explores the desire, anguish, joy, grief, happiness, sorrow, loves, hates, insight or its lack that constitute the primary universe inhabited by human beings – the universe of social and emotional experience. Key questions about mind – its relation to brain; the nature and source of consciousness; how the motion picture in colour and sound that plays inside our heads arises from the electrochemical activity of cells packed in there – remain as hard to answer as ever. Indeed, if anything, in some respects *harder* to answer, because knowledge of the brain closes down a number of options for thinking about the source of mental life that beforehand seemed to offer the explanatory closure that people are always eager to have. And, at the same time, the more we know about the brain, the more apparent the complexities and limitations of our thinking about mind become.

Of the three regions of discovery discussed in this book, advances in neuroscience are the most immediately consequential in practical terms. They are already being applied clinically and in other ways, even though the science itself is still at an early stage of empowerment by its newly available technologies of investigation. Until these technologies were developed, the science of the brain, and the psychological and philosophical understanding of the mind, have between them made very little, and very slow, progress – in truth: scarcely any progress at all. That is because of the complexity of all three targets of enquiry – the brain, the mind, and their connection.

There has to be caution in thinking about neuroscience, for the best of reasons: that the progress being made in it is so great, and happening so rapidly, that it is premature to take as definitive the portrait of brain function it currently paints. There is nevertheless much to consider regarding what it has already found. The set of questions it prompts, most especially about what its findings imply, multiply as rapidly as the science itself advances. As with almost everything else, context and background are important; in the pages that follow therefore I consider the background of thinking about brain and mind, survey the new technologies of neuroscience and what they are revealing, discuss the question of what they say about mental life, and consider the implications – not least some ethical ones – of what neuroscience might make possible.

First it is necessary to have a map of the landscapes that abut and overlap in these debates: *neuroscience, psychology, neuropsychology, cognitive neuroscience, neurology*, and the *philosophy of mind*.

These different labels denote different, though connected, foci of interest. The most comprehensive of them are 'neuroscience' and 'psychology', which each embrace a wide range of subject-matters. *Neuroscience* denotes the study of the nervous system and principally the brain, from every relevant perspective of anatomy, physiology, and the biochemistry and biology of molecules, cells and their development, both in normal and pathological modalities. The primary targets of study are neurons and their connections, which involves not only direct examination of the technologies that can image them as precisely as possible but modelling their interconnected activity mathematically and understanding the psychological correlates of that activity.

Psychology is the study of mind and behaviour across the range of phenomena connoted by both terms: perception, reason, memory, learning, motivation, emotion, intelligence, personality, relationships, the development of these capacities, problems that arise in connection with them and treatments for those problems, and the use of social, neurological, pharmacological, and forensic methods in research and applications. The subject has a number of specialist sub-areas, such as developmental psychology, social psychology, clinical psychology, and more.

Neuropsychology is one of these specialist areas. It is the study of how behaviour and mental life are produced and mediated by the nervous system and principally the brain. (The endocrine system, which produces hormones, is also relevant.) In association with *neuropsychiatry* it has a clinical emphasis in addition to the research project of understanding how mental life and behaviour are neurologically based, how disorders of cognition, behaviour, and mental life in general can arise from neurological disease and injury, and how they might be treated.

Cognitive neuroscience focuses on the brain and sensory pathways of the nervous system in order to understand how they mediate and process perception, memory, attention, language capacity, decision-making, and emotion. Whereas clinical neuropsychology and neuropsychiatry address pathologies of these functions, normal functioning has to be understood in order to provide the contrast with pathological or absent function; as cognitive neuroscience addresses what the brain and nervous system normally do, the neighbouring studies seek to understand what has gone wrong when problems arise, and how to remedy them.

Neurology is the medical practice that specializes in disease and injury in the brain and nervous system. Until quite recently it was said that the cleverest doctors gravitated to neurology because of its interest, even though there was little they could do to help. The interest remains, but the possibilities of care and cure are growing all the time.

At the centre of all these activities lie a *commitment* and a *core enquiry*. The commitment is to the proposition that the brain is the centre, cause, operating system, and seat of consciousness, mind, and mental life. This is why 'neuro-' is prefixed to 'psychology', 'psychiatry' and the '-sciences' of the brain and mind in general in these labels. The core enquiry therefore is the descriptive and analytic aspect of neuroscience itself, aimed at a comprehensive understanding of the brain.

This commitment is a powerful one. No credibility attaches to alternative views that decouple mind and brain as separate substances; no credibility attaches to theories that locate the seat of consciousness and mental life anywhere other than in a brain. I call it a 'commitment' rather than an 'assumption', because the evidence in its favour, and everything associated with accepting it as true, is so immensely powerful; indeed too much so to permit the weaker notion of being merely assumed rather than established to an intellectually coercive degree – which it is.

To understand this commitment's power, we need to see what the alternatives to it are. First, note that a sceptic or severe critic would say that neuroscience fully exemplifies at least five of the problems that beset enquiry: the Pinhole Problem, the Map Problem, the Hammer Problem, the Meddler Problem, and the Metaphor Problem. The first three are connected. The human brain is often said to contain 100 billion neurons (the best method of counting so far suggests 86 billion, give or take a few billion) with as many as a *trillion* connections between them, tiny and densely packed together, in constant flux with many neurons growing or dying and very many of the connections constantly changing. Even with the best current technologies available, we are looking at all this through a pinhole. This means that we are working with extremely small-scale maps of brain regions, and that the most recent and powerful techniques for imaging brain activity in real time (using functional magnetic resonance imaging, fMRI, and other techniques) are crude – the severest sceptic might say they are expensive and high-tech versions of phrenology, given that (for example) fMRI's one-millimetre resolution is like seeing Everest from space rather than the cutlery in the kitchen drawer, which is the level at which we need to see. And, as this is our instrument for seeing what is going on inside the skull, we take what it shows us as being what there is to see.[1]

These first three criticisms amount to saying that, because we are at such an early stage, despite the gleaming equipment and high science at work in neuroscience laboratories, we are in danger of jumping to conclusions – a version of the Closure Problem. This has to be admitted; but to recognize that neuroscience is at a relatively primitive stage compared to what it will become, and is rapidly becoming, is not to impugn it. At most these criticisms amount to cautions or reservations about what can so far be concluded. They also underestimate the extent and significance of what has already been learned and what can already be done on the basis of it.

The criticism drawn from the Metaphor Problem has more to it. The metaphor at work in neuroscience is that of *computation*, and it is so compelling that it goes without challenge. It is the latest in a long line of metaphors for explaining cognition and mental life, and this fact – indeed the fact that a metaphor is doing a great deal of explanatory heavy lifting in this field – at very least requires that it be investigated and its appropriateness justified. At least one recent theory of consciousness repudiates the computational metaphor altogether.

Thinking about the mind has always relied on metaphors and similes because there is a direct relationship between ignorance and the resources of would-be explanatory comparisons we employ to make up for it. At the beginning of the Early-Modern Period – the sixteenth century – the favoured metaphor for how the brain works was either clockwork or hydraulics, both of which were regarded as marvels. Clockwork had been around for centuries (see the remarks on technology in Part I) but had become miniaturized, for all the world like a little brain with its tiny cogs and springs, working away without a human or animal operating the mechanism once it had been set going. The metaphor was indeed applied by many to the entire universe, the deity being nominated 'the divine artificer'.

Water flow had also long been in use as a source of energy for various contrivances including clocks. In the sixteenth and seventeenth centuries hydraulic devices made statues move and speak, to the wonder of visitors to places like the gardens of Saint-Germain-en-Laye, where Descartes himself had been inspired to form his view of the brain and nervous system as mechanism, and of animals as machines without consciousness (because 'lacking a soul').[2] It was however in this period that the machine metaphor for the brain – though requiring the input of *mental* activation to get going, with mind conceived as a different substance – recommended itself alongside its use for much else in nature's workings.

In the nineteenth century the metaphor came closer to home; discoveries about electricity and its galvanic effects on nerves were the first clue, the metaphor of the telegraph and later the telephone system shortly afterwards completing the picture. William Godwin, father of Mary Shelley, witnessed a display of galvanism performed on a corpse in London sometime in the 1790s (Luigi Galvani's book about electrical stimulation of the muscles of corpses, *De viribus electricitatis in motu musculari*, was published in 1791): an interesting source for the Frankenstein story. Some of those present were so terrified by the corpse's movements when stimulation was applied that they fainted. The telephone metaphor persisted well into the twentieth century because of its aptness: the telephone exchange seemed a good model for the network of nerves visible under the microscope.

But the introduction of the computer in the 1950s provided an even more powerful and persuasive metaphor, not only or even mainly because of its structure but because of its operation: computation. Computation is the carrying out, in sequential steps according to a rule or algorithm, of a calculation or procedure aimed at an outcome of a given kind – in algebra, solving an equation; in the brain regions, coordinating vision and motor control, computing (see how readily the term supplies itself) the distance, angle, muscular force, required finger extension and retraction, and so on, for reaching out to pick up a cup from a table. The metaphor of computation imports a family of powerful concepts, among them 'feedback', 'code' 'algorithm', and 'information'.

The chief problem with the *computer* metaphor is that it is not clear whether, or to what extent, the brain is a digital device. It does not use binary logic or arithmetic as such. But nor is it an analogue device, though the continuously varying flow of stimuli both internally and from the brain's sensory interfaces with the world might suggest that it is. Instead, the brain appears to handle information by estimating and statistically approximating, and, because it is non-deterministic (not invariable and automatic), it cannot rerun its handling of information without the chance of variation. Excitation and inhibition in the firing of neurons is either/or, like an electrical circuit, and that is binary; but the net outputs of weightings in excitation and inhibition are analogue in character. So it appears that brains operate in a way that is both and neither binary and analogue. If this is right, then however *computation* is carried out by the brain – how it processes and deploys information – the required model for it, so critics argue, has yet to be found.[3]

The brain itself becomes a metaphor and a model. *Neural networks* are organized collections of units operating according to successively applied algorithms to send forward sets of mutually weighting signals to yield an output, which the system can be trained – more accurately: can learn – to produce by repeated runs.[4] Connectionist models in turn use neural networks and parallel distributed processing to mimic brain activity. The power and utility of this reciprocal modelling – brains likened to connectionist networks, neural networks to brains – adds to the persuasiveness of the view that 'computation' is not a metaphor after all but a correct description of the brain's activity. Yet this fails to persuade proponents of alternative views, such as Roger Penrose.[5]

A final point at this juncture concerns the most important point of all, namely, the neuroscientific commitment itself: the commitment to the view, as the fundamental premise of the enterprise, that the mind is wholly explicable in terms of brain physiology. I shall show that, even though there is nothing in the universe that is not ultimately a matter of physics, the brain is not the whole story of the mind, and that there is still another neuro-prefixed enterprise needed in the picture, which one might call *neurosociology*, aimed at registering the role of the social environments of brains. This is an overlooked and potentially key point. There is a case for saying that a mind is a relational entity, which cannot be understood only in terms of what happens within a cranium. On this view a mind is the product of interactions between the brain and other brains and the physical environment. The neurosciences are apt to treat mental phenomena on a 'narrow-content' basis; its assumption is that everything to know about the brain is in the brain, forgetting the rich implications of the brain's being essentially an interactive device, for which the inputs and outputs, the fact of being plugged into social and physical worlds, are key to what it produces in the way of cognitions and mental life, and the source both of its thriving and of some of the kinds of damage and deficits it can experience. This aspect of the question relates to the last of the abutting and overlapping fields mentioned, namely, the *philosophy of mind*.

There are lessons to be learned about all these matters from the reasons why, before neuroscience as we now have it, thinking about the mind and its relation to the body took the forms it did. This – where the frontiers of knowledge about mind and brain used to lie, and why – is the topic of the next section.

1. Mind and Heart

Despite the advances in scientific and especially neuroscientific understanding, it can occasion no surprise that the majority view in the world today still is that mind – somewhat blurrily conceived as, or as intimately associated with, spirit or soul – is not the same thing as brain and body. This is a basic assumption of religious world-views in which an afterlife figures, because by hypothesis the person and her consciousness and memories must survive bodily death in order to translate to the non-physical dispensation that is heaven, purgatory, hell, or whatever destinations are believed to await different conditions of the deceased.

Sadly for the democratic impulse, majority-held metaphysical views are more likely to be wrong than right, and as it happens the kind of belief just sketched has anyway not been universally held, though now standard among members of the historically young religions of Christianity and Islam. Early Christians believed, as did votaries of the Jewish sect from which their views were drawn, in bodily resurrection rather than a disembodied spiritual afterlife; it was only after some four centuries that Platonic (in the form of Neoplatonic) ideas about an 'immortal soul' separate from the body entered into Christian thinking, largely because the Second Coming had not happened as expected, and the bodies of 'saints' (martyrs and believers who had died) were found to have rotted in their graves, contrary to the implication of Christian belief that those who are holy shall not see 'corruption' – hence the great disappointment of Father Zosima's devotees in *The Brothers Karamazov* when his corpse begins to decay rapidly after death.[1]

Indeed Plato's view of the existence of the intelligent part of souls separate from bodies was not widely held even in his own day; in the dialogue *Phaedo* or *On the Soul* Socrates's interlocutors are sceptical about his arguments for the soul's immortality. This raises questions about what was understood by the 'underworld' as a destination for the dead in Greek mythology – though one thing we know about it is that it was not a desirable place to be; the heroic impulse was governed by the desire to survive in others' admiring memories, not in a place of posthumous reward. In Chinese ancestor worship, in mummification and elaborate grave goods accompanying corpses in Egyptian burials, and presumably in the grave goods burials of prehistoric peoples, belief in some form of continued existence is indicated, though it is unclear whether it implies a non-corporeal existence, or whether the material goods provided indicate belief in some form of posthumous materiality (mummification suggests this).

These remarks show that the question of what kind of thing the mind is has not been open and shut at any time in history. Nor is the question of *where* it is during a physical lifetime either. In fact the question of the mind's location is as alive today as ever, and not just among religious believers. Amazingly, the two principal rival theories about the mind's location, independently of its being or not being physical, namely whether it is associated with the brain or the heart, continued right into the modern period, one of the last and most distinguished espousers of the heart theory being no less than the discoverer of cardiovascular circulation, William Harvey.[2]

The idea that the heart is the seat of the mind seems easy to ridicule, until one sees why it was espoused even by the most advanced and brilliant scientist of his day, Aristotle. In adopting this view he was taking sides in a debate that had been in progress since well before his participation in it. Moreover his participation was intended in part to effect a rapprochement between the two sides of the argument, because he saw the brain and heart as constituting a single mutually interacting system.

The usual starting point for this discussion is the Presocratic philosopher Alcmaeon of Croton. Alcmaeon was born about 510 BCE and is thought to have been a pupil of Pythagoras. Croton, a Greek colony in southern Italy, was one of the major centres of medical studies in the Greek world, influenced by Pythagorean thought but empirical in its approach, as testified by Alcmaeon's anatomical research; he reportedly described the Eustachian tube (the pharyngotympanic tube connecting the back of the nose to the middle ear) and the anatomy of the eye and optic nerves, including what he took to be their juncture at the optic chiasm in the brain (in fact the chiasm is the point at which the optic nerves cross – 'chiasm' means 'crossing' – on their way to the primary visual cortex at the back of the brain).

What we know of Alcmaeon comes from a work by Calcidius, translator into Latin of, and commentator upon, Plato's late dialogue *Timaeus*, the only full text by Plato known until the later medieval period. Calcidius's primary interest was Plato's cosmology, but he mentions Alcmaeon's work in passing, citing him as the first person to undertake dissection for anatomical study. It is not known whether Alcmaeon dissected humans or other animals, but either way it provided him with his reason for locating the seat of mental life in the brain.

For Alcmaeon the joining of the optic nerves explains why the eyes always work in concert. He called the optic nerves 'light-bearing paths' into the brain, and thought the eye itself contains light, as evidenced by the fact that pressure on or in the eye, such as a blow or a sneeze respectively, or disease or damage in the optic nerves or retina, will cause phosphenes, sensations of light such as a flash, or 'seeing stars'. The view that the eye contains light was only finally abandoned in the eighteenth century CE.[3] A number of other Presocratic philosophers shared Alcmaeon's view, chief among them Democritus, who in turn influenced Plato. Democritus and his teacher Leucippus – the Atomists – held that everything consists either of atoms or the void in which the atoms move. Atoms come in different grades of fineness: the light, swift, perfectly spherical atoms constitute the psyche, distributed throughout the body but clustering mostly in the head. Somewhat less refined atoms gather mainly in the heart, constituting it as the seat of emotion, while atoms of still less refinement cluster in the liver, explaining why it is the seat of the appetites, including hunger and lust. This tripartite division of the soul's functions – intellect, emotion, and appetite – is adopted by Plato, including their localities; in the *Timaeus* he says, 'the gods, imitating the spherical shape of the universe, enclosed the two divine courses in a spherical body, that, namely, which we now term the head, being the most divine part of us and the lord of all that is in us: to this the gods, when they put together the body, gave all the other members to be servants.[4]

Hippocrates and his school were emphatic that the brain is the seat of thought and emotion: 'the source of our pleasure, merriment, laughter and amusement, as of our grief, pain, anxiety and tears, is none other than the brain. It is the organ that enables us to think, see and hear' – and it is to blame not only for sleep disturbances, forgetfulness, and eccentricities but for epilepsy also, the 'sacred sickness', which Hippocrates scathingly said had nothing to do with the sacred or supernatural but was an affliction caused by an inability to drain phlegm from the brain.

The Hippocratic school did not engage in anatomical dissection but relied – on the whole, rather well – on clinical observation and experience, so, although it was good at describing the manifestations of diseases, matters were otherwise regarding their causes. Hence the view about phlegm, which relies on the 'blood, phlegm, yellow bile, black bile' four-Humours view whose balance, or lack of balance, is taken to explain health and sickness. Yet, despite lacking knowledge of underlying anatomy and physiology, the Hippocratics were categorical in rejecting the claim that the heart has anything to do even with emotion, as the quotation shows.[5]

The brain-theory tradition running from Alcmaeon to Hippocrates was the minority view. The much older and more widely held view, that the seat of the mind is the heart, was assertively held by Aristotle, who dismissed the brain theory as 'fallacious'. He cited strong empirical reasons in support of the heart theory, as follows. It is clear from experience that emotion affects the heart; it beats rapidly in fear or excitement, anger or arousal; it is slow and steady when one is calm. One feels nothing in the brain in any of these states. The heart is the source of blood, which is required for sensory experience, and it is warm, which is indicative of life in higher creatures. The brain is relatively bloodless, and cold, and has no sensation. The heart itself can feel pain, but the brains of living animals can be cut without the animal showing any pain or discomfort. Through its vascular system the heart is connected to all the muscles and sense organs; lacking blood, the brain cannot be so. The heart is essential for life, but the brain is not, as we see from the fact that almost no animal is without a heart, but there are many without a brain. The heart develops first in the foetus, and is the last organ to stop working at the end of life; the brain develops after the heart, and can stop working before the heart does. That the eyes, ears, nose, and mouth are carried aloft in the head is for convenience, affording a higher vantage point; their proximity to the brain does not indicate that the information they convey is destined for the brain. In any case sensations occur all over the body and require a central point where they can be integrated into what Aristotle called the 'common sense'. That is why the heart is at the centre of the body, as befits its importance.[6]

But this is not, said Aristotle, to relegate the brain to unimportance. On the contrary, that the heart is a hot organ requires that it be balanced by a cool organ, so that it can 'attain the mean, the correct and rational position. Thus does the brain, which is cold by nature, temper the heat and tumult of the heart.' The brain is enclosed merely in bone, unlike the heart in its thick coat of muscle, bone, and wraparound lungs, and that is why the brain is colder. It dissipates heat readily, cooling the blood. We see that in cold weather wearing a hat retains some of the heat that would otherwise by radiated away by the brain. If the brain grows too hot, it becomes congested and produces phlegm, the source of epilepsy. But the key point is that the brain is a vital adjunct to the heart, keeping it at the right temperature so that it can perform its task of thinking and feeling properly. The large size of the human brain thus correlates to the superior nature of human intelligence compared to all other creatures, because it is an excellent radiator. Aristotle was also inadvertently on the right track in blaming mental illness on malfunction of the brain, though in his view because it was not efficiently cooling the heart.

One reason often overlooked in discussions of Aristotle on the mind is that the concept of mind in Greek philosophy is not quite the concept of mind we have, but is *psyche*, in Latin *anima*, the force that among other things *animates*, or makes things alive, and most significantly for Aristotle is the principle of motion and change that distinguishes animate from inanimate things. Motion and change were important concepts for Aristotle, who charged a number of his predecessors with failing to explain how these phenomena can occur, and who was consciously opposing the Parmenidean view – as illustrated by the paradoxes of motion devised by Zeno – that change and motion are impossible.[7] For Aristotle, the warm pulsing heart contrasts with the cold inactive brain as the best candidate to be the seat of the principle of activity in all its forms, including thought and feeling.

A biologist far in advance of his time, Aristotle dissected a total of forty-nine animals varying in size from a snail to an elephant, and, in the course of doing so, examining brains and noting the meninges, hemispheres, and ventricles. Indeed in some of these cases he vivisected his experimental subjects. It might therefore seem very surprising that he did not recognize the true function of the brain. This has been attributed to the fact that he was not a medical man, and therefore had no experience of observing and treating head-injury patients exhibiting symptoms of mental deficits of some kind, perhaps noting that injuries in the same region of the head produced similar deficits in different patients. That by itself would have been a large pointer in the right direction.

Charles Gross is right to stress Aristotle's positive influence in the development of science in general and an understanding of the brain in particular – despite the foregoing – in the centuries after the conquests of Alexander and the Hellenization of Egypt and the eastern Mediterranean world. Alexander's friend from boyhood who became Egypt's ruler, Ptolemy I, founded a great institution in Alexandria, the Musaeum – 'seat of the Muses', a university and research institute – at which scores of state-funded scholars studied, taught, and researched. The men who advised Ptolemy on setting up the Musaeum were pupils of Aristotle's leading follower and successor, Theophrastus. Ptolemy had tried to lure Theophrastus himself to Alexandria to head the Musaeum, but Theophrastus sent his pupils Strato and Demetrius instead, and they founded what was in many ways a continuation of Aristotle's school, the Lyceum. Aristotle's example in practical science, especially dissection, was followed at the Musaeum, resulting in a flourishing of anatomy and notably neuroanatomy in the work of Herophilus of Chalcedon and Erasistratus of Ceos. They were the first anatomists known to undertake systematic and extensive dissection of the human body. Tertullian claims that between them they vivisected more than six hundred condemned prisoners. One of their chief interests was the brain; Herophilus was the first to distinguish the cerebrum and cerebellum, to identify the association between the optic and oculomotor nerves, to appreciate the inner structure of the eye itself, and to recognize the difference between the intracranial nerves and blood vessels.[8] Erasistratus described the valves of the heart and recognized it as a pump, and distinguished motor from sensory nerves and traced both kinds to the brain. Both were empirically satisfied that the brain is the seat of the mind.

Tertullian expressed the outrage of Christian thinkers at the Alexandrian practice of human vivisection. It was defended by the Roman historian Celsus, who said, 'It is not cruel, as most people maintain, that remedies for innocent people of all times should be sought in the sacrifice of people guilty of crimes, and only a few such people at that.'[9] This *kind* of justification is used in defence of experimentation on live animals today. On this point, if on few others, one is inclined to agree with Tertullian. As Gross observes, 'Vivisection of humans was never systematically practised again [until the Third Reich]' – he might have added vivisection by Japanese experimenters in its captured territories during the 1930s and 1940s – 'Even the dissection of human cadavers disappeared in the West until it was revived in the new medieval universities, and then only for forensic, not medical or scientific, purposes.'[10]

An interesting suggestion by Gross is that the reason for Alexandria's scholars' acceptance of human dissection was the Egyptian practice of mummification, which required removal of the brain and other organs from a corpse before embalming could take place. Another reason is that the Greek ascendancy in Egypt meant that scruples about the subject population were few, hence the use of prisoners as vivisection subjects. Moreover in the Hellenic world in general it had become normal to regard corpses without much sentimentality, as merely the abandoned frame left by the person who had animated it. Whatever the reason or complex of reasons, anatomy flourished at Alexandria for a time, but it was another four centuries before further major advances occurred in it, this time in the work of Galen.

Galen became an accomplished anatomist on the basis both of dissection and vivisection of animals – it was forbidden to dissect human cadavers in the Rome of his day, so to understand human anatomy he dissected their closest analogues, monkeys; in particular, Barbary apes (macaques). Inevitably the restriction to animals meant that he made mistaken inferences to human anatomy, something not fully realized until the work of Vesalius in the sixteenth century. But his genius compensated for much.

A prolific researcher and thinker, Galen wrote over five hundred dissertations on medicine and philosophy, including ethics, and was equally emphatic in his admiration for Hippocrates and Plato and his disagreements with Aristotle and Erasistratus. As with so much else of the literature of antiquity, only a small part of Galen's corpus has survived; many of his works are said to have perished in a library fire in Rome in 191 CE. What survived proved highly influential in Arabic and European medicine later. Indeed Galen was *the* medical authority until the Renaissance.

Although he admired Hippocrates, Galen did not restrict himself to observation of symptoms for his research but took scalpel in hand to investigate their physiological basis. His interest extended beyond anatomy and physiology to pharmacology (collecting and prescribing medicinal plants) and psychology (recognizing and describing psychosomatic disease). He was an inventor of surgical instruments also, for use both in dissection and operations. He was deeply versed in the medical theories of his predecessors, influenced in particular by the Hippocratic view of the four Humours, one of the implications of which is that not only will imbalance among them cause physical illness but that an habitual preponderance of any one of them will affect the character and its psychology: thus some people are sanguine (preponderance of blood), some choleric (preponderance of yellow bile), some melancholic (preponderance of black bile), some phlegmatic (preponderance of phlegm).

A key concept in Galen's theory is *pneuma*, literally 'breath' but with the meaning also of 'spirit'. It is inhaled into the lungs, whence it passes to the heart, liver, and brain, and thence to the rest of the body, transformed by the heart into *vital spirit* (*pneuma zotikon*), which causes the body's warmth (its life), and by the brain into *psyche* (*pneuma psychikon*), the mind or soul. Psyche occupies the brain's ventricles and from there activates the nerves spread throughout the body, transmitting movement and receiving sensation. The brain is the seat of all the cognitive functions of thought, memory, imagination, volition, and sense, and Galen calls it the *hegemonikon*, the master or controller.

Through his dissection studies Galen identified ten of the twelve *cranial nerves*, the *corpus callosum*, which connects the two cerebral hemispheres, the brain's ventricles, structures in the hippocampus and midbrain such as (respectively) the *fornix* and *tectum*, the *recurrent laryngeal nerves* that branch from the *vagus nerve* (Cranial Nerve 10) to the muscles of the larynx, the blood supply of the brain, and the spinal cord and its significance in movement and sensation, a result of observing different levels of paralysis and anaesthesia in spinally injured gladiators below the level of their injuries, and the same in his vivisection at different levels of the spinal cords of monkeys. He noticed that sectioning the spinal cord at the fifth cervical vertebra (in the neck, in humans about level with the larynx) results in paralysis and anaesthesia of all limbs but not immobilization of the diaphragm. Semisection of the spinal cord results in loss of voluntary movement on the same side as the section (*ipsilateral*) and loss of sensations of temperature and pain on the opposite side (*contralateral*). Paralysis or muscle weakness on one side of the body – *hemiplegia* – in association with facial palsy is the result of a lesion in the contralateral hemisphere of the brain, implying involvement of the cranial nerves; lack of facial palsy indicates that the lesion is in the spine. He made an opening in the skulls ('trepanning') of head-injury victims to drain intracranial haematomas (accumulated blood from bleeding inside the skull) and to alleviate pressure. In all, he was a remarkable anatomist and medical scientist.[11]

Galen's admiration for the *Timaeus* prompted him to mirror Plato's 'tripartite soul' arrangement in his specification of the seats of the pneumas in his physiology – mind in the brain, emotion in the heart, and appetite in the liver. 'Mind', as noted, consists of reason, memory, perception, imagination, and will. Any disharmony, *dyskrasia*, between the three seats of pneuma will cause mental illness, just as imbalance of the four Humours will cause physical ill-health and, in lesser degree, different emphases of character. He described a variety of mental illnesses and affections of the brain: mania, delirium, phrenitis, paraphrenia, coma, catalepsy, epilepsy, and varieties of dementia. And he was careful to distinguish mental illness from passions such as love and ambition, even when these become too consuming, saying that they require counselling – we might now say, psychotherapy – rather than medical treatment. In this regard he borrowed the Stoics' advice on how to achieve *ataraxia* – 'peace of mind' – namely: to face with courage what you cannot control in the world around you, and to master the appetites, fears, and desires within you.[12]

Galen's writings were the bible of medicine until the sixteenth century CE. Human dissection was not performed until the end of the thirteenth century, a gap of some fifteen hundred years since the researches at the Alexandrian Musaeum, because of doctrinal proscriptions both in Christianity and Islam, though in the Muslim world scholars such as Avicenna were instrumental in preserving and transmitting the surviving parts of the corpus of Greek science and philosophy. One of the earliest revivers of experimental anatomy was Mondino de' Luzzi of Bologna, known as Mundinus, who followed Galen in locating the cognitive powers in the brain by nominating their ventricles as the source of the *pneuma psychikon*. But neither Galen's authority nor his later followers were sufficiently convincing to put a stop to the controversy about the mind's location, as to whether it lies in the heart or brain. In the late sixteenth century Andrea Cesalpino, an Aristotelian, still argued that 'the heart is not only the origin of all the veins but also of the nerves.' Descartes in his *Principles of Philosophy* noted that the disagreement about where to site the meeting place of mind and body (his choice was the pineal gland in the brain) was alive and continuing at his time of writing in the first half of the seventeenth century, contemporary with Harvey.

Note, however, that the claim that the brain is the seat of mind is not the same as the claim that mental phenomena are identical with, or are produced by, brain activity. Most people continued to be dualists about mind and body, their question not being 'which part of the body is responsible for mental phenomena' but rather 'with which part of the body is the – separately existing – mind associated, or through which it acts'. The most materialistically inclined of thinkers, until the Enlightenment mostly to be found in pre-Christian antiquity, conceived of mind, *psyche, anima,* as consisting of *pneuma* or very refined fluid, so that those who thought that mind-stuff is transmitted via the nerves from whatever is its point of origin – whether heart or brain – were not thinking as contemporary neuroscience does about the organ responsible for mentation. Thus one sees the famous anatomist Samuel Soemmerring in his *On the Organ of the Soul* (1796) attributing mind to the fluid found in the ventricles of the brain, while his contemporaries Karl Friedrich Burdach and Johann Friedrich Meckel both nominated the brain itself as the 'organ of the soul', the former saying it was the entire brain, the latter attributing the 'primitive functions of the soul' to the lower brain and its 'higher powers' to the upper brain – this apportionment being approximately on the right track, unsurprisingly for an anatomist who based his views on embryology and comparative studies.

Even into the early nineteenth century some researchers thought that the nerves are tubes along which vital spirit or fluid flows – they had at first little understanding of the role electricity might play, still less of the electrochemical propagation of action potentials (by means of reversal of the polarity of sodium and potassium ions through the wall of the axon), which is how impulses are conducted, and the purely chemical transmission across most synapses by neurotransmitters.

But the use of electrical stimulation to study brain and nervous-system function was not long in coming. Luigi Galvani's frightening experiments on the corpses of criminals in the late eighteenth century had shown the way. A number of nineteenth-century physiologists experimented on animals living and dead, much being learned and more hypothesized from electric shocks to the feet of decapitated frogs, and from vivisection experiments on a variety of creatures including dogs and monkeys. One of the leading figures in these developments was David Ferrier, who drew a precise map of motor functions in the cortices of the two latter animals, corroborating stimulation of the relevant areas, which demonstrated activity, by then lesioning them to demonstrate the loss of function that followed. He applied his map of the macaque brain to human brains – with rather approximate results – and the map was used in clinical practice and neurosurgery. Fortunately for patients, medicine is an evidence-based proceeding, and neurologists quickly learned to use the map as a guide rather than a floor plan in developing ways of determining lesions or tumours by observing their effects on behaviour.[13]

Proper understanding of nerve cells began in the second half of the nineteenth century with Camillo Golgi's development of a technique for hardening and staining brain tissue to clarify it under the microscope. His technique was used – and improved – by the celebrated neuroanatomist Santiago Ramón y Cajal, whose drawings of neurons are still used today. His work was instrumental in replacing the idea of the nervous system as a 'reticulum' – a network of fibres, a tangle of which was all that anatomists could see before Golgi – with that of neurons as discrete cells. Research with instruments of even higher resolutions than were available to Cajal shows that the neurons of the brain and nervous system come in many different types.

Yet, even as neuroanatomy was reaching the mature stage as a science that it now occupies, and the quarrel between 'cardiocentric' and 'encephalocentric' theories of mental location or connection was being firmly resolved in the latter's favour, the question of mind and its relation to body remained an issue, at least in the sense that a need was felt to show why dualistic views are untenable.

'Mind–body dualism' is the metaphysical thesis that mental phenomena and physical phenomena – minds and bodies – are two different *kinds* of thing. The *locus classicus* for a statement of the dualist view is Descartes's *Meditations on First Philosophy*. In it Descartes argued that everything that exists in the world consists either of material substance or mental substance, where 'substance' is a technical term in philosophy meaning 'the (or a) most basic and fundamental kind of existing stuff'. Descartes defined 'material substance' as 'what occupies space; extended stuff', and 'mental substance' as 'thought; thinking stuff'. The idea that mind and matter are *really distinct* things (in the literal sense of 'really', namely 'in reality') is supported by a methodological claim Descartes relied upon: 'the fact that I can clearly and distinctly understand one thing apart from another is enough to make me certain that the two things are distinct [in reality]' – a questionable claim.

By describing matter and mind as *essentially* different – different in their essence, in what makes them the kind of thing they are – Descartes raised the apparently insuperable problem of how they interact. How does a physical event like stubbing one's toe result in the mental occurrence of pain? How does the mental event of thinking 'It's time to get up' cause the physical event of rising from one's bed? Descartes proposed that mind and matter interact in the pineal gland in the brain, choosing it because it is a unitary structure, unlike the twinned structures around it, and was centrally placed, handy to serve as the site from which the whole network of nerves radiates and to which they return. His successors were quick to see that this would not do as an explanation of how mind and brain interact, for it simply hides the problem inside a conveniently located small organ, with no account of how the interface works.

These successors resorted to some heroic solutions to the interaction problem. Their strategy was to accept dualism but to deny that mind and matter interact, the appearance of their doing so being the result of the hidden action of a deity. Different versions of this solution were proposed by the French philosopher and Oratorian priest Nicolas Malebranche and the German philosopher and mathematician Gottfried Wilhelm Leibniz. Malebranche thought that a deity makes mental and physical events match up whenever a match is required, thus: the deity detects a mental sensation of hunger, follows it by prompting a physical movement to the kitchen for preparation and consumption of a peanut-butter sandwich, following that in turn with a mental sensation of satiety. This doctrine is known as 'occasionalism': on every occasion on which a mental or a physical occurrence requires a physical or a mental correlate respectively, the deity provides one. Possession of infinite powers by the deity is evidently useful in such a scenario, given the number of correlations required at every moment. As it would involve the deity in some highly questionable mind-to-body and body-to-mind matchings, doubtless even a deity might, in such straits, come to regret creation.

Leibniz likewise thought that interaction does not occur, and that a deity set the physical and mental realms going, in exact parallel with one another, at the moment of creating the universe, to make it *appear* as though mind and matter interact. This is accordingly known as 'parallelism', and it incurs the price of commitment to strict determinism – for otherwise the parallels between the realms would fail. Familiarly, determinism raises problems about free will, moral responsibility, and conceptions of human nature.

Obviously enough, neither of these efforts to circumvent the problems raised by dualism is very palatable. The majority of philosophers after Descartes's time have taken the only plausible alternative to be a form of monism ('mono' meaning 'one'), that is, commitment to there being only one kind of substance. There are three main possibilities: there is only matter, or there is only mind, or there is a neutral substance of which both mind and matter are expressions or effects. Each of the three has had proponents – Spinoza, Berkeley, and William James are representative figures for the latter two versions[14] – but it is the first option, namely the reduction of mental phenomena to a material basis, that has been most influential, despite being the one least consistent with theism.

Descartes himself, after conceding the futility of the pineal gland suggestion, abandoned the effort to find a satisfactory way of dealing with the problem of interaction, but did not give up dualism itself. One reason was his reluctance to abandon theistic orthodoxy, given the inconsistency with it of what he thought to be the exclusive alternative, materialist monism. But there were other reasons, which fed more generally into the widespread assumption of the truth of dualism. The chief was that the properties of mental and physical things are completely different: whereas physical things have positions in space, weights, velocities, and (if macroscopic) colours and odours, mental things, such as ideas, memories and hopes, have none of these properties. It follows from views such as those Descartes held, in which conscious existence can survive bodily death, that mental phenomena have no dependence on physical phenomena for their existence, and we see from plenty of examples (rocks, sticks, raindrops) that most physical things display no connection, still less dependence, on mental things – thus rendering the mental and physical two-way independent. And, as it happens, in a slightly different version, the unlikeness of mental and physical things remains a sticking point in thinking about the mind even today, in the problem of *qualia* (on which, more below).

A much more recent act of heroism in response to the problem of mind was to refocus the question of the relation of mental to physical phenomena by translating talk of all things mental into talk of *behaviour* – that is, by reducing mental concepts such as pain, emotion, and desire to descriptions of people's observable behaviour and dispositions to behave. This solves the problem by getting rid of reference to mental phenomena altogether, at the same time having no need to supply a theory of the brain as the source of those phenomena. Behaviourists laid stress on the fact that behaviour is publicly observable; interpreting mental phenomena in these terms obviates the need to rely on reports of introspection as the data of psychological enquiry, and therefore of anything inner and subjective. So 'X is in pain' is replaced by 'X is bleeding, wincing and groaning' – thus giving us, in the behaviourist's view, an objective and perspicuous account of the meaning of 'pain'.

The leading figures in promoting this view were the psychologists B. F. Skinner and J. B. Watson, and versions of their views were developed by the philosophers Gilbert Ryle and W. V. O. Quine. There are differences between the views of these thinkers, but they all faced the same crucial difficulty, which is that behaviourism does not succeed in eliminating reference to the fundamental mental phenomena of belief and desire that figure in almost all *explanations* of behaviour. To describe a man's body as moving into a shop and reappearing with a packet of biscuits in one hand would not be an explanation of his action. His beliefs about the availability of biscuits in the shop, his desire for biscuits, his intentions about what to do with them, and so on, are necessary for such an account. Something essential is left out if we try to analyse, say, 'X desires a biscuit' into 'If such-and-such circumstances obtain, X goes into shops and reappears with packets of biscuits' – unless the reference to beliefs and desires has been covertly smuggled into the 'such-and-such circumstances' clause.

In setting out his views in his best-known work, *The Concept of Mind*, Ryle coined the phrase 'the ghost in the machine' to characterize pithily the kind of view Descartes and dualists in general hold, and that the scientific and philosophical consensus had come to reject.[15] In rejecting it, they had replaced the interaction problem that dualism prompts with a *black box* problem, given that the workings of the brain were wholly opaque and the way mental phenomena are caused by it, or are instantiated in it, or supervene upon it, was simply unknown and could not even be guessed. Behaviourism, and a successor theory called 'Functionalism', were efforts to find ways of characterizing and understanding mental phenomena without having to have a theory of how the brain operates. Functionalism is the suggestion that we should understand mental states such as pain and desire in terms of the *roles* that pain and desire play – that is, we can identify pain as the link in a causal sequence relating certain kinds of inputs with certain kinds of behaviour. Unlike the behaviourists, functionalists do not *reduce* 'pain' to 'observable winces and groans', but describe it as what makes functional sense of the connection between such and such circumstances and such and such behaviours. Like behaviourism, the theory does not rely on a commitment to some way of understanding how the functional states arise inside the black box; it restricts attention to the inputs going in and the behaviour coming out, leaving the whirring of the machine to be whatever it is.

But shying away from any effort to understand the inside of the black box is obviously unsatisfactory, and efforts to investigate how brains are related to the mental phenomena associated with them were already under way, and had been so for some time.

Contributors to brain research in the nineteenth century differed over the question whether brain functions are specialized by locality, or are global. In the early nineteenth century the phrenologist Franz Josef Gall promoted the theory that different faculties or functions are localized in different parts of the brain; and, even if his effort to identify the localities by feeling the bumps on people's skulls now seems absurd, in principle his idea about localization is no different from the view held by many neuroscientists now. The French physician and physiologist Jean Pierre Flourens showed by experiments on rabbits and pigeons that the principal divisions of the brain are functionally dedicated: perception and judgement in the cerebral hemispheres, balance and coordination of movement in the cerebellum, autonomic functions such as respiration and circulation, essential to the maintenance of life, in the brainstem. Unable to find specific localities for memory and reasoning, he concluded that they are produced by global brain activity.[16]

Flourens's method of investigation is known as 'lesion studies', that is, correlating mental function with injury to areas of the brain. In that respect it is similar to Ferrier's work, described above. Another French physician, a younger contemporary of Flourens, who made a significant discovery by the same technique but this time in relation to human brains, was Paul Broca, who identified an area of the frontal lobe (the *inferior frontal gyrus*), which is specific to the *production* of speech. Later in the century the German psychiatrist and neuropathologist Carl Wernicke identified another region of the cortex, further back on the upper part of the temporal lobe (the *superior temporal gyrus*), specific to the *comprehension* of speech. Both centres are localized in the left-cerebral hemisphere in 90 per cent of right-handed people and 70 per cent of left-handed people. The two areas are connected by a bundle of fibres called the *arcuate fasciculus*. In both cases it was injury or disease to the respective regions of the brain that alerted Broca and Wernicke to their function.[17]

Until the advent of brain-imaging technology, lesion studies – damage and disease in humans, induced lesions in animal studies, phenomena observed in patients undergoing brain surgery – were the principal means of exploring the human brain. One of the most famous examples of a brain injury and its effect on mental life is the case of the railway-construction worker Phineas Gage. In September 1848 Gage was a foreman in a group of workers using explosives to clear rock for the laying of railroad track in Vermont. A three-foot-long iron rod was blown through his head, at an angle upwards through his left cheek and eye socket and out through the top of his head, destroying a portion of his brain's frontal lobes. Gage was momentarily dazed, but was conscious when carried to a cart, in which he sat for the ride back to his hotel in the nearby town of Cavendish, then walked upstairs to his room. While being examined by Dr John Harlow, the local physician, he vomited because he was swallowing quantities of blood draining internally from the injuries inside his head, and the effort of doing so 'pressed out about half a teacupful of the brain through the exit hole at the top of the skull, which fell upon the floor'.[18] His physical recovery from the injury was at first difficult because of infections to the wounds in his head, but it was achieved through Harlow's skilful and assiduous care. Harlow published the first of his two accounts of the case in the *Boston Medical and Surgical Journal* of December 1848.[19]

Gage lived for a further twelve years, described by those who knew him as 'no longer Gage', very different in personality from the way he was before the accident – more aggressive, less emotionally stable, given to uttering profanities, no longer the steady, reliable, shrewd individual he had been before the accident. Towards the end of his life he seems to have regained some degree of stability of character, but had also developed epilepsy, dying during an epileptic seizure in May 1860 at the age of thirty-six.[20]

The case of Phineas Gage was a powerful indication of the relation between brain and personality, and it added fuel to the debate between the localizers and globalists on the question of brain function. Because of the sensationalism that surrounded Gage's story, which became famous in the newspapers as the 'American Crowbar Case', it has proved difficult to arrive at an accurate account of the nature, extent, and duration of the mental changes that followed the physical injury. His pre-injury character was talked up to make the post-injury changes seem more dramatic, and so, because in the remaining years of his life Gage was something of a fairground wonder and worked in a variety of places, including Chile (as a stagecoach driver), forming a reliable clinical picture is not possible. The commendable Harlow himself made efforts to trace Gage's progress, and even secured Gage's skull after the unfortunate man's death, but the evidence he gathered is anecdotal.[21]

The examples set by Broca and Wernicke – which, incidentally, were initially greeted with scepticism by some in the scientific community, but which were already being pursued independently by others in search of an understanding of consciousness, as we shall see in Section 3 below – resulted in a growing number of lesion studies indicating localization not just of language capacity but of memory, emotion, vision, and motor control. In the 1950s the role of the hippocampus in memory was identified, and surgical interventions to help sufferers from severe epilepsy – by severing the *corpus callosum*, the structure that connects the two halves of the brain, to prevent seizures spreading from one half to the other – revealed that (putting matters at the minimum) the left hemisphere is usually dominant for language and reasoning, the right hemisphere for spatial ability.[22]

The Nobel Prize in Physiology or Medicine was awarded to neuropsychologist Roger Sperry for his split-brain studies in 1981. Beginning in the 1950s Sperry studied monkeys, cats, and humans, and found that, when the brain's two hemispheres are separated, they act as independent centres of mental life. He had been intrigued by the lack of any apparent dysfunction in people who had undergone a division of the corpus callosum, whose sheer size, consisting of a thick bundle of neural fibres, suggested that it must be an important structure – and yet severing it was not interfering with patients in any obvious way. Beginning from the recognized laterality of brain function – each hemisphere governs the opposite side of the body and visual field – he designed experiments that fed information to just one hemisphere, and thereby discovered the hemispheres' mutual independence. For example, he showed a patient's right eye a word denoting an object, and asked the patient to say the word; the patient could do so readily. When he showed the image to her left eye, the patient could not utter the word, though if asked to draw the object it named, she could do so.[23]

If well in other respects, split-brain patients retain their memory and social skills, and function normally. It seems that they cannot learn new skills requiring independent action by limbs on opposite sides, such as playing the piano, but the obvious deficits are few. More obvious deficits occur with more localized lesions, associating those areas with the functions that are disturbed or lost as a result. Before non-invasive imaging of the brain became available, lesion studies offered the only possibility of drawing a map of the brain to identify these associations.

However, this assumption – that lesions show that particular cognitive functions are localized in particular anatomical structures – is problematic, at least for the reason that it is hard to establish *invariable* one-to-one correlations between structures and functions. For one thing, individual brains differ. For another, some functions are carried out in a distributed manner across the brain, and the plasticity of the brain means that neuronal structures can develop and change in response to novel challenges. A fixation on the modularity or localization of function is in danger of overlooking these points. When injury or disease affects the brain, it often involves more than one region; some areas of the brain are more susceptible to stroke damage than others, and the damage can be extensive, disturbing or disabling many functions. On the other hand, the redundancy of the brain means that small strokes can occur without obvious deficit. In any case, it would be wrong to assume that when damage occurs to a particular region, intact regions remain the same; it is often observed that lost functionality can be restored by other parts of the brain assuming responsibility for it.

And there is also the fact that an otherwise healthy brain region can cease to function because a distant region whose connection to it is necessary for its operation has been damaged, or the connection to it severed; so inference from locality to loss of function and vice versa might mislead.

These criticisms of the lesion method have been standard since Broca, but matters have been vastly improved by the advent of imaging technologies: single-photon emission computed tomography (SPECT), sometimes combined with CT (computed tomography X-ray), which is quicker; positron emission tomography (PET) – 'tomography' is a method of creating images by slices or sections using one of these forms of penetrating radiation; near-infrared spectroscopy (NIRS) measuring cerebral oxygenation; electroencephalography (EEG) measuring electrical 'brain wave' activity; magnetoencephalography (MEG) and transcranial magnetic stimulation (TMS), the first recording and the second inducing magnetic fields in the brain; and magnetic resonance imaging (MRI). The last of these in particular has enabled great strides in the study of the brain and its functioning. The MRI concept is as simple as it is compelling: a subject is positioned in an fMRI (functional MRI) scanner and offered various stimuli, including tasks, pictures, sounds, invitations to remember or imagine, and so on, and blood flow is monitored to see which regions attract more blood when the stimulus is presented.

The simplicity of the concept belies the indirectness of the method. It works as follows. When neurons fire, they incur a metabolic cost. Not having their own stores of glucose and oxygen, activated neurons have to summon them from the brain's blood supply. In the region where this occurs, a change results in the proportion of oxy-haemoglobin to deoxyhaemoglobin, as oxygen is transferred to the neurons. Oxygenated and deoxygenated blood are magnetically different: the former is diamagnetic (is repelled by a magnetic field), the latter paramagnetic (is attracted by a magnetic field). An increase in paramagnetic response indicates an increase in deoxygenated blood, showing that the neurons in the region are using more oxygen. The fMRI scanner detects and records this variation. What it is imaging by this means is called the BOLD response – 'blood oxygen level dependency', or 'haemodynamic' response, exploiting the quicker uptake of oxygen by activated neurons relative to more quiescent ones. The raw data thus collected is a three-dimensional array of thirty thousand *voxels* (a voxel is a 3D pixel), captured every two seconds at a spatial resolution of about a single millimetre.[24] There are hundreds of thousands of neurons in each voxel. A neuron fires within one hundred milliseconds of being presented with a stimulus; the

BOLD response takes about six seconds. The temporal resolution of fMRI is, accordingly, very slow in comparison with neuronal activity.[25]

fMRI does not measure the absolute amount of metabolic activity in neurons, only the difference between the levels of such activity before and after a stimulus is offered to the subject. Nor does it measure the intensity of neuronal activity. What causes the BOLD signal is unknown. Action potentials? Synaptic activity? Inhibition? And what is being observed is not a causal connection, just activity whose co-occurrence fairly close in time suggests to observers a connection of some kind between it and the offered stimulus.

A problem fMRI faces is that there is a difference in magnetic response between brain tissue and air, causing fluctuations or distortions in signal that can make it seem that a BOLD response is lacking in certain regions of the brain otherwise thought to be significant for cognitive function. Moreover, subjects in an fMRI scanner are subjected to tremendous amounts of noise, which might interfere with their responses, and they are also subjected to a powerful magnetic field that is constantly switching, the assumption being that brain activity in its presence is what it would be in its absence, or at any rate is not affected by it in ways that make observations unreliable. As with all observation and experimentation, what comes to mind is the Meddler Problem.

Apart from these technical questions, critics further point out that there are interpretative limitations on the data fMRI provides. Even if a lot of data correlates a particular region with a particular function, it does not establish whether the region is *necessary* for the function (lesions studies, by contrast, can do this), or is being activated as a by-product of activity in another region. Nor does it say anything about the contribution being made by regions that are constantly activated, no increase or other change in their activity therefore being observable.

Reservations aside, fMRI is a powerful tool. One of its greatest advantages is that allows the *healthy* brain to be studied, which by definition lesion studies do not permit. It is non-invasive, the spatial and temporal resolutions it offers – limitations aside – are considerably better than most other techniques, and it is a resource of great utility to clinicians and researchers alike, having proved itself handsomely in clinical settings in neurology, neuropsychology, and neuropsychiatry, and having produced a rich harvest of data and insight in cognitive neuroscience research.

Although neuroscience is still at an early stage of the new technology-empowered impetus it has received, its spectacular advances have achieved two things: it has settled the question of the seat of the mind, and it has stimulated research into cognition, consciousness, and a new understanding of the nature of mind. These are the topics of the next three sections.

2.   The Cognitive Brain

The internet readily provides detailed atlases of the brain and its structures, so the following is schematic only, to provide orientation for what cognitive neuroscientists are hypothesizing about brain activity and its relation to perception, thought, memory, and emotion.[1]

A standard trope about the brain is that it contains 100 billion neurons with 100 trillion connections between them. The Brazilian neuroscientist Suzana Herculano-Houzel found a way of estimating the number of neurons in the brain to arrive at a more accurate figure. She did it by turning the brain into a homogeneous liquid, a 'soup', and counting the stained nuclei in a sample under a microscope. The number she found is approximately 86 billion, give or take 8 billion.[2] The rounded-up impressionistic figure is therefore not too far off the higher estimate. On either count, the brain contains a very large number of neurons that, along with their accompanying glial cells, makes for a densely crowded intracranial space. The comparison is standardly made with the number of stars in our galaxy: 250 billion ('give or take 150 billion': at these numbers the idea of approximation is itself very approximate). The point of bandying these numbers is to emphasize the tremendous complexity of the brain and the tiny size of its vastly multitudinous components, and thus a different but no less blurry kind of approximation affecting investigation of brain function.

It used to be thought that there are ten times as many glial cells as neurons in the brain; the soup work described above suggests that the ratio is 1:1. *Glial cells* ('glia' is Greek for 'glue') constitute the electrically inactive supportive and protective packaging around the neurons. Over half the brain's neurons are crowded into the cerebellum, a ball of tissue lying centrally underneath the cerebral hemispheres as part of the hindbrain. 'Cerebellum' means 'little brain', a misnomer from this point of view but not in point of size: the cortex of the brain has 80 per cent of the brain's mass, though containing 'only' between 14 and 16 billion neurons.

A human brain weighs about 1.5 kilograms (3.3. lbs.). At an average volume of 1,274 cubic centimetres a male brain is about 150 cubic centimetres larger than the average female brain.[3] With its somewhat spongy texture, convolutions, and wetly glistening appearance, the brain – like most of the body's organs – is not an especially aesthetic object, but it is among the most fascinating things in the universe.

One can visualize the brain as a jet pilot's helmet with earphones on each side. (We also have to imagine a tennis ball under the back of the helmet – this is the cerebellum – which slightly spoils the image.) The two cerebral hemispheres, separated by a deep longitudinal fissure but joined across it by the thick bundle of neural fibres that is the corpus callosum, each have four general sections. The earphones – imagine them as elongated ellipses lying horizontally – are the *temporal lobes*, positioned just about level with the top of one's ears. Remove them for a moment and contemplate the body of the helmet itself. The frontal lobe constitutes about 45 per cent of the cerebrum; behind it and separated from it by the *central sulcus* – a fissure running transversely across the brain – is the *parietal lobe*, which occupies about 35 per cent of the cerebrum; and behind it again, at the back of the brain, is the *occipital lobe*, occupying about 20 per cent of the cerebrum. The familiar convoluted appearance of the brain consists of hills and crevasses, the hills called *gyri* (singular *gyrus*) and the crevasses called *sulci* (singular *sulcus*). The convolutions or folds increase the surface area of the brain, making optimal use of the space inside the cranium.

Before reattaching the earphones – the temporal lobes – to the helmet, look at their inner surface and the surface of the cerebrum against which they nestle. The gaps between these respective surfaces are the *lateral sulci*, one on each side, and hidden within their folds is an exceedingly important part of the brain, the *insular cortex*, to which a number of functions are attributed, including consciousness itself.

The cortex consists of a number of layers, each consisting of columns and microcolumns of neurons arranged perpendicularly to the cortical surface. A distinction is drawn between the *neocortex*, the upper six layers of the cortex, and the *allocortex*, the lower four layers. To the neocortex are attributed the higher functions, such as language, motor control, and sensory perception. To the allocortex, where structures of the limbic system such as the hippocampus are located, are attributed memory, emotion, motivation, and the sense of smell.

The cerebrum is connected to the spinal column by the *brainstem*, which consists of three principal structures: the *midbrain*, *pons*, and *medulla oblongata*. The first of these plays a multiplicity of roles, taking part in the processing of visual and auditory information, motor control, sleep and wakefulness, and control of body temperature. The pons (Latin 'bridge') carries messages between the cerebral cortices and the cerebellum and the medulla oblongata (Latin 'elongated pith'). This latter is a centre for autonomic activity such as breathing, blood pressure and heart rate, and involuntary activity such as vomiting and sneezing.

All vertebrate brains are arranged in three sections: *forebrain*, *midbrain* and *hindbrain*. Each section has its own fluid-containing hollows, called *ventricles*. The forebrain is the cerebrum; the midbrain and hindbrain form the brainstem and cerebellum. In humans the cerebrum is very enlarged – this is the jet-pilot helmet with earphones – and it is regarded as the site of thinking and the complex processing of information. To the frontal lobe of the cerebrum are attributed reason, judgement, problem solving, control of emotions and behaviour, and the personality. The occipital lobe is the primary visual centre ('primary' because it delegates some of its functions to neighbouring regions). The temporal lobes at the sides of the cerebrum are involved in language and sound processing, while the parietal lobe in the midst of the cerebrum integrates sensory information and plays a part in spatial and motor processing. Tucked deep in the temporal lobes are the *amygdalae* (singular *amygdala*, Latin 'almond'), which are almond-shaped clusters of cells associated with emotion, chiefly negative emotion – fear, anxiety, aggression – and involvement in memory and decision-making.

Such, in rough outline, is a description of the brain and some of the attributed functions. Cognitive neuroscience aims to be more circumstantial in identifying and understanding the relationships between structures and functions, and therefore the functions themselves. The level of detail at which this investigation proceeds is not only the brain *regions* but their constituents and the connections between them: *neurons* and *synapses*. The grand endeavour of seeking to map the trillions of connections among neurons is the Human Connectome Project ('connectome' by analogy with 'genome').

Neurons and their mutual connections are therefore the primary target of interest. They are remarkable little machines. Each consists of a cell body from which protrude an *axon* and a number of *dendrites*. The cell body, also called the *soma*, contains the nucleus and other organelles, the nucleus containing most of the cell's DNA (that is, all of it other than mitochondrial DNA), which combines with various proteins to form chromosomes. Cell bodies constitute the grey matter of the brain. Axons and glial cells are the white matter. Axons are wrapped in a sausage-string-like sheet of fatty tissue called *myelin*, which insulates the axon and speeds up the passage of the action potential – the electrical impulse – along it, in effect by causing the impulse to jump from gap to gap in the sheath. The gaps are rather poetically called 'the Nodes of Ranvier', after the French medical scientist Louis-Antoine Ranvier, who first described them in the early twentieth century.

The action potential jumps along from node to node – gap to gap – by means of oppositely charged ions of potassium and sodium changing place across the axon wall, reversing polarity at that point. Axons can end in a number of projections called axon terminals, or *telodendria*, which align across a gap with the ends of dendrites. These gaps are the synapses, and at this point the arriving signal ceases, in most cases, to be electrical and becomes chemical, the axon terminals releasing neurotransmitters across the synaptic gap to be picked up by the dendrites on the other side and thus to stimulate another signal onward to their own cell bodies. (Some axons connect synaptically with other axons and some dendrites with other dendrites.) In the usual case, a neuron sends signals through its axon and receives them through its dendrites. Some of the dendrites act to excite, others to inhibit, signals; the sum over the excitations and inhibitions determines whether the neuron will send a signal further onwards along its axon.

Action potentials do not vary in amplitude but in number, that is, in how many 'spikes' – all of the same size – are propagated through a neuron. The degree of density of the spikes encodes the information being transmitted. Imagine a version of Morse Code in which each sound is the same in loudness as every other (which is indeed the case in Morse signalling) but instead of patterns of dots and dashes you have just dots and the silences between them: the longer the silences, the lower the intensity; the shorter the silences, the higher the intensity. The code consists in the number of dots in a given period of time. A crude analogue would be dots per second: one dot per second is *A*, twenty-six is *Z*, and so on for all between (ten for *J*, etc.). In Arthur Koestler's autobiographical novel *Darkness at Noon* prisoners communicate by tapping on the pipes running from cell to cell in a version of this, using a five-by-five grid – five horizontal rows with five letters in each row – and assigning a pair of numbers to each letter of the alphabet: Row 1, Column 1, identifies the letter *A*, Row 2, Column 3, identifies the letter *H*, Row 3, Column 5, is the letter *O*, and so on. This shows that very complicated information can be transmitted by simple means, which is what the combined action potentials of neurons achieve.

As each axon is communicating with numbers of dendrites, its signal is being transmitted to as many other neurons; and each neuron is receiving signals from many other neurons. It is calculated that each neuron connects with about ten thousand others. Consider that each of these ten thousand is connected to ten thousand; consider the total of many tens of billions of neurons there are in layers and layers of the brain; consider the unresting activity of the brain even in sleep – and the sense of tumultuous interconnectivity and activity becomes overwhelming.

There are three main kinds of neuron: *sensory*, *motor*, and *interneuronal*. Sensory neurons carry information from the organs of sight, sound, smell, taste, and touch to the brain, there to be processed, interpreted, and reacted to (painful sensations of touch, as when one is burned by a flame or pricks a finger on a needle, pass to the brain through the spinal cord, which itself deals with the required involuntary reaction of pulling the affected member away, even before the brain registers the pain or the act). Motor neurons take messages from the brain and spinal cord to muscles, organs, and glands. Interneurons are the connecting neurons within a brain region, forming neural circuits that carry out various functions.

In this sketch, and not least in the last sentence of the preceding paragraph, it will appear that the brain is again being characterized as if it is modular, with specific regions supporting specific functions. This is the implication that can with some readiness be drawn from lesion and imaging studies, as mentioned above. But, as also mentioned above, matters are probably not so straightforward. To picture the brain as a network of networks, with functions being more distributed than localized but with hubs of networks playing important roles in mediating those distributions, might come closer to the truth. This is one of the ideas at work in the Human Connectome Project, launched with significant funding in 2010.[4] The goal of the project is to map the connectivity of the brain to a scale of millimetres rather than individual neurons (there are fifty thousand neurons in a cubic millimetre), which as a first step is more modest than what will eventually be the target: to map the one hundred trillion connections within the brain. This latter presents a vastly more ambitious target than the Human Genome Project faced.

Seeing the brain as a connectome does not mean giving up the idea of functional specialization altogether, however, because the alternative is not complete globalization but the idea, just hinted, of networks of networks in which the distributing hubs play something like the role of local specialization for a given type of function. Schematically, it might be that a network apt for the type of task that is useful in a variety of applications can be called upon to put its talents into combination with others from other networks, all summoned by a coordinating hub, the effect of their combination being the performance of a specific macro-task. On another occasion its talents might be combined with those of a different set of networks for a different macro-task. Smaller networks can thus be recruited to play roles in various larger networks as occasion and necessity demand. Variations on this theme now command something of a consensus among neuroscientists.[5]

There are some factoids about the brain that would be useful in that game where items of information are greeted with alternate cries of 'Woe!' and 'Hurrah!' For example, about 10 per cent of our cortical neurons die between the ages of twenty and ninety, which is one neuron every second. But it appears not to be true, as once thought, that new neurons cannot be born in the brain; they have been observed to do so in the dentate gyrus in the temporal lobe, part of the hippocampal circuit that plays a role in new memories and finding one's way in new environments. The brain ages, and, despite its strong protective casing of bone, is a vulnerable organ; but it is 'plastic', meaning that its wiring changes constantly, and in drastic cases such as injury or disease the contributions of a given region can be taken over by other regions. Best of all, brain plasticity remains throughout life.

Cognitive neuroscience has made considerable strides in investigating the brain's functioning in sensation, attention, movement, memory, speech, decision-making, emotions, and how we understand the written word and numbers. To understand something of how this is done, consider just two different but connected cognitive functions: vision and memory.

It is a matter of utmost wonder how vision occurs. In normal conditions a two-dimensional pattern of irradiations of the retina at the back of one's eye – photons striking the retina and stimulating its constituent rods and cones to fire – prompts a sequence of events whose outcome is a Technicolor movie of a three-dimensional world arrayed in depth outside and around one's head. We seem to see through our eyes as if they were windows to a world outside us. But in fact the world is inside the brain, and mostly at the back of the brain; and everything we experience – buses, spiders, teacups, and the winter wind – consists of electrochemical-action potentials.

The optic nerves leave the eye from about the middle of the retina, making a blind spot – the *fovea* – which we do not notice because the brain helpfully fills in the missing bit from the surrounding information. Impulses sent along them from the retina cross at the optic chiasm and there diverge into at least ten separate pathways through the brain. The dominant pathway travels to the occipital region at the back of the brain, the location of the *primary visual cortex*, also called the *striate cortex*, or VI. This path detours through a part of the thalamus with the forbidding name of *lateral geniculate nucleus*, or LGN. The thalamus is a general centre for processing sensory inputs, and the LGNs – one in each hemisphere – sort the incoming optical information into colour and movement aspects, preparing it for later phases of processing further along. The later phases deal with edges and contrasts of light and dark, these being significant for analysis of depth and movement in the visual field.

Because the left side of the brain deals with information from the body's right side and vice versa, the assumption is commonly made that the left eye sees only the right half of the visual field, and vice versa for the right eye. In fact each eye sees both halves of the visual field, as you can easily test by closing one eye, *pace* part-occlusion of the opposite visual field by your nose. What happens is that the right half of the retina in *each* eye sees the left visual field, and likewise the left half of each eye sees the right visual field. However, the data from half of each retina coming from the opposite half of the visual field goes to the LGN in the hemisphere on the same side as itself – thus, data from the left halves of each retina, receiving input from the right visual field, goes to the LGN in the left hemisphere. So both LGNs have visual data from both eyes.

Recordings from single neurons in the brains of cats reveal that cells in the primary visual cortex, VI, are specialized for detecting orientations of edges and lines. This discovery earned a Nobel Prize for physiology for the scientists who made it, David Hubel and Torsten Wiesel of Harvard University. The discovery was accidental at first: they noticed a cell in a subject cat's VI area activating when an angled crack occurred in one of the slides they were showing it. Further experiments confirmed that individual cells can be made to fire by showing the subject a line whose orientation excites that particular cell. Although the individual cells combine to achieve an overall representation bottom-up, they receive feedback top-down from higher levels of analysis dealing with more complex perception of surfaces and shapes. It is as if the higher levels of perception, elsewhere in the occipital region and in the distribution of further visual processing to the temporal and parietal lobes – for example, when the data is being analysed as representing a shape – signal back to the orientation-detecting cells in VI for confirmation or additional data, telling it (so to speak) what to check more particularly.

The ten or more pathways of optical input have developed during the course of evolution, with more sophisticated pathways not replacing but supplementing more primitive ones. One route leads from the eyes to a part of the hypothalamus that makes use of information about day and night. Nocturnal mammals would, of course, only venture out if they knew it was night; somewhere in the remote ancestral past of primates this facility was of crucial survival value. Perhaps, in the modified form of home-time and sundowner time, it still is. Another route links perception of a sudden flash of light, or change in intensity of light, with involuntary movement of the eyes and body, perhaps as an avoidance response to an unexpected challenge, or to summon attention to something in the periphery of the visual field that could turn out to be significant – the movement of a snake-like shape, say.

The existence of a multiplicity of visual pathways, each subserving more particular purposes for which vision is of value in evolutionary terms, probably explains one of the more extraordinary phenomena of cognition: *blindsight*, or cortical blindness. People who cannot see – who are not conscious of seeing – can nevertheless respond to information from what would be their visual field if they were normally sighted. In experiments such people are asked to indicate where an object is, and do so accurately, though reporting that they cannot see it. The opposite of this condition is the remarkable Anton–Babinski Syndrome, in which completely blind individuals claim that they can see, though unable to respond to anything in what would be their visual field if they were sighted.[6] The obvious and most plausible account of blindsight, once other possibilities such as partial vision or residual light-dark contrast are excluded, is that there are pathways of optical input that bypass conscious awareness. There is no doubt that visual awareness is highly selective – most people now know about the famous example of the gorilla walking through a group of ball players, unnoticed by observers who have been told to count the number of times the ball is passed from one player to another – and this adds the thought that unconscious visual experience might occur in other and more ways besides.[7] If one can ignore some visual data, one might also be using visual data without registering it, perhaps through pathways that can or do dispense with conscious monitoring.

Vision is of course far from being a merely passive sense; it exists to be employed. Information processed through the LGNs and the occipital region is shared around the latter and also forwarded to the temporal and parietal lobes in what are known respectively as the *ventral stream* and the *dorsal stream*. The ventral stream to the temporal lobes involves the capacity to recognize objects, among them particularly significant ones such as faces, and therefore connects with memory also. The dorsal stream to the parietal lobes is deployed in attending, acting, and moving. Because these functions are dispersed more widely over the regions in question, damage there can result in selective visual deficits – for example, a sufferer might lose colour vision but still be able to see movement, or lose the latter ability while still seeing the world in colour.

Localization is evident in the case of colour vision; an area of the occipital lobe known as the lingual gyrus and labelled V4 is thought to be the *main* centre for colour vision, at least in the human brain; damage to it results in the world appearing in shades of grey to those afflicted. This is known as *achromatopsia*, a rare condition because there are two V4 areas, one in each hemisphere, only one of which is likely to be damaged by stroke or injury at a time. V4 is identified in fMRI scans by its activation when 'Mondrians' (coloured squares, named for the artist) are presented in the visual field. Colour is obviously important information in evolutionary terms, given that the brain not only has a dedicated colour region but has equipped it to maintain 'colour constancy', that is, the recognition of a colour as the same under different conditions of lighting. Food, poisonous substances, mating invitations, and danger signals are colour-coded in many cases; getting them right can be a matter of life, death, and the survival of one's genes.

Damage to a small area known as V5 can result in the loss of ability to perceive motion. For sufferers, everything appears as a series of stills; a passing airplane manifests staccato-like and motionless at different successive points in the sky. Wranglers on a dude ranch in the US south-west claim that this is how horses monitor their environment, taking photos alternately on each side of their heads with their laterally monocular vision, and why – as prey animals, on constant lookout for predators – they will shy away from the side where they think something new has appeared when they photograph it again. Wranglers on dude ranches are not sentimental about horses, and generally are not neuroscientists. Those who admire these animals for their nobility and beauty, as well as those who are equine neuroscientists, attribute a slender field of binocular vision to them, and 360 degrees of view with both eyes combined – though apparently rather blurred, lacking in depth perception, and red–green colour-blind. The horse comparison is interesting, because these claims draw heavily on the relevant neuroscience with no subjective reports possible to corroborate them outside the fiction of Anna Sewell, and show how much can be determined by that means and inferences from behaviour alone.[8]

An important use of vision is object recognition. To understand this, identifying the brain structures involved has to be coupled with the psychoneurology and psychology of learning and memory. Objects are unfamiliar to the extent that they match nothing in memory, but the way visual processing works can nevertheless allow for classification and hypotheses about what an unfamiliar object might be. This requires distinguishing figure from ground, recognizing similarities with previously experienced objects, grouping saliences in the patterns of data presented, and venturing an interpretation on that basis. In seeing any object, familiar or otherwise, one has to integrate its parts into a whole and to maintain 'object constancy' from different angles and distances in different lighting conditions. The region with a significant role in this is the *inferotemporal cortex*, the bottom curved surface of the temporal lobes.

In turn, an important aspect of object recognition is face recognition. From earliest infancy faces matter and attract our attention, and different models have been proposed of how this works. The main suggestion is that the *fusiform gyrus*, also in the inferotemporal cortex, is dedicated to face recognition and analysis of facial expressions. It is accordingly known as the fusiform face area, FFA. Researchers have found lower densities and lower numbers of neurons in the FFA cortical layers of autism spectrum disorder patients, which correlates with lower activation of this region in fMRI scans when they are presented with face images. Both the lower density of neurons and observed hypoactivity in scans imply lower connectivity. Researchers hypothesized that if this is how things are in the fusiform gyrus, it might also be so in other areas of the brain correlated with reduced functionality in autism – for example, the amygdala, involved in emotion. Studies revealed a lower density and connectivity of neurons there too.[9]

The FFA responds less to other stimuli than it does to faces, this time supporting the view that cognitive functions are localized more than they are distributed, even if some degree of distribution is involved. There is evidence that the FFA, present in both temporal lobes but more active on the right side, works to make definite decisions about faces – *categorical perception*, as it is known – so that if a composite picture blending two well-known faces is presented, the FFA ignores the ambiguity and chooses one or other of the individuals outright.

The social importance of facial recognition adds further weight to the suggestion that the brain goes in for face-specialized visual processing. An inability to recognize faces, *prosopagnosia*, or 'face-blindness', can be inherited in about 2.5 per cent of cases, but is most frequently the result of brain damage, as in strokes. As one would expect on the evidence mentioned, the damage occurs in the occipital or temporal lobes and most often in the FFA.[10]

Experimental subjects in fMRI scanners asked to imagine seeing something 'in their mind's eye' show activation of V1. This finding settles what had been a debate about which cognitive functions are recruited for visual imagination. An earlier theory had been that visual imagining is basically semantic, that is, draws on the same resources as word understanding and memory. Memory is an obvious component even in the visual imagining of fantasy scenes, by providing elements to be recombined into fantasy objects. But asking a subject to picture a scene while being scanned in a fMRI machine provides clear evidence that the brain's visual centres activate. Further confirmation comes from stroke patients who cannot visualize aspects of a scene – colour or shape – if the damage they have suffered occurs in a region associated with that aspect of visual processing. The 'fractionation' of processing is demonstrated by the phenomenon of patients who can visualize an object perfectly well but cannot recognize it when they see it.[11]

The picture that emerges is that the simplest act of seeing in normally sighted people – as when you look up from this page at the scene around you – is an enormously complicated set of events in which a flood of electrochemically encoded data is passing from your eyes through multiple channels to – principally – the back of your brain and then forward again to the temporal and parietal lobes, at each stage of its journey being processed and analysed in preparation for the work to be done at the next stage, at the same time recruiting memory and capacities for comparing, contrasting, inferring, and evaluating. The information is being used for more than seeing; it is part of the fundamental data for appropriate responses to the environment if such are called for, and for planning; and it contributes to balance and control of movement.

But there is more to it still. Think of a sportswoman catching a ball; much of the visual processing and motor coordination required for that simple act is unconscious, worked out by the neural circuits chiefly in her occipital and parietal lobes and their control of muscular activity and skeletal positioning. She watches the ball, she sees it flying through the air to her proximity; but she does not consciously compute the trajectories, muscular coordinations, and the like that enable her to catch it. Now think of a different kind of case – of someone seeing a small movement in space that nevertheless is powerfully charged with social meaning: say, a man seeing his wife touch another man's hand in a way that suggests secret intimacy. Here the visual data triggers a cascade of abstract semantic interpretation and inference of a socially learned kind. Vision is a *cognitive* process: we do not see, we always *see as*; it is highly interpretative, the shapes, colours, orientations, and contrasts detected by the successive neural stages of the process have meanings well beyond their basic content. Some of this is describable in the language of neurology, some of it in the language of psychology; the great question concerns a translation-relation between them: the *how*, or *whether*, of a translation-relation. Think of the ballplayer and her jealous husband: her catching the ball can be described in wholly neurological terms. Her husband's interpretation of her flirtatious touching of another's hand is readily done in the language of psychology, not so readily in the language of neurology. This perennial dilemma remains to be solved.

Memory was mentioned more than once in preceding paragraphs, and a moment's reflection shows how central and crucial a function it is. It is required for almost every cognitive act we perform. It is constitutive of our identities and personalities. Its loss is destructive of our capacity to relate to the world, and to be ourselves. Age and a variety of brain injuries compromise memory, making it possible to locate the regions principally involved, and – as always with brain research – to aid the search for medical remedies.

One of the most famous names in memory research is Henry Molaison, who died at the age of eighty-two in 2008.[12] The neuroscientific literature knows him as H. M. At the age of seven H. M. had a bicycle accident, which was later thought to be the reason why, at the age of ten, he began to suffer epileptic seizures. Over the next decade and a half these became so severe that he, his family, and doctors agreed that surgery was the only remedy. H. M.'s neurosurgeon at Hartford Hospital in Connecticut was William Scoville, who had previously operated experimentally on psychotic patients, removing portions of their temporal lobes. Scoville surmised – erroneously, as it proved – that the source of the seizures was H. M.'s hippocampus, which he separated from the surrounding allocortex using a cauterizing blade, and then sucked out with a vacuum device. H. M. was awake during the procedure, seated in an operating chair.

H. M.'s hippocampus appeared atrophied, and the neighbouring tissue damaged; Scoville removed it (not all, as it turned out much later), and with it the amygdala and a portion of the medial temporal lobe known as the *entorhinal cortex*, one of the hubs for neural circuitry involved in memory and time perception as well as spatial navigation. The surgery reduced the frequency and severity of H. M.'s epileptic attacks but did not stop them. The chief result was that H. M. could not thereafter form new memories – this is *anterograde amnesia* – although his *working memory* and *procedural memory* were intact. Working memory, sometimes mistakenly thought of as 'short-term memory' but similar to it, is the capacity to hold a certain amount of information in mind for a limited period in order to get current tasks done. Procedural memory is memory of how to do various things – tie one's laces, brush one's teeth – without consciously thinking about them. H. M. also had some retrograde amnesia, loss of most pre-operation memories in the two years before the operation, together with moderate memory loss relating to the decade beforehand; but memories of his childhood up to his early teens were unaffected (he was aged twenty-seven at the time of the operation).

The history of H. M.'s memory dysfunctions, the results of scanning studies when these became possible, and post-mortem examinations of his brain have provided a wealth of evidence and suggestion for understanding memory. The first and most obvious point to ponder arises from the capacities he lost and retained, indicating that different memory functions have different brain locations. Working memory, procedural memory, long-term memory, and new memory formation all appeared to rely on different areas for their support. The hippocampus seemed quite definitely to be associated with new memory formation and with learning (H. M. had difficulty acquiring new semantic information after the operation). But scanning and post-mortem evidence introduced some ambiguity. For one thing, the operation had caused more damage elsewhere in the brain than had been realized, while at the same time leaving about half of the hippocampus in place. For another, a lesion was detected in the frontal lobe that had not been detected before. It turned out to be a little harder to localize function than at first thought.

But these studies have not impugned localization altogether; on the contrary, the pendulum continues to swing back and forth on the subject. Examination of H. M.'s brain and the results of investigations of other patients with similar damage to their medial temporal lobes strongly indicate that these regions are involved in long-term memory, but not in working memory or word memory, nor in the ability to acquire new motor skills. Remarkably, although unable to acquire new memories – more accurately: new 'episodic' or everyday autobiographical memories – H. M. was able to acquire new *implicit* memories, that is, non-conscious ones, like tying shoelaces or riding a bicycle, by 'repetition priming', which is the improvement of a response by repeated exposure to a stimulus. This too showed that, whereas episodic long-term memory is located in the medial temporal lobes, other memory capacities are located elsewhere. This includes topographical memory – years after his operation H. M. moved to a new residential home, and was able to draw a floor plan of it, showing that the spatial processing capacity associated with the *parahippocampal gyrus* (an area of grey matter surrounding the hippocampus) was unimpaired. In fact the portions of the hippocampus and its neighbourhood that survived Scoville's scalpel and vacuum cleaner appear to have enabled H. M. to acquire a few scraps of postoperative memory also, including information about leading public figures such as the US president of the day.

An interesting finding is that H. M.'s retrograde memory was fine for childhood memories but increasingly less reliable for periods leading up to the time of his surgery. This phenomenon is noticed in the elderly and those with common types of dementia, and it suggests that long-term memory of childhood is not stored in the medial temporal lobes. One suggestion this prompts is that these lobes have an important role in consolidating memories formed later, but that what is learned early – including semantic memory – has a different formation and location.

In addition to the hippocampus at least four other brain locations are associated with memory: the fornix, thalamus, mammillary body and frontal lobe. Injury or disease in any of these areas can cause different kinds of memory impairment. Distribution of neural sites prompts one to think not only of different kinds of memory but different uses of it, not just recalling information such as a telephone number or events such as a traffic accident one saw but carrying out procedures and finding one's way around an environment. It also might explain the usefulness of forgetting, which may make certain kinds of remembering more efficient, and it might explain why we quite often do not remember something we were told or read, namely, that we did not process the information properly on first encountering it – snatching at it too hastily, not really being interested, being distracted, and the like. In the case of forgetting for efficiency reasons: remembering where you left your spectacles an hour ago is served by forgetting where you put them yesterday or last week (unless you always put them in the same place).

Can one choose to forget something? That seems hard, but in experimental settings it appears to be possible to induce forgetfulness. The work on this prompts a degree of scepticism, however, when one reads about fMRI studies seeking to identify brain regions that respond to instructions about remembering and forgetting.[13] Remembering relative to forgetting is ascribed a location in the hippocampus; forgetting relative to remembering is ascribed a location in the right dorsolateral prefrontal cortex. These areas are not obviously connected, so possibilities are sought: could a structure close to the dorsolateral prefrontal cortex, say, the anterior cingulate cortex, so affect an intermediate structure, say, the entorhinal cortex, that it inhibits the flow of information to the hippocampus? The technical terms themselves generate an air of plausibility.[14] One recalls the 86 billion neurons and the highly indirect nature of investigation of very large-scale clusters of them – and looks forward to the onward progression of the science.

If it is hard, though possible, to forget something deliberately, it seems all too easy to be made to 'remember' something that did not happen. False memory and – a more common occurrence by far – distorted memory are interesting and important phenomena. Experiments have shown that people can be induced to form a memory of something that did not happen by prompting strong associative links. One experiment involved getting subjects to remember words like 'bed', 'rest', 'awake' but not the word 'sleep'. The subjects were quite confident that they had heard the word 'sleep' in this array, the experimenters describing it as 'a powerful illusion of memory'.[15]

A controversy was stirred by 'recovered memory therapy' in which clients of this practice recovered a suppressed memory, or came to believe that they had done so, the recovered or confabulated memories typically of a traumatic event, such as sexual abuse in childhood, this being given as the reason for the memory's suppression. The therapeutic technique involved hypnosis, guided imagining, the use of drugs, and the interpretation of dreams. Given that people are in general highly suggestible, not just in being led to think a certain way by charismatic or forceful personalities but (for example) by assertions and conspiracy theories endorsed by others on social media, the potential for getting someone to believe strongly that they are 'remembering' something is great. One result was a number of court cases, the condemnation of 'recovered memory therapy' by the American Psychiatric Association and the Royal College of Psychiatrists in the UK, and the refusal of insurance companies in some jurisdictions to cover therapists engaged in the practice. At the same time the occurrence and ill-effects of abuse are all too real. This led a Royal College of Psychiatrists committee to say, 'Given the prevalence of childhood sexual abuse, even if only a small proportion are repressed and only some of them are subsequently recovered, there should be a significant number of corroborated cases. In fact there are none.'[16]

Damage to the brain's *orbitofrontal cortex* – the region sitting just above the eye sockets (orbits) – has been linked to confabulation, the confident production of false memories. Some theorists think that they might consist of parts of genuine memories that have been disassembled and recombined. 'Confabulation' is taken to be different from lying, in that the liar states as true something he knows to be false, whereas the confabulator may have no awareness of the falsity of the supposed memory.

The complexity of memory, imagination, belief, and confabulation makes it hard for neuroscience to suggest ways in which true and
false memories can be distinguished by noting activation of different
parts of the brain respectively. But one observation suggesting the
opposite is that if a subject is presented with words describing a memory or a 'memory' to a single hemisphere, a response is noted in the
other hemisphere if the memory is genuine but not if it is false.[17] This
has prompted the interest of those who repose faith in lie-detector
('polygraph') technology, widely used in the United States but dis-
approved of in most other jurisdictions. Could lying be reliably
detected? The testimony secured on the basis of current lie-detector
technology has played its part in condemning some to Death Row;
the sophistication of neuroscientific theory and technology would
reinforce the practice.

Across the range of cognitive functions and psychological phenomena –
and the derangements of both by injury or disease – that neuroscience
explores, the ones most susceptible to investigation are the sensory
pathways and the functions whose loss or disruption is readily observable: memory, speech, emotion, and certain kinds of behaviour
expected to be rationally responsive to cues in the social and physical
environments. As the foregoing shows, correlations are what neuroscience observes – not causes, and therefore not explanations – but
correlations that are highly suggestive of explanations, so much so
that in some cases clinical applications can be derived from them.

Neuroscience's contributions are less direct, at least currently,
when it comes to the larger complexities of human life: the effects on
relationships and long-term behaviour of desire, affection, responsibilities attached to roles or jobs, long-term planning, and the like – and
this includes understanding how bad experiences leave palpable psychological traces but no detectable injury to the brain (at least at the
level of structure currently accessible). These phenomena are much
harder to understand in terms of action potentials in brains alone.

Nevertheless, the assumption neuroscientists make, that in a future
perfected state of their science everything that relies upon brain
activity will be understood in terms of brain activity, is one they *have*

to make. At the very minimum it has to be a regulative principle of the enterprise that this is its goal. Neuroscience therefore embodies a strong form of *reductionism* to the effect that all mental phenomena will ultimately be understood in neurological terms. Some critics of reductionism – who are apt to insist on defining reductionism as 'seeing nothing in the pearl but the disease of the oyster' – think that this is in principle not possible. They might think this because they are proponents of theories about emergent properties, in which psychological phenomena supervene on neural activity and cannot exist without it, but which cannot be explained in terms of the neural activity alone without some remainder – for example, what being in a given state *means* or what it *feels like* to be in that state. This latter, especially, is a key point in the problem of consciousness, discussed in the next section.

The debate here is a key one: on the one hand, there are those who think that psychological language – which uses terms denoting such concepts as desire, belief, hope, love, need, longing, joy – bears the same relation to the terms of neuroscience as the long-ago language of 'demonic possession', as an explanation of illness, bears to the language of today's medical science. The neurophilosophers Patricia and Paul Churchland are of this view.[18] Others argue that intentional concepts – the belief/desire concepts just mentioned – are indispensable, because they constitute the 'theory of mind' by means of which we interpret and anticipate others' behaviour.[19] This applies to non-human behaviour too: the statement that one's cat wishes to be fed as an explanation of why it is mewing and patrolling the corner of the kitchen where its dish is always placed has an explanatory simplicity and informativeness that only experts would think they see in an fMRI scan of its brain.

One can accept this while at the same time acknowledging that the idea of a 'theory of mind' has been powerful in understanding (for example) childhood cognitive development and – differently but relatedly – autism, in neither case excluding the valuable possibility that neural circuits can, as suggested above, be identified that will help in understanding how they work, so that where problems exist they might one day be remediable. The common-sense observation that both approaches have their uses and, in addressing the same or related phenomena, might be mutually intertranslatable one day is attractive. Subject to the observations in Section 4 below, it seems the right approach to take. Critics will iterate their view that of the problems that beset enquiry, neuroscience clearly – indeed, they will say, to a marked degree – exemplifies all five of the Hammer, Map, Lamplight, Meddler, and Parmenides problems. If your fundamental theoretical concept is neuronal-action potentials, everything will reduce to them (Hammer, Parmenides), and, as you are looking at the only thing you can see – very large structures relative to the size of their components – your map is exceedingly small scale, and you are drawing it with devices that might affect the landscape itself (Lamplight, Map, Meddler). All this has to be acknowledged; but neuroscientists know this, and take it self-critically into account – the scientific method is in large part a matter of anticipating and countering these very barriers to enquiry. Recognizing that enquiry confronts such problems is not an invitation to give up but to work round them.

3.   Neuroscience and Consciousness

The great, and true, commonplace about consciousness is that it is simultaneously the most familiar and the most mysterious thing in the universe. It is the most familiar because we experience it, intimately and immediately, in all our waking moments and, in somewhat stranger forms, many sleeping ones. We also experience its distortions when drunk or drugged, or infatuated. We therefore know what it *feels like to be conscious* pretty well.

It is the most mysterious, because we have little idea of what it is, and no idea of how it arises from brain activity – some will add: *if* it does so.

Until the recent past there were broadly three kinds of view about consciousness. One was to attribute it (once again: the old standby) to an immaterial principle, a soul or mind, inhabiting or in some way colluding with the body. Another was to think that there is a physical principle of some kind, a refined form of matter, perhaps a liquor or vapour, that is sent from the brain (or heart, see above) to pervade the body, animate it, and sense its contacts with the surrounding world. The third option is to ignore the problem altogether and accept things as they are. Counting heads in the world today, one would find the third option to be the most common, though the first will be invoked when a demand is made to think about it – an instance of the Closure Problem.

The first two options carry various implications. The first of them is standardly connected to belief about conscious existence apart from body, as, for example, after bodily death, and is consistent with the idea of conscious existence in things other than humans – animals, plants, even indeed mountains and rivers, or (as panpsychists hold) the entire universe – though many who think that consciousness does not require a bodily connection also deny that it exists in connection with anything lower in the scale of existence than humans (so, in the medieval conception of the Great Chain of Being, all the levels above humans – angels, archangels, and so forth, are conscious). In the presence of so great a mystery as consciousness there is no control on speculation.

The consensus today, backed powerfully by empirical enquiry, is a version of the second option, with the adjustment that the physical basis consists not in liquor or vapour pouring through nerves conceived as tubes but in the electrochemistry of the highly complex interconnectivity of neurons in the brain.

The word 'conscious' was brought into common English usage by the philosopher John Locke in the second edition of his *Essay Concerning Human Understanding* (1691). He did not coin the term – some etymologists identify earlier uses – but he gave it currency. It is an anglicization of the Latin word *conscius*, 'knowing with', from the verb *conscire*, 'to be privy to' (*con* 'with' + *scire* 'to know'). Locke used the term in his attempt to explain how a person can be self-identical over time, that is, can be the same person at a later time in her life as she was at an earlier time.[1] This is the intriguing problem of 'personal identity'. What justifies our commitment to the idea that you are the same person at age fifty as you were at age twenty, in the sense that at fifty you are entitled to the money you started saving at twenty – it is your money; you are the same person as the one who opened the savings account – while at the same time you are probably 'quite a different person' at fifty (staid, mature, settled) from what you were at twenty (flighty, promiscuous, confused)? Does this run together two different ideas, one about the continuity of your body (it has just got older and fatter but it is the same body, *pace* cell replacement, amputations, and so forth) and the other about the continuity of your personality? Which is *the* 'you'? If both can change in their different ways, what sense does it make to say 'you have changed' unless there is a 'you' who has done the changing – a 'you' who has stayed the same enough to be recognized as having changed – and what is the thing that has stayed the same enough for this to be said?

Locke's answer was that personal identity consists in 'consciousness of being the same person over time'. This consciousness consists in self-awareness, memory, and a self-regarding interest in the future.

You are, so to speak, 'privy to yourself'; you know yourself to be you, and this is a function of memory. His view challenged theologians because it set aside the assumption, until then accepted, that the vehicle of personal identity is an immortal soul, and it prompted disagreement among other philosophers, because it seemed to get matters the wrong way round: it made memory the basis of identity, whereas surely[2] personal identity is the basis of memory – for how can this memory be *my* memory if I am not the same person who had the experience that originated the memory?[3]

The debate sparked by Locke, along with other ideas emerging in the seventeenth- and eighteenth-century Enlightenment, prompted a revolution in thinking about mind and the self. I discuss aspects of this in the next section. For present purposes one notes that Locke's idea that consciousness consists in 'being self-aware as a centre of continuing experience' is only part of the story about consciousness at best, and indeed is too restrictive; for one can be, and often is, conscious without being self-aware, as when one is absorbed in listening to music or working intensely at something. In such cases it takes a visitor from Porlock to recall one to oneself, to become self-aware again.

Nevertheless *awareness* has to be an essential part of the story, and the answer to 'Awareness of what?' admits of two general answers: of an object of attention, and of the felt quality of being in that state of awareness. The first answer relates to the *intentional* character of consciousness, namely, that conscious states are typically directed upon something – at very least (to take a minimal case) on one's mantra or breath in meditation, though there the aim is to achieve a sort of unconscious consciousness that is objectless; hard for all but adepts to do. The second answer relates to *qualia*, the experienced properties of being in a given state of consciousness: what it *feels like* to be in pain, to drink a cold beer on a hot day, to see a red dress or hear a trumpet. Emotional states have phenomenal properties too – what it *feels like* to be happy, sad, excited, tranquil; though here the intentional object, generally an emotion-prompting thought, person, or state of affairs, can be diffuse – as when one wakes on a holiday morning, under no pressure, feeling relaxed and content; perhaps the recognized absence of an immediate object can itself be the intentional object.

The experience of qualia is called 'phenomenal consciousness', and in view of the fact that an experiencer can be aware of his qualia, phenomenal consciousness counts as 'access consciousness' also; the subject can introspect them. Where a subject is conscious – extemporizing on the piano, say – but so thoroughly absorbed as not to be *self-*conscious, is he not accessing his state of consciousness? Surely, we wish to say, the pianist is experiencing qualia; how else would he be able to modulate through the keys as he extemporizes? He hears the notes, and there is 'something it is like' to hear musical notes – so he is accessing his own consciousness without being conscious of doing so. This suggests the need for a further distinction, one between *accessing* and *monitoring* one's own consciousness – the latter being consciousness of one's consciousness, self-reflexive awareness of what one is experiencing – for purposes of reasoning, planning, directing one's activities, communicating with others, and the like.

In neuroscientific investigations of consciousness, introspection by human subjects is an important adjunct to locating brain regions that might be significantly involved; identification of these could be important in determining whether someone is conscious but unable to report the fact, as in locked-in states. Stimulating a region of the human brain and associating it with the subject's reports, and in the case of both human and animal brain stimulation observing associated (caution prevents saying 'resultant') behaviour, establishes strong and often precise correlations. These observations help to narrow the 'explanatory gap' between what can be known about brain states and what occurs in subjective experience. For, although it was recognized that we would still not know how brain events and subjective experiences are *connected* despite establishing correlations, it was thought worthwhile to try to understand something about the correlations nevertheless – for a chief example, something about the neurological conditions in which consciousness is respectively present and absent. This is a more modest aim, but one that might be a first step to achieving the goal of eventual reduction.

For a time in the nineteenth century enquirers were vexed by the question of where in the brain and central nervous system correlations with consciousness might be located. It was, in effect, an update of the mind–body interaction problem but without dualism – so in fact more of an upgrade than an update. Experiments with decapitated frogs that moved their legs away from stimuli made some enquirers think that consciousness resides in the spinal cord as well as the brain. Variations on that theme – noting that decapitated frogs do not try to escape water that is becoming ever hotter – suggested that the spinal reflex was unconscious after all, and that the seat of consciousness must be somewhere in the brain itself, though it was disputed whether it was purely in the cortex, or in the mid- and hindbrain too. By the end of the nineteenth century the consensus had settled upon the thought that it is somewhere above the spinal cord anyway.[4]

In the first half of the twentieth century neuroscience was significantly advanced by the work of Karl Lashley and his successors at the Yerkes Primate Center, sited originally at Orange Park in Florida and now at Emory University. Lashley coined the term 'engram' to denote a physical change in the brain after learning, and studied learning and memory in animal subjects, though not always arriving at the correct conclusion (he thought that the V1 occipital region is the seat of memory). He and his successors at Yerkes observed important correlations between behaviour and lesioned brain areas, especially in the temporal lobes, which demonstrated that they have a role in learning and memory. Among other things their work showed that visual processing is not confined to V1, and that 'working memory' is supported in the prefrontal cortex. These were significant advances.

One approach to investigating the connection between brain regions and conscious experience was to alter consciousness and see which parts of the brain were implicated. Accordingly, the Chicago University psychologist Heinrich Klüver administered mescaline to rhesus monkeys and observed their behaviour. Mescaline is a psychedelic drug with effects similar to LSD and psilocybin ('magic mushroom'), derived from the peyote cactus and long traditionally used by Native Americans in Mexico. Klüver took it himself, enabling him to describe some of the visual effects, one of which is seeing 'cobweb figures' and other recurrent shapes.[5] Klüver noticed that under mescaline's influence his monkeys behaved somewhat like epileptic sufferers, and invited the neurosurgeon Paul Bucy to collaborate with him in exploring the brain regions this suggested, namely the temporal lobes, a common originating site of epilepsy's electrical storms. Bucy produced lesions in the monkeys' temporal lobes on both sides, dramatically altering their behaviour; they exhibited what Klüver and Bucy called 'psychic blindness', or loss of discrimination, ceasing to be afraid of snakes and people, eating substances they would previously have avoided, becoming sexually promiscuous, and seeking to mate with animals of other species as well as their own.[6]

The 'Klüver–Bucy Syndrome' came to be recognized in human patients with bilateral damage to their medial temporal lobes and amygdalae. Docility, 'pica' (eating inappropriate things), hyperorality (the desire to explore objects with the lips and mouth, as infants do), hypersexuality, and a diminished capacity to recognize familiar things are symptomatic of the condition and were what Klüver and Bucy saw in their monkey subjects. Further studies of the temporal lobes' roles followed, notably by Brenda Milner, who contributed substantially to understanding learning and memory in her studies of epilepsy sufferers who had undergone temporal lobectomies.[7]

The growing evidence from lesion studies, split-brain, and amnesiac patients, and then the impending arrival of technologies offering minimally invasive or non-invasive ways of investigating brain function, prompted a resolve to make consciousness itself the target of scientific study. A consolidating moment occurred in the early 1990s in papers by Francis Crick and Christof Koch, who argued that, because so much is known about how vision is processed in the brain, the visual system could be a useful place from which to embark upon an empirical search for consciousness.[8] Despite the intractability of what the philosopher David Chalmers called the 'Hard Problem of Consciousness', namely, explaining how the experience of qualia arises from brain activity (the 'easy problem' is describing the role of brain regions in perception, learning, memory, and emotion), Crick and Koch advocated searching for 'the neural correlates of consciousness' as at least a first step.[9]

The Crick–Koch call to arms was made at just the time a well-known professor of psychology, Stuart Sutherland, was announcing in *The International Dictionary of Psychology* (1989) that 'Consciousness is a fascinating but elusive phenomenon: it is impossible to specify what it is, what it does, or why it evolved. Nothing worth reading has been written on it.'[10] In the decades since, that situation has changed dramatically.

For neuroscience to provide an explanation of consciousness requires effective psychological and computational models for interpreting neural data, so that conscious states can be identified as such, and – more importantly still – so that their *content* can be specified. This latter would be the really big breakthrough. The lesser goal of establishing correlations is achieved when third-person observation systematically aligns brain states either with behaviour or with first-person introspective reports, or of course both, as mentioned above; but an *explanation* of the key aspect of the Hard Problem, namely *how consciousness arises* from neural activity, would be attained when third-person observations are sufficient by themselves to identify not just the occurrence of conscious states but their content and how they have been caused.

A start on the first goal – of being guided by the neural data in distinguishing between conscious and non-conscious mental states – is provided by the fact that the existence of non-conscious processing of representations is observable, as in blindsight: subjects can respond to visual data while reporting that they cannot see. This makes it meaningful to look for what else is going on that makes states conscious when they are so, the 'extra' that distinguishes conscious from non-conscious sensory processing – in this example, the differentiator between sighted and blindsighted responses to visual cues.

Various theories have been advanced. The 'Global Neuronal Workspace' model, as described by its first proposer Bernard Baars in 1988, is that 'Consciousness is accomplished by a distributed society of specialists that is equipped with a working memory, called a global workspace, whose contents can be broadcast to the system as a whole.'[11] The 'specialists' are computational units, not brain regions; the content of a short-lived working memory is sent to all the specialist units so that they can jointly contribute to processing that content. On this view consciousness is an integrative function, a 'brainweb' consisting of 'a massive parallel distributed system of highly specialized processors'.[12]

In their development of the Workspace Model, Stanislas Dehaene and his colleagues focus on the multiplicity of neural networks active in the brain, unconsciously processing information in parallel. In their view the information becomes conscious when the neural circuits that represent it are 'mobilized by top-down attentional amplification into a brain-scale state of coherent activity that involves many neurons distributed throughout the brain'. They call the neurons that are connected to each other over long distances 'workplace neurons' and think of them as making information globally available to the participating networks.[13] In both the Baars and Dehaene versions of a brain 'workspace' the key idea is that consciousness is a function of integration at a global level of activity. Note that both versions say what consciousness is *functionally* and, in general terms, that it is caused by the whole brain or large regions of it, but not *how* the qualia experience itself — colour, taste, the felt quality of sensation — arises. In the quotation from Dehaene, mention is made of 'top-down *attentional* amplification' — which a critic might say looks like smuggling in what the model is supposed to explain.

An alternative, more localized model is 'Recurrent Processing' in sensory areas of the brain connected to each other and sending information back and forth in feedforward, horizontal, and feedback sweeps of information. The feedforward connections are those that go from earlier to later phases in the processing of information, as noted above in connection with visual input, where data processed in V1 is distributed onwards to temporal and parietal regions for finer-grained interpretation as shapes and surfaces. Feedback occurs when the higher regions return data to those lower in the chain; horizontal connections are those within cortical regions. Recurrent Processing involves the two latter types of connection in several levels, the processing at the highest levels constituting consciousness of the content. As in the Workspace Model, integration is key, the difference being that this model does not attribute consciousness to the whole brain but only to the neural circuits involved. The chief proponent of this view is the Amsterdam University neuroscientist Victor Lamme.[14]

A different approach is offered by 'Integrated Information Theory', which takes consciousness itself as the starting datum and seeks to infer what properties a physical system must have in order to give rise to it. To constrain what these properties must be, the essential properties of consciousness itself must first be specified. These are described as 'axioms', whereas the essential properties of any supporting physical system are described as 'postulates' merely, because such systems are inferred and need not be biological. The five essential properties of consciousness are that it is *real, structured, specific, definite*, and *integrated*. Whatever physically underlies it has to be causally adequate to these properties. (There is a mathematical model for information integration that illustrates how the structure can satisfy the requirement to produce integrated and definite outcomes.) In taking the experience of qualia as its starting point, the theory explicitly *defines* subjective experience as 'a system's capacity to integrate information', and accordingly treats it as 'a fundamental quantity, just as mass, charge or energy are. It follows that any physical system has subjective experience to the extent that it is capable of integrating information, irrespective of what it is made of.'[15]

Although thought-experiments are used in setting out this view, a considerable body of empirical evidence now exists to support the general themes discernible in the theories just sketched – namely, that consciousness consists in the integration of information achieved by different neural circuits interconnecting, 'talking to each other', in ways that can be observed; for example, using transcranial magnetic stimulation (TMS) and electroencephalograph (EEG) readings, one can see a wide and enduring distribution of cortical responses to a TMS impulse (which gives the brain a magnetic nudge or tap) in an awake subject, when in a sleeping subject the response to the same impulse is limited and brief.[16] This technique offers a way of measuring the degrees of consciousness in various states, such as wakefulness, rapid eye movement (REM) sleep, deep sleep, and coma, with clinical possibilities for detecting consciousness in vegetative and locked-in states, and for communicating with sufferers by tracking neural responses to questions and other stimuli (such as responses to medical

treatments).[17] TMS has also been used to interrupt top-down feedback in the visual system, preventing subjects from being aware of
the visual input being received by lower levels of the system.[18]

In short, the evidence makes it plausible to generalize the thought
that unconscious processing of sensory information is restricted to
the relevant sensory areas, whereas conscious processing is distributed throughout the cortex through the interconnections among
otherwise individually specialized neural circuits.[19]

This empirical data also supports the idea that consciousness is
intentional. Experiments show how the brain contributes to experience by filling in gaps, introducing colour contrasts into the visual
field where there are none but when imagining that they are there
helps in judging perspective or distance, interpreting sensory data
according to expectations, making a 'best guess' about what sensory
data indicates about the physical environment. This has confirmed
a suggestion made more than a century ago by the psychologist
Hermann von Helmholtz that the brain is a prediction device, a probability machine, drawing inferences from the data it receives to
construct a theory of how things are in the world around it and using
its complex feedback mechanisms to check, adjust, and minimize
error in its predictions.[20]

The idea that the brain engages in high-level integration of information being traded around its component neural systems is no
surprise, because one thing that would result is weighting of the
probability calculations offered by the systems involved, to the obvious survival and reproductive benefit of the brain's owner. But does
this have to involve *consciousness*? One can imagine a being whose
neural circuitry is extremely sensitive and responsive for survival
reasons, without requiring of it that it have self-reflexive awareness
of the fact that the sensory inputs, and the responses to them, are
occurring – and, still less, a need to know subjectively what it feels
like to have them. Indeed one thing neuroscience makes clear is
that *most* such processing occurs in all animals, including humans,
without awareness of it anyway – that they benefit from being unconscious, for we know that being aware of what we are doing when, for

example, we play a tennis shot we do it less well. The question of what consciousness adds, which is closely linked to the question of why it exists, is not answered by the theories just sketched. They do not resolve what consciousness researchers call the 'Zombie Problem': that all animals respond selectively to the impacts upon them of their environments, pursue goals, avoid dangers, and do everything they do in ways that could in principle be explained without having to impute any degree of consciousness to them. Thus – again – the idea that consciousness consists (from the neural point of view) in high-level information integration says nothing about why it occurs, still less *why it feels a certain way* for that integration to be taking place; and it says even less still about *how* it occurs physiologically.

The idea of computation is implicit in all these theories. Nobel laureate Roger Penrose rejects computational models and argues instead that different thinking is required for consciousness to be explained. To reject computational models has significant implications, because upon their basic principles turn two important ambitions. One is the development of artificial intelligence via machine learning; the other is the Human Connectome Project. If Penrose is right, AI would never be able to shake off the 'A', however smart it became, and the Human Connectome Project's aim of exhaustively representing mental operations and properties in terms of neural interconnectivity would be incomplete, if not impossible.

Penrose begins by rejecting the applicability to mind of the notion of 'computation' itself. As the term suggests, computation is what computers do (more accurately, it is what mathematically idealized computers called 'Turing Machines' do). It consists in the ordered running of specified procedures. In Penrose's view, even the most sophisticated computational models cannot simulate consciousness, because the latter has something fundamentally non-computational about it. He takes this to follow from Kurt Gödel's proof in logic that no set of rules for proving propositions in some formal system can ever be sufficient to establish all the true propositions of that system. Penrose takes Gödel to have shown that no set of proof-rules can ever prove all the propositions of, say, arithmetic that humans can know to be true. From this he further takes it to follow that human thinking is not a form of computation.

What, then, is the alternative? Penrose argues that it is to be found at the level of quantum events in the microstructure of the brain. A number of other people have made similar claims, not least because mysteries are good things to hand-wave at when one is short of explanations, and quantum phenomena certainly appear mysterious. But Penrose is a distinguished physicist with major contributions to quantum theory, mathematics, and cosmology – most famously, perhaps, his work with Stephen Hawking on black holes – so, among all those who appeal to the quantum level as a possible site for consciousness to emerge, he has some of the best credentials. His proposals have, however, been met with a near-universal chorus of disagreement.

In collaboration with the anaesthesiologist Stuart Hameroff, Penrose suggests that a process called 'Objective Reduction' occurs in the particle constituents of microtubules in the brain to give rise to phenomena both of consciousness and free will. Microtubules are large molecules of a protein called tubulin that serve as the scaffolding of cells, giving them their shape and engaging in a number of cellular activities, such as molecular assembly and cell division. Hameroff proposed that electrons in the microtubules could form a condensate in which quantum effects occur, giving rise to what Penrose calls *objective-reduction collapses* of the wave function, which in his version of quantum theory are simultaneously non-computational and non-random.[21]

Controversy over the Penrose–Hameroff theory includes the assertion that the brain is 'too wet, warm and noisy' for wave-function collapses to play any role in neural events. Neurophysiologists dispute the accuracy of the theory's account of cellular structure in neurons and their relationship to glial cells.[22] Recall that the charge against Descartes was that he had hidden the mind–brain connection inside the pineal body in the brain. That charge could be iterated here, with microtubules serving as the latter-day hiding place; but at least in the Penrose–Hameroff suggestion something a little more circumstantial is offered, namely, the role played by quantum states of electrons in the atoms constituting microtubules. This is not much of an advance: a quantum choreography of wave-function collapses still does not say how it is we subjectively experience colours, fragrances and the rest. If the theory turns out to be pointing in the right direction, even if not correct as it stands, one consequence is that consciousness might be not be confined to animals: for quantum states underlie everything we know about the physical universe.

In debates about the Cartesian legacy of the mind–body problem mentioned in Section 1 above, certain 'heroic' solutions to it were described. Given the difficulties implied and apparent in the foregoing, the problem of consciousness itself invites heroic solutions. One – flying in the face of what would seem the most obvious of facts – is the 'Eliminativist View', that consciousness does not exist. Another – this one adopted by David Chalmers, because the Hard Problem seems intractable without reaching out for more imaginative and radical solutions – is to accept the possibility (just hinted in the preceding paragraph) of some form of panpsychism, the universal and fundamental ubiquity of consciousness, thus – as in Tononi's view – starting with consciousness as the basic fact instead of trying to determine how it arises as a special and unusual case from certain kinds of non-conscious matter.[23]

The Eliminativist View is associated, not quite accurately, with Daniel Dennett.[24] He does not say that it does not exist, but he does call it an 'illusion', more precisely, a 'user illusion', like the icons on the screen of a computer. He makes use of two resources in advancing this view. One is the idea of the brain as billions of little non-conscious elements working in concert to process the information it receives from sensory pathways. The brain's components are not conscious; by composition therefore the brain as a whole is not conscious; what its operations give rise to is an illusion of consciousness. The other resource is the fact of perceptual illusions and magic tricks that make things *seem* to be what they are not; Dennett makes good use of optical illusions to illustrate the way the brain makes things up, fills in blanks, sees motion where there is none, makes different things seem the same and the same things seem different. He applies these two ideas in support of the idea that consciousness is a marginal and relatively insignificant feature, perhaps (but he does not say this in these terms) an epiphenomenon, or by-blow, of what really does the work of mind: the billions of little non-conscious elements.

Dennett's first resource – the billions of little elements idea – is uncontroversial at least for most neuroscientists and philosophers. Accepting it does not, however, make the Hard Problem of Consciousness go away, as the foregoing shows. The relevance of an appeal to illusions and magic tricks in solving the Hard Problem is unclear. Yes, the brain fills in, interprets, adds, subtracts and deceives, sometimes because it is useful (in a telephone conversation a proportion of what someone says is lost, but the brain supplies it) and sometimes because the way the visual systems work makes them susceptible to being tricked. If anything the latter case speaks *for* the existence and utility of consciousness: being aware, knowing, that one is the subject of a visual illusion could be more than interesting or entertaining, it could be life-saving. Being *unaware* – not representing to oneself that what one is seeing is not what is really happening – could be disastrous. One answer to 'Why does consciousness exist?' might therefore be that it provides opportunities to correct and override what non-conscious mental operations are saying. But, in any case, if 'consciousness is an illusion', to whom or what is it an illusion?

The idea that consciousness provides opportunities to correct and override – and, by implication therefore, monitor – what non-conscious mental operations are saying in turn elicits thoughts about choice, volition, and conceptions of mental life that play a major role in our ideas about human nature and its moral and social dimensions. Neuroscience might disclaim being in a position to talk – yet – about these kinds of phenomena, but at least one thing is worth mentioning in connection with them: that the historical evidence of change, sometimes dramatic, in moral outlooks and social arrangements suggests that these things are not determined by the evolved neural mechanisms of the brain, which is why the idea that mental and social phenomena, as products of consciousness, are reducible to sums over billions of non-conscious events provokes scepticism. We can account for the role that experience plays in individual and social change; why not also recognize persuasion, discovery, serendipity, accident, the vectors created by the interaction of many social and historical forces colliding?

Dennett's view of mental phenomena as issuing from the concerted activity of large numbers of unconscious subunits is very similar to a number of others of the same general tenor. One is the 'Society of Mind View' put forward by Marvin Minsky in a 1986 book of that title, in which he describes mind as consisting in the connections between many individually mindless simple parts, which he calls (misleadingly, since they are mindless) 'agents'. The individually mindless components of Robert Ornstein's theory constitute mind as 'a squadron of simpletons'.[25] The views of Dennett, Minsky, and others all emerged at about the same time – which was, significantly, the period when the computer metaphor was at its most persuasive.

In the theories canvassed here, the concept of *emergence* has played no part, though it is consistent with them. Empirical evidence for emergent properties might come from investigation of how the higher feedback mechanisms – those already identified as integrating information in neural circuits – function differentially in response to challenges that have different kinds of meaning for the subject; that is, responses that require *thought* – analysis, reflection, bringing knowledge to bear – which is in short to say: where *mind* is in play, in the sense of what we describe most economically, and understand most clearly, when we employ the concepts of our everyday 'theory of mind' – the familiar way we interpret others' behaviour and intentions, an informal theory about beliefs, desires, feelings, memories, and reasons, which we take it are to a significant extent known to the agent himself or herself. In neuroscience, vague references to the 'mind-brain' sometimes occur as an acknowledgement – or as a side-stepping of the issue – that ultimately what we wish to understand is mind in this 'theory of mind' sense: mental life, thought, knowledge, reason, feeling, the passions and anxieties. In addressing the brain as the basis of mental phenomena, neuroscience is proving markedly successful in identifying the neural correlates of cognition. But the mind in general, and consciousness in particular, is yet to be integrated into that picture more fully.

4. The Mind and the Self

The problem alluded to at the end of the previous section can best be explained as follows. Imagine two people observing a series of events on a field. One of them is a physicist. She describes the events in terms of mass, momentum, frequencies of radiation, the principles of mechanics. The other is a sociologist. He describes the events as a football match. They use two different languages for two different purposes. Both are accurate in their own terms. The great question is: what is the relation between the two languages?

Reductionism says that the language of sociology can be and will be translated without remainder – without leaving anything out – into the language of physics one day. The problem is the 'without remainder' claim. The events have a meaning to the football match's participants and spectators that it is hard to imagine, at present, being captured in the language of physics. Reductionists respond by saying that translation will proceed as follows: sociological concepts will be redescribed in the language of neurology, in which representations of meaning are identified as complex events in brains, whose constituent neurons will then be described chemically, the chemical descriptions ultimately being reduced to the terms of a quantum, string, or successor theory.

Broadly two alternatives offer themselves. One is that this reductionist claim is true. The other is the emergentist claim that levels of organization of phenomena may have properties that other levels do not have, and that vanish as one descends, or ascends, to those other levels. To illustrate their thesis, an emergentist might cite the property of 'being alive'; an animal has this property when all its component anatomical parts are correctly connected and physiologically functioning, but the individual limbs and organs do not have this property when the animal is dismembered.

Consider a spectator at the imagined football match. Such

concepts as 'goal', 'foul', 'penalty', 'referee', 'off-side', 'captain', 'centre forward', and the relationships between them – the referee's recognition of a foul and awarding a penalty, his doing so incorrectly, the significance of the relative number of goals scored by each team, the role of a football team captain – are indeed represented neurally in his brain. At any moment in the game he has the whole thing 'present to his mind', as the phrase has it, understands what is going on, cares about it and the outcome, recognizes the possible implications of each occurrence in the unfolding sequence of events. The question of how this complexity is instantiated in, and managed by, his brain is an empirical question, which could be answered once the connectome is mapped. Another and very different question concerns how this complexity has meaning and importance for the owner of the brain. Indeed the terms of the implied relationship, the meaning and the thing it is meaningful *to* or *for*, are unequal from the point of view of resources we have with which to understand them. We say meaningful to or for *a person* (and sometimes we say *to a family, to a nation*, or any collective of persons with interests *as* a collective), and this makes sense. But the 'meaning' itself – the significance, import, value, difference-made – is less easy to characterize.

In order to get some purchase on this, let us turn the telescope round and look through its other end – not from the neurological but from the mental end.

One way to think about minds is to broaden the perspective from what is exclusively inside heads to include the physical and social environment that surrounds heads. This idea is prompted by the thought that what we know when we understand a concept has, in most cases, to involve a connection between a brain event and something in the world. For an obvious example: to understand the concept of a tree, and to be able to distinguish between trees and other things – dogs and buildings, say – the relevant physiological occurrences inside the head have to stand in a determinate relationship with trees and non-trees outside the head – even if indirectly: someone who had never encountered a tree in perceptual experience might have indirect knowledge of them through books and pictures. But a less obvious aspect of having a concept of trees is that whenever we think or talk of trees, the relationship between what is happening inside our heads and trees outside our heads has to remain in some form, in order for our discourse correctly to be about trees rather than some other thing. Nothing mysterious or magical is implied by this; it just means that to explain the thought of a tree, as distinct from a thought of anything else, reference to trees out there in the world is unavoidable. You could not 'individuate' the thought of a tree in someone's mind – tell it apart from thoughts about other things – without reference to trees, and without reference to one or other route (mainly perceptual) by which the thought could be a thought for that thinker.

The idea that thought is thus essentially connected to the outside world is intended to illustrate the more general idea that 'mind' is not solely describable in terms of brain activity alone, but must be understood as a relationship between that activity and the social and physical environment external to it. Philosophers give the name *broad content* to thoughts that can be properly described only in terms of their thinkers' relations to these environments, and some argue that there can be no such thing as *narrow content* – that is, thoughts specifiable independently of their thinkers' environments but just in terms of what is going on inside the skull. If it is right that there can be only broad content, the implications are very great; it means that understanding minds involves much more than understanding brains alone: it involves understanding language, society, and history too.

This is an idea that seems obvious enough when exploring anything to do with culture and creativity. In seeking to understand an artist, say, we naturally look to the influences, the historical circumstances, the experiences that contributed to shaping him or her. Yet, in thinking about how brains produce mental phenomena, we tend to downplay the significance of the environment of the head in which the brain is housed. Consider: even in connection with the psychophysiology of perception, the nature of the distal end (outside the head) of the perceptual relationship is specified only minimally, and the equipment at the proximal end (inside the head) – optic nerves, the visual cortex – is the main focus. Of course, to understand vision it is necessary to have a detailed specification of the neural activity in which it consists; and perhaps there is no more to be said about what stimulates the retina's rods and cones and the firing of the optic nerves than that incoming light propagates at such-and-such frequencies. But understanding *vision* is not yet understanding *visual perception*. For recall that we never just see but always *see as*; concepts are always deployed in visual experience; and these concepts concern what exists beyond the lens of the eye, out there in the surrounding world. Recognizing this is the motivation for saying that understanding mind, as opposed to the brain that gives rise to it, requires a broad-content approach.

To say in this way that minds are not only brains is not – to emphasize the point again – to intend anything non-materialistic by this remark; minds are not some ethereal spiritual stuff *à la* Descartes. What it means, rather, is that minds are the product of interactions: between brains and other brains, and between brains and the natural environment. As essentially social animals, the brain–brain interactions, and the complex social reality they jointly constitute, are probably the most significant environment for most people. People are nodes in complex networks from which their mental lives derive most of their content. That an individual mind is, accordingly, the product of interaction between many brains is not something that shows up directly on an fMRI scan. The historical, social, educational, and philosophical dimensions of the constitution of individual character and sensibility are vastly more than the electrochemistry that an individual brain generates by itself without external input.

It follows that, although neuroscience is an exciting and fascinating endeavour in the process of revealing a great deal about brains and the way aspects of mind are instantiated in them, it cannot teach us *everything* of what we would like to know about minds and mental life. Yale psychologist Paul Bloom put his finger on the nub of the issue when he commented on neuropsychological investigations into the related matter of morality. Neuroscience is pushing us in the direction of saying that our moral sentiments are hard-wired, rooted in basic reactions of disgust and pleasure. Bloom questions this by the simple expedient of reminding us that morality changes. He points out that those who were reading his article 'have different beliefs about the rights of women, racial minorities and homosexuals compared with readers in the late 1800s, and different intuitions about the morality of practices such as slavery, child labour and the abuse of animals for public entertainment. Rational deliberation and debate have played a large part in this development.'[1] As Bloom notes, widening circles of contacts with other people and societies through a globalizing world are formative in this; for example, we give our money and blood to help strangers on the other side of the world. 'What is missing, I believe, is an understanding of the role of deliberate persuasion.'

Contemporary psychology, and especially neuropsychology, ignores this dimension of the debate, not through inattention but because it falls outside its scope. Experiments on decision-making, for example, focus on simple time-defined choices, and they record brain activity assumed to constitute the decision before the subjects report the decision themselves. Leave aside the fact that there are questions about what this research really shows, and note that even if portable headset fMRI scanners could track correlations and time-delays in brain activity associated with thinking about a marriage proposal or which college to apply to, the correlations would not constitute an account of the decision process itself. Put this together with the thought that mind is more than brain, and the scale of the task in understanding one feature of mental life – making choices – becomes yet more apparent. The complexity of understanding brains and minds – and the qualities of wisdom, wit, intelligence, perceptiveness, maturity, ability, and more (and on the other side: resentment, bitterness, prejudice, hostility, hatred, and more) exemplified by some of the latter – does not entail that questions about these topics are permanently unanswerable. Rather, the endeavour itself forces us to think afresh about what questions we are asking and what phenomena we are investigating. On the great questions of mental lives and their character and qualities as we live them, we are more likely to learn about them from literature, history, and philosophy than from neuroscience. This is not to downplay the importance of neuroscience, far from it; but it is to suggest that enquiries at the level where mental, moral, and social, as opposed specifically to neural, phenomena are at issue, the connection has to be made.

This connection is indeed made, according to some, by the new field of *neurophilosophy*. As earlier sections showed, questions about the relation of mental states to brain states, not least in respect of subjectivity, consciousness, and representation, have been central to the philosophy of mind ever since dualism and non-materialist monisms (various idealisms and 'neutral monism') were rejected as serious possibilities. The proponents of neurophilosophy hold that a raft of further questions – about morality, intention, free will, selfhood, and rationality – are now amenable to investigation with the greater empirical depth offered by neuroscience's research technologies. The armchair speculations of traditional philosophical enquiry, they say, can yield place to something with a more solid basis and some surprising and suggestive findings already to hand. Even before fMRI studies on decision and volition began to suggest that these are preconscious processes, it was known that people whose brain hemispheres had been separated by commissurotomy seemed to have two sometimes competing centres of selfhood; and studies of brain chemistry have provided insights into the nature of mental disturbance, emotion, and social bonding.

These are good points. It is therefore not to be either sceptical or critical of this project to say that, nevertheless, a sense of proportion has to be maintained about its philosophical promise. For when one thinks about persons, their characters, what they know and believe, the frameworks of concepts that organize their views, their attitudes and responses to the world, and the way they give weight to competing reasons for action in it, the neurophilosophical approach is only part of the story, because in principle it cannot be the whole story. The reason has already been given: minds are more than brains in the sense of having to be understood 'broadly'. In saying that individual minds are more than the individual brains that instantiate them, one is once again describing them as the product of continuous feedforward and feedback interactions with parents, teachers, their communities, and physical environments. Put the point this way: to say that a mind is a brain plugged into two kinds of environment, social and physical, is to say that a brain that is not thus plugged in is not the seat of a mind. A description of the brain that leaves out the context of its functioning and how it acquired the nurture part of its character is not the description of a mind.

There is a simple empirical observation that lends credibility to this perspective. It is that brains are needed only by living things that move and sense their way about. A map of areas of the human brain responsible for movement, when metamorphosed into a model of a human, shows huge hands and a huge mouth; far more of the motor cortex is devoted to these areas than to any other in the body. Another empirical correlation is that the human brain consists of layers of its own evolutionary history, with more primitive structures (shared with other animals) overlaid by a large cerebral cortex responsible not just for sensory experience and movement – and the complex interactions between them – but the mental operations of thought and reason, the use of language, and the degree of social complexity that a human typically navigates, a complexity perhaps not distinguished in *kind* but certainly in *degree* from other animals.

An important corollary to thinking about mind as an irreducible category – for that is the implication of the foregoing remarks – is the role played by a concept of the *self*. John Locke's concern about personal identity has an interesting dimension to it: that it focuses upon the notion of a *person*, chosen in order to avoid making the principle of continuing identity dependent on body, which grows, changes and ages, and which might be so affected by, say, an accident that the memories, character, and aims associated with that individual might vanish and be completely replaced, even though the body is the same. Recall what was said of Phineas Gage in just such a case. Locke's choice of the concept of a 'person', rather than that of a 'self', was deliberate. A *person* is a *forensic* entity, that is, a being that is the bearer of moral and legal rights and responsibilities. A normal adult human is a person, a commercial company is a person, a baby is not a person (it has rights but no responsibilities yet), and a demented human has ceased to be a person except by the courtesy our concern and compassion prompt us to accord him. These are matters of definition. For Locke, self-awareness of being such a thing requires memory, and memory entails continuity of that self-awareness; hence his choice of 'person' as the entity that persists. It follows – and he accepted the entailment – that if a person ceases to remember earlier phases of life, he is no longer that person; the chain of his identity has broken.

The theologians who placed all three properties of identity, self-hood and personhood in a 'soul' were, as noted, offended by the suggestion. Much of the discussion that followed Locke's proposal set aside the forensic aspect of personhood and thought wholly in terms of a *self*, somewhat question-beggingly because it reintroduced the idea of a metaphysical entity not much different from a soul, and therefore easy to reject – as Locke's eighteenth-century successor David Hume did, in asking whether you would find a 'self' among your current perceptions and feelings if you looked within yourself. His answer was: no, you would not; you would find only a temporary bundle of the aforesaid perceptions; and he took this to be an empirical refutation of the idea of a self, which he therefore described as a useful fiction allowing us to imagine that we remain self-identical over time. His theory is accordingly known as the 'Bundle Theory of the Self'.

Hume's views about the self and personal identity were published fifty years after Locke's theory. In between there was a great controversy over the matter, not just between Locke and the theologians (especially a bishop called Edward Stillingfleet) but in the reading public at large. It became such an issue that the *Spectator* magazine in 1712 demanded that there should be a conference of 'all the wits of the Kingdom' to decide the question of what personal identity consists in.[2] Shortly afterwards a group known as the 'Tory wits' ('Tory' then had different connotations[3]), including Jonathan Swift the satirist, Alexander Pope the poet, John Gay the playwright, John Arbuthnot the queen's physician, Viscount Bolingbroke the statesman, and others, decided to spoof the debates of the day by writing a book called *The Memoirs of Martinus Scriblerus*, the hero Martinus being an enthusiast who dabbled in all the debates but not very intelligently in any. A section of the book is devoted to the question of personal identity in a way that would today be regarded as in poor taste, involving a pair of Siamese twins, one of whom Martinus marries – hence the dilemmas that arise – but that marshals the arguments brilliantly.[4]

The longer-term legacy of the selfhood debate has arguably been consequential. Despite Hume's outright rejection of the concept, a significant feature of Romantic attitudes to artistic ownership of creativity's outcomes can be viewed as turning on a full-blooded embracing of the idea of selfhood. Consider the idea of 'genius': originally a genius was a being who leaned over one's shoulder and whispered inspirations ('in-spir-ation' means 'breathing-in') into one's ear. The Romantic conception takes the genius into the artistic self and identifies the two. The opening lines of Swinburne's *Hertha* could serve as a motto for the creative self: 'I am that which began; out of me the years roll; out of me, God and man.' The idea of the self had, one might say, been democratized; everyone was or had a self, was fully an individual, a status once attributed chiefly, if not only, to great heroic figures who stood out from the crowd.

It is not a great distance from the idea of selfhood to the idea of its having unreflective, unaware depths. An indirect influence in this respect stemmed from the thinking of an older contemporary of Locke, the philosopher Baruch Spinoza. In the final two books of his great work *Ethics*, respectively named 'Of Human Bondage' and 'Of Human Freedom', he speaks of how we can be held in thrall by unclear, half-formed, imperfectly realized ideas, unconsciously or only half-consciously grasped, from which we free ourselves when we understand those ideas and ourselves clearly, and see the truth. It is an interesting speculation how much the idea of the self and the individual, conscious and otherwise, in the 'Masters of Suspicion' of the nineteenth and early twentieth centuries – Marx, Nietzsche, and Freud – owes itself to the idea of selfhood, individuality, and the person since Locke.

All this is mentioned because in discussion of the mind one also has to address the question of the self – of the sense of self, self-awareness, the perspectival nature of experience in which each individual is at the centre of a universe in space, time and interpretations of the personal and social meanings of what happens in that experience. To the extent that neuroscience explores the basis and nature of mind, it is committed to saying something about selfhood also. Indeed it has to: the experience of 'being in the world' is a key dimension of conscious mental life, and it relates not only to perception, proprioception (the internal awareness of one's body, comprising both what is happening in it and what one is doing with it), and intentionality, but to the contexts of information being processed by the individual in these respects.

The concepts of 'integration' and 'emotion' both spring to mind, and connect with each other, in thinking about how to unpack the idea of the self. In the computational theories of consciousness sketched above, integration of information plays a key role, and this is reprised in theories such as Michael Gazzaniga's of the self as an 'Interpreter'. For other theorists such as Antonio Damasio consciousness has its source in the (initially ill-defined) *feeling* of being a self. Both the idea of a self-monitoring conscious function and the idea of an emotional seat of self-awareness are high level from the perspective of neuroscience's present targets of research, but they look less intractable from theories hospitable to emergent properties, and less so still from what might be called *operational* theories in which the concept of the self, understood in Gazzaniga's or Damasio's terms, plays a powerful explanatory role. Whether the emergentist or operationalist perspective (they are not incompatible) persuades, at some point a completed science of the mind has to include an account of why a sense of selfhood is such a central feature of experience.

As the title of his book *The Feeling of What Happens* suggests, Damasio takes consciousness to consist in a distinctive kind of feeling: the 'feeling of feeling'.[5] Consider its emergence developmentally. Feeling that we are feeling is a primitive level of selfhood, a strong, persistent if at first vague awareness of occupying what we later regard as our own uniquely first-person perspective. The self and its objects – the things that prompt emotional responses in us – come to constitute a relational model of the world; at this point consciousness has advanced from a feeling of feeling to a feeling of knowing. For Damasio this offers a handle on the central phenomenon of conscious selfhood: the sense we each have of being the owner and viewer of a movie-within-the-brain that is our own aware experience, representing a world to us of which we are the centre.

So much is familiar; the interesting part is that neuroscience helps

to flesh out the account. A phenomenon Damasio says he found striking in the data on pathologies of consciousness was that some patients
can be awake and aware of their surroundings enough to interact
with them, yet in non-conscious ways, showing that consciousness is
not the same thing as mere wakefulness or sensitivity to stimuli. To
understand the extra dimension that is consciousness one has to
identify how it would confer survival advantage; otherwise higher
mammals would not have evolved it. Damasio's suggestion is that the
appropriate utilization of energy, and the protection of the organism
from harm, which are chief goals for any living creature, are much
enhanced by an organism's ability to place itself in a map of its environment, and to make plans and judgements about the best courses of
action in relation to it. Creatures that are automata – although aware
and sensitive to their environment – might do this well enough but
not as well as genuinely conscious creatures.

Damasio accepts the evidence from the kind of neurological and
neuropsychological data, described above, indicating that there is a
degree of localization of mental capacities and that much mental processing happens at non-conscious levels, but he adds the claims that
the same evidence supports the idea of different levels of consciousness and that consciousness is not one but many things. Accordingly
he distinguishes between *core consciousness* and its associated primitive
sense of self, and the higher-level phenomena of *extended consciousness*
and its subject, which he calls the *autobiographical self*. On this view
consciousness is not to be identified with the cognitive functions of
language, memory, reason, and attention, nor viewed as constituted
by them, but regarded as more fundamental because presupposed by
them.

An aspect of Damasio's previous work that is an element in this
theory is that emotions are fundamental both to consciousness and
reasoning.[6] Deficits of consciousness in brain-damaged patients are
always accompanied by deficits in emotional capability. Damasio also
discovered that brain damage that destroys the capacity to feel certain emotions can similarly result in impaired reasoning; just as too
much emotion interferes with logic, so does too little. But it is the
direct point about the relation of emotion to consciousness that is

most intriguing, for to locate the origins of consciousness in feeling is to say that emotion lies at the basis of thought and personal identity.

Gazzaniga developed his theory of the 'Left-brain Interpreter' from work done with the Nobel-winning neuropsychologist Roger Sperry – described above – on people who had undergone commissurotomy procedures separating their cerebral hemispheres.[7] Prompted by observations of the kind described above, Gazzaniga conducted experiments with, in the first instance, three patients – one of them both before and after commissurotomy had been performed – and derived his conclusions from these observations and the later work they inspired. An accurate account of them, and by extension of the conclusion expected to follow from the full reductionist potential of neuroscience, is provided by the opening summary of an interview conducted by Shaun Gallagher with Gazzaniga in the *Journal of Consciousness Studies*: 'Psychology is dead. The self is a fiction invented by the brain . . . Our conscious learning is an observation *post factum*, a recollection of something already accomplished by the brain. We don't learn to speak; speech is generated when the brain is ready to say something . . . We think we're in charge of our lives, but actually we are not.'[8]

On this view, the fiction of a self is an emergent property of the highly modularized nature of brain activity. 'Highly modularized' means that there are very many localized functions operating at different levels in the brain, and all the work of cognition and emotion is done by them, without conscious input. The brain is in charge; the Interpreter, located in the left hemisphere, provides post facto justifications.

If you were to have asked me why I had jumped, I would have replied that I thought I'd seen a snake. That answer certainly makes sense, but the truth is I jumped before I was conscious of the snake: I had seen it, I didn't know I had seen it. My explanation is from post hoc information I have in my conscious system: The facts are that I jumped and that I saw a snake. The reality, however, is that I jumped way before (in a world of milliseconds) I was conscious of the snake. I did not make a conscious decision to jump and then consciously

execute it. When I answered that question, I was, in a sense, confabu-
lating: giving a fictitious account of a past event, believing it to be
true. The real reason I jumped was an automatic nonconscious reaction to the fear response set into play by the amygdala. The reason I
would have confabulated is that our human brains are driven to infer
causality. They are driven to explain events that make sense out of
the scattered facts.[9]

The Interpreter's aim of making sense can lead to problems; it generates what is called the 'narrative fallacy' – the misleading interpretation
of events as stories that have an orderly cause–effect structure – and
not infrequently prompts misjudgements, as when a losing streak at
the gambling table makes one think that one's luck must surely be
about to change.

Gazzaniga's conclusion is that the Interpreter's activities as a post
facto rationalizer mislead us into thinking that we are a unified self that
is in command, or could be in command, of our choices and decisions,
our lives and our life story. The brain, in its multitude of independent
modules, does everything in advance of our knowing what it is doing.
The Interpreter can be tricked into various misconceptions and mistakes, both experimentally and in real-world situations, among the
latter by the lesions and diseases its brain is vulnerable to, adding to the
evidence that it is 'only as good as the information it gets', as Gazzaniga
puts it, from the systems constituting the brain.

On the face of it, there is an incoherence at the heart of this account.
Think of the gambling example: the Interpreter, imposing its idea of
logic, thinks that it cannot go on losing forever, and that a long losing
streak must 'surely' be about to change; and so the person goes on
gambling – that is, *chooses* to go on gambling. If some non-Gazzaniga
theory that there is a controlling self somehow instantiated in the
brain were true, this choice would have the effect of directing relevant modules in the brain to buy more chips, shake the dice in the
cup, toss them on to the craps table, and so on. But in the Gazzaniga
theory, although the Interpreter is fooling itself into thinking that its
losing streak must end, it – *it* – is not choosing to go on gambling;
neither it nor anything else is doing any choosing; there is only

non-conscious brain activity. So why is the brain continuing to gamble? If the brain does what it non-consciously does, it does not matter what the Interpreter thinks, either way; the Interpreter is an epiphenomenon. So what explains the apparent *result* of Interpreter mistakes if the Interpreter is not having an effect on brain activity, and, through it, its actions in its social and physical environments?

The chorus of neuroscientific reductionism – which, to repeat, may well be true – is that 'we' are each deluded into thinking that 'I' exist, for there is no 'I', only the illusion of one. As in the question prompted by Dennett's views, *what* (since we cannot ask *who*) is being deluded into thinking it exists? If there is one thing salvageable from Descartes's views, it is that something has to account for the (illusion of?) the persuasiveness of his *cogito*, 'I think therefore I am.' *What* falsely thinks it is? Moreover: if there is no 'I', then there is no 'them' either; we do not inhabit a world of persons, of agents, but of automata – zombies, not to put too fine a point on it – and conceptions of human nature, agency, morality, responsibility, value and meaning are fictions likewise; to think otherwise, as indeed we do, is to live according to a massively and systematically erroneous theory about ourselves and the world.

Everything Gazzaniga says about the high degree of brain modularity, the largely unconscious processing it carries out, and the emergent Left-brain Interpreter that all the brain's modules feed with information is almost all certainly true. The single point on which a critic might disagree is that the Interpreter is merely epiphenomenal. Everything Gazzaniga says is consistent with the idea that by means of the brain's architecture the emergent property of interpretation is causally effective in its feedback on the brain's modular activities. After all, the empirical data shows that higher levels of visual processing feed back to lower levels and act upon them. Gazzaniga-type theories in effect say that feedback stops at a certain point – goes no higher than upper-level modules – and somehow does not happen at the highest level: the Interpreter level. If anything, this goes against what is also empirical data – the data we have about ourselves (Damasio's 'feeling of being a self') and the data underlying the theory by which we navigate our interactions with other humans and animals, namely, that they are conscious and intentional agents too.

Enumerating the investigatory problems (Pinhole, Map, Meddler, etc.) that affect neuroscience, as they affect almost all forms of enquiry, provides a salutary reminder of this important science's youth. However, the investigatory problem that most affects youthful enquiries is the Closure Problem – in its standard form, the desire to reach a conclusion, to have a completed explanation or story, to tidy up and sign off; but in this connection it might better be described as *jumping to conclusions*. The neural correlations of cognitive functions are being identified with ever-greater precision in ways that already offer clinical applications and much hope. Adjusting for the Ptolemy Problem in that connection – 'It works, but is it true?' – the dramatic reversal implied for our view of human nature, that we are not agents but zombies, gives pause for thought. Yes, it might be true; if so, what are we going to do about it? Nothing? Live the lie we tell ourselves, and continue to punish and reward ourselves and each other, educate our children in the hope of making intelligent and mature beings out of them, struggle with our consciences, read, learn, and listen to debates, thinking that we thereby inform ourselves and improve the choices we make – all an illusion because the brain is already doing what we think 'we' do?

  The alternative is to challenge neuroscience with the task of really understanding consciousness and selfhood, either to explain them or to explain them away, but in either case to give an account of the empirical fact of the deep and persistent experience of both. This is no more than a continuation of the Crick and Koch challenge to make consciousness itself the target of scientific study. In the decades between their issuing that challenge and the date at which these words are written, the chief tendency has been to eliminate consciousness or at least its influence; if a single cautionary note emerges from the discussion, it is that this might be a case of jumping to conclusions.

It is no longer a science-fiction possibility that technologies of neuroscientific investigation might be used in morally questionable ways.

A new generation of lie-detector techniques is a real possibility, and the prospect of identifying the *content* of conscious states is no longer remote; scanning of visual-systems activation already provides researchers with indications of content.[10] It is an entailment both of the assumptions and the aims of neuroscience that in a perfected state of itself it will enable more direct – that is, not via standard sensory pathways – communication with brains (desirable in the case of locked-in patients), and, further, the corollary of this: the ability to specify the content of cognitive states, or, in popular parlance: 'mind-reading'. Total loss of the privacy of one's thoughts is one thing; the prospect of techniques of controlling thought, introducing thoughts and memories, extinguishing existing memories, altering personality, controlling behaviour, and any of these activities with malign as well as benign purposes in view, must also be contemplated; none of this lies beyond the bounds of conceivability, given what is already known, and what capabilities already exist, in neuroscience.

Elsewhere I introduce a notion under the label 'Grayling's Law', which states that *Anything that CAN be done WILL be done, if it is of advantage to whoever can do it.* (The corollary is that *things that can be done will NOT be done if it incurs costs to those who can prevent it* – which explains why insufficient action is taken over anthropogenic climate change, preventable diseases in poor regions of the world, and the like.)[11] This Law entails that problematic developments such as autonomous-weapons systems, genetic modification of foetuses, uses of AI that breach civil liberty desiderata, mind-control neuroscience, *will happen* because they cannot fail to be of advantage to private and public agencies in various and obvious ways. Of the three areas of advance in knowledge surveyed in this book, neuroscience is the one that most, and most nearly, invites moral questions. In spite of the iron inevitability of my troubling Law, as we see from the example of the sterling but fundamentally ineffective efforts to control lethal autonomous-weapons systems by human rights groups and international agencies such as the UN, there will be debate and anxiety about neuroscience when its advances begin to reach a tipping-point of applications. Such debates can never begin too early.

Conclusion: The View from Olympus

Humankind has passed from the certainties of faith to the uncertainties of knowledge, from faith to ignorance via knowledge – a new, knowledge-filled ignorance, a surprising and paradoxical state of affairs because so much has been learned, and so much mastery acquired as a result of the great quest for knowledge – and yet, like climbers on a mountain, the higher we go the further away we see that our ignorance stretches; we see that the frontiers of knowledge themselves lie unmappably beyond the horizon.

The sheer extent of ignorance revealed to us by our giant strides in knowledge suggests that we are only at the beginning of a journey. If humanity can survive these first steps – there are no guarantees it will; we are still bedevilled by too much primitivism in our thinking and feeling: we still go to war, quarrel among ourselves, believe nonsense, waste our short lives on trivialities – the way the world, time, and mind will seem to others further along in humanity's story are inconceivable to us now.

Modern times began in the sixteenth century, when belief was abandoned in the idea that it is possible for a single individual to know everything there is to know. The idea of 'Renaissance Man' was the idea of a person who could see coast to coast from the top of the epistemic mountain – the view from Olympus, so to speak. Now there are very few who would claim to be able to see that view, even in generalist terms. Specialism, necessary for genuine advances in knowledge to be made, places us in silos. An historian who studies the Bronze Age is likely to know very little about quantum physics, the quantum physicist very little about Bronze Age history. Humanity's great strides in knowledge have been achieved at the price of losing a conspectus, a sense of place in time and meanings, a sense of human focus amid the nonhuman and sometimes inhuman immensity of things.

The aim of education should be to equip us with a thorough knowledge of a specialism for which we have aptitude, and at the same time a good general literacy in science, history – including the history of ideas – and arts. Higher education has undergone a remarkable reversal. It has gone *from having general literacy as its goal*, leaving special expertise to form itself later as the outcome of individual interest and experience, *to inculcating special expertise as its goal*, leaving general literacy to form itself later as the outcome of individual interest.[1] The reversal occurred without a halfway house, like a switch being thrown. In educational jurisdictions where both general literacy and specialism are valued and their mutual fruitfulness understood, the idea of liberal arts at the undergraduate level and specialism as a postgraduate acquisition was once persuasive; impatience, expense, and the exigent needs of economies in search of foot soldiers for their technological commerce are squeezing out the idea that education and intellectual maturation need to go together, and that both should be replaced by the single endeavour of *training* instead. A corollary of this is the desire to shorten higher education to two years' duration, halving it or more than halving it from what was its norm through the greater part of its thousand-year existence.

If the thrust to specialism, and training rather than education, goes so far that the connections between different fields of enquiry become invisible, and in particular if the 'Two Cultures' gap between science and the humanities, identified by C. P. Snow in the mid twentieth century, grows even greater than it already is, there will be a real danger that human affairs will become unmanageable. Take a single example: AI and the sophistication of computer technology (including its miniaturization) in weapons systems, especially autonomous-weapons systems, are already being combined and developed apace.[2] In a dispensation where no connections are made between the impact of technological developments on the one hand, and social, political, legal, moral, and humanitarian considerations on the other hand, dangerous mismatches can occur.

Understanding how much we know, and how great is the ignorance exposed by our increased knowledge, is valuable in helping to keep the connections in view. It is helpful in other ways too. In the past, in the age of certainties – mainly theological – certainty could be murderous. It was thought that if I am right and you are wrong, especially about the greatest questions concerning reality and the safety of our souls, your wrongness is dangerous, and has to be dealt with. But if we are all paddling the boat of enquiry together in an ocean of ignorance, our perspective changes – for the better.

This point might be put in different words: words about truth. In the Introduction I put this as a question: '*The Truth Problem*. Given that empirical enquiry gives us defeasible probabilities, what are the standards (such as the sigma scale in science) that can be regarded as satisfactory short of certainty? Does this imply that we have to treat the concept of truth pragmatically, as a (possibly unattainable) goal of enquiry upon which, in the ideal, enquiry strategically converges? Where does this leave the concept of "truth" itself?'

The answer is embedded in the question. The concept of truth is the concept of an idealization, towards which enquiry strains its every sinew, and by which we measure the degree of confidence we place in our findings and proposals. This has an important implication: that when we think of knowledge as our best and most rigorously supported belief, we are in effect thinking of *rationality*, of what it is *rational* to believe. 'Rational' means '*ratio*-nal', that is, *proportional*; it is about the ratio or proportion of our beliefs to the evidence we have for them and the soundness of the reasoning we apply to it. This is why it is *irrational* to believe that there are fairies at the bottom of the garden. All the evidence for fairies, such as it is, comes from stories, legends, and other people's beliefs. This applies to quite a few too-influential ways of thinking, not least religion.

There is an asymmetry between rational beliefs (taken as true when the supporting evidence is very strong) and irrational beliefs (taken outright as false when, in addition, we find that premising them or acting on them leads to a high incidence of poor outcomes). This is that we find that rational beliefs tend to be highly coherent with each other – mutually consistent, and often mutually supportive, and when coherence fails (as between quantum theory and general relativity) the matter is not allowed to stand. Irrational beliefs, on the other hand, tend to be independent of each other and two or more can be, and often are, held together or simultaneously with rational beliefs, even when inconsistent. An example of the first case is when it is believed that ghosts can pass through walls – not interacting with matter – and can do one physical harm – thus interacting with matter. Pointing this out is how I tried to dissuade my own offspring, when children, from fear of the supernatural. An example of the second case is holding the belief that the deity is omnipotent and wholly good, inconsistently with recognizing that natural evils such as childhood cancers exist; the inconsistency implies a failure either of omnipotence or goodness on the deity's part – or, more rationally still, the nonexistence of such an entity; though the usual solution to this problem, known in theology as the 'Problem of Evil', is that suffering, including children suffering from cancer, serves some greater long-term good opaque to us now. The carpet of divine inscrutability is always a good place to sweep difficulties under, and the kind of manoeuvre in which it consists is a mark of irrational belief in its own right.

The twelve problems that beset enquiry – the Pinhole Problem and the rest – set the terms of enquiry; they are what variously have to be worked round, taken into account, dealt with, accepted, understood or – best of all – solved. They define the nature of enquiry. Recognizing their existence goes hand in hand with the new world of enquiry that has lately generated such huge amounts of knowledge and such a huge – even larger – understanding of our ignorance. In the age of certainty, before the scientific revolution of the sixteenth and seventeenth centuries, when the stalled quest of classical antiquity was at long last resumed, these problems of enquiry were scarcely thought of. They are the children of knowledge and its correlative ignorance, and they are what help, and will help, knowledge in its explorations of that ignorance.

Knowledge brings the ability to do things; the ability to do things can create moral dilemmas. These latter can be made worse by the new ignorances that new knowledge brings with it. Of the three domains of new knowledge surveyed here, neuroscience is the one that carries seeds of good and bad things with it – both very good, and potentially very bad. Knowing what knowledge portends gives us a chance to reflect.

A final point: enquiry is exhilarating. As the human past looked up at us from the eyeless sockets of australopithecine skulls, as preclassical antiquity emerged from the *tells* of Mesopotamia, as the secrets of nature opened to the mathematics and particle colliders of physics, as the differently oriented stripes of a visual field registered in the cells of an occipital cortex, the sense of frontiers being crossed gave those involved an insight into what motivates humankind's better endeavours: the unparalleled excitement of discovery. The great thing about a frontier – so different from a boundary, a border, a wall – is that it is an invitation to travel across it, and travel onwards. And to travel prepared.

Appendix I: Figures

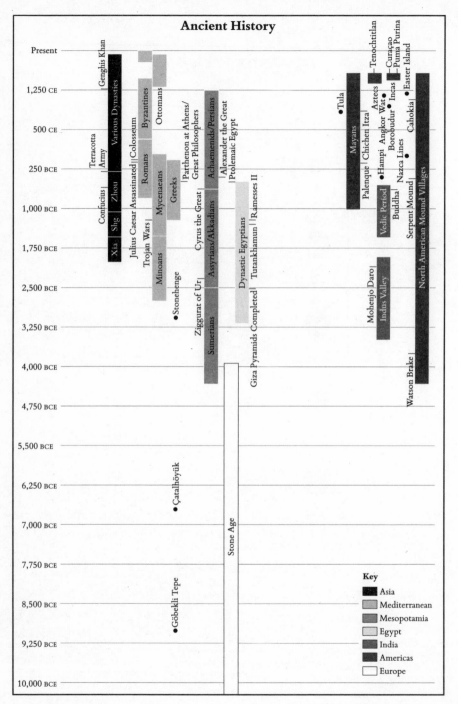

Figure 1

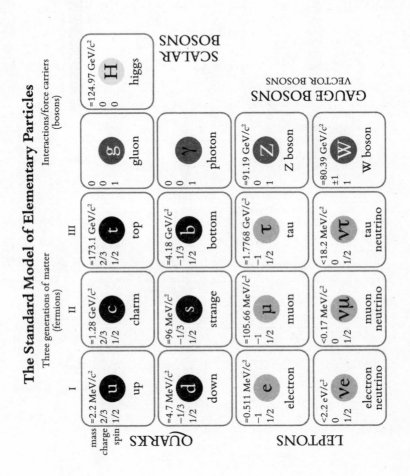

Figure 2

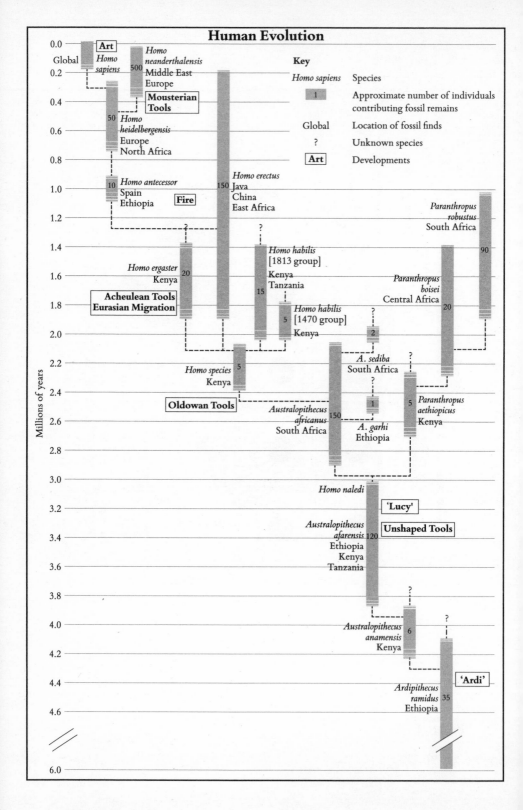

Figure 3

The Brain

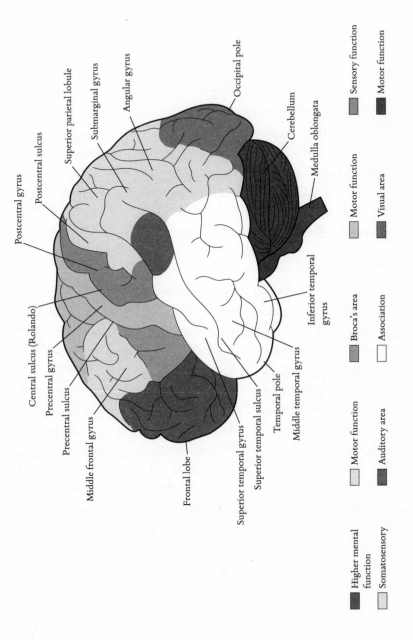

Central sulcus (Rolando)
Precentral gyrus
Precentral sulcus
Middle frontal gyrus
Postcentral gyrus
Postcentral sulcus
Superior parietal lobule
Submarginal gyrus
Angular gyrus
Occipital pole
Cerebellum
Medulla oblongata
Inferior temporal gyrus
Middle temporal gyrus
Temporal pole
Superior temporal sulcus
Superior temporal gyrus
Frontal lobe

Higher mental function
Somatosensory
Motor function
Auditory area
Broca's area
Association
Motor function
Visual area
Sensory function
Motor function

Figure 4

Appendix II: *The Epic of Gilgamesh*

The *Epic* tells of the imperious Gilgamesh, who had established himself as a lion in the eyes of his subjects, a handsome and ferocious warrior and a powerful ruler, 'Supreme over other kings, lordly in appearance, he is the hero . . . the goring wild bull . . . a mighty net, protector of his people.' Nevertheless he had begun to distress his people because of his arrogant behaviour, not least his assertion of the *droit de seigneur* that entitled him to sleep with any bride on the first night after her marriage. No girl was safe from him. The people complained to the gods, one of whom, Aruru, took some clay and moulded a man from it to be a check to Gilgamesh. This was Enkidu, whom Aruru placed among the wild beasts to be brought up by them. Enkidu's 'whole body was shaggy with hair, he had a full head of hair like a woman, his locks billowed in profusion . . . He ate grasses with the gazelles, and jostled at the watering holes with the animals.' He was seen by a trapper, who was struck with fear by Enkidu's wildness and great strength. The trapper hastened to tell his father about the savage man he had seen, and his father advised him to go to Uruk and report the matter to Gilgamesh.

When Gilgamesh heard about Enkidu he told the trapper to take the harlot Shamhat to the watering hole where Enkidu drank, there to entice him to have sexual relations with her, thus taming him. 'Go, trapper, take the harlot Shamhat with you. When the animals are drinking at the watering place, have her take off her robe and expose her sex. When he sees her, he will draw near to her, and his animals, who grew up in his wilderness, will be alien to him.'

The plan was put into action. The trapper and Shamhat waited at the watering place, and when Enkidu and the beasts came to water's edge the trapper said, 'That is he, Shamhat! Expose your sex so that he can take in your voluptuousness. Do not be restrained – take his energy! When he sees you he will draw near to you. Spread out your robe so he can lie upon you; perform for this savage the work of womankind. His animals, who grew up in his wilderness, will become alien to him, and his lust will groan over you.' Shamhat displayed her body to Enkidu, and spread out her robe, and 'took his energy . . . he lay upon her, she performed for the savage the task of womankind, his lust groaned over her, for six days and seven nights Enkidu stayed aroused, and had intercourse with her until he was sated with her charms.'

The plan worked. When the animals saw his 'utterly depleted body', they ran off, and Shamhat, telling Enkidu that he was beautiful, encouraged him to accompany her to Uruk, telling him of the charms of the city and of the glory and prowess of Gilgamesh. Meanwhile Gilgamesh had had a series of dreams presaging Enkidu's arrival, which his mother interpreted for him, telling him that he and Enkidu would become loving friends: 'There will come to you a mighty man, a comrade who saves his friend, his strength is mighty. You will love him and embrace him as a wife, and it is he who will repeatedly save you.'

Shamhat taught Enkidu to eat bread and drink beer, and in general prepared him for entering civilization by having him live with some shepherds whose flocks he helped to guard against wolves and lions. He shaved and rubbed himself in oil, becoming presentable; and at last entered Uruk.

He did so just as Gilgamesh was about to exercise the *droit de seigneur* at a marriage feast over which the bridegroom had gone to great expense. Enkidu, told of this, was outraged at the thought of such a violation of the nuptial bed, and hurried to the place of the feast, blocking the way into it and thereby refusing Gilgamesh entry. The two fought; when it became clear to Gilgamesh that he was not going to beat Enkidu, they stopped fighting, and their bond of friendship was instantaneous: 'They kissed each other, and became friends.'

Missing text means that we do not know why Gilgamesh and Enkidu resolved to journey to the forest to cut down the great Cedar Tree guarded by the monster Humbaba, other than that the feat would bring them renown; for Humbaba was fearsome, his 'roar is a flood, his mouth is fire, his breath is death; he can hear any rustling in his forest a hundred leagues away! Who even among the Igigi gods can confront him?' The two heroes had special weapons made, and their journey to the forest was long, with Gilgamesh again being haunted by dreams that seemed ill-omened, but that Enkidu interpreted positively. They reached the forest, where Gilgamesh fought and defeated Humbaba. The overthrown Humbaba pleaded with Gilgamesh to spare his life, saying that he would be his servant from then on; but Enkidu persuaded Gilgamesh not to spare the monster, but to 'grind up, kill, pulverize and destroy him!'

After felling the giant Cedar, 'whose top touches heaven', they floated it down the river to Nippur, there to be made into a gigantic door, taking the severed head of Humbaba with them. When they had washed the dust from their bodies and matted locks, and had oiled themselves, Gilgamesh put on his crown, and appeared so handsome that the goddess Ishtar said to him, 'Come, Gilgamesh, be my husband, grant me your lusciousness, be my husband and I will be your wife, I will have a chariot harnessed for you of lapis lazuli and gold, with wheels of gold and horns of electrum, and it will be harnessed with great storming mountain mules!' But Ishtar's history of taking and then disposing of husbands in unpalatable ways made Gilgamesh reject her, after listing her ill-treatment of her many consorts – to her great anger: she hastened to her father, the god Anu, to complain about Gilgamesh, saying that he had 'recounted despicable deeds and curses against me!' and demanding that Anu give her 'the Bull of Heaven, so he can kill Gilgamesh in his dwelling'. She threatened that if Anu did not give her the Bull of Heaven she would 'knock down the gateposts of the Netherworld, I will smash the door posts, and leave the doors flat down, and let the dead go up and eat the living!'

Anu gave Ishtar the nose-rope of the Bull of Heaven and she led it down to the Euphrates, where its snorts opened pits in the ground, into the first of which a hundred young men of Uruk fell and were swallowed up, and into the second of which two hundred young men of Uruk fell and were swallowed up. Enkidu grappled with the Bull, shouting instructions to Gilgamesh where to strike with his sword, between the nape of the neck and the horns; when the Bull fell dead they ripped out its huge heart and presented it as an offering to the god Shamash. The genitals of the Bull Gilgamesh cut off and threw in Ishtar's face. Ishtar summoned all the 'cultic women, joy-girls and harlots and set them to mourning over the genitals [the 'hindquarters'] of the bull'.

But killing the Bull of Heaven, and felling the Cedar Tree, exacted a penalty: one of the two friends had to die in compensation for these deeds. The gods in counsel decided it had to be Enkidu, who therefore fell ill. His illness was protracted, and Gilgamesh anxiously watched over him, until the day that, even as he was speaking to Enkidu on his sick-bed, he saw that his beloved friend had died: 'What is this sleep that has seized you? You have turned dark and do not hear me!' And then he saw that 'Enkidu's eyes do not move, he touched his heart, but it beat no longer.' In an agony of grief Gilgamesh shaved his head and tore off his fine clothes, pacing up and down 'like a lioness deprived of her cubs . . . he cried bitterly, roaming the wilderness' – and then, when he had mourned over the dead body of his beloved comrade for eight days, he saw a maggot drop from Enkidu's nostril: and he was filled with horror at the thought that he too was mortal, and must die.

The fear of death so gripped Gilgamesh that he resolved to travel to find his remote ancestor Utanapishtim – 'Utanapishtim the Faraway', who had been granted eternal life by the gods – in order to get the secret of immortality from him. The journey was long and difficult, and Gilgamesh was soon in a sorry state of dirt, emaciation and disarray, so that those he encountered would not believe that he was the legendary Gilgamesh until he had explained to them his quest and the reason for it. At last he was able to make his way across the Waters of Death to where Utanapishtim and his equally immortal wife lived. When Gilgamesh explained why he had come, Utanapishtim told him how he had himself acquired immortality. When Anu and all the gods apart from Ea decided to wipe out mankind in a great flood, Ea told Utanapishtim to build a ship and take his family and animals into it. Utanapishtim built a great ship, as big as a field, with six decks, and stocked it with oil, beer, wine and butchered meats, and brought animals on board, and his family; and the immense flood came and covered the mountains: 'From the horizon a black cloud arose . . . the land shattered like a pot . . . all day long the south wind blew, submerging the mountains in water, overwhelming the people like an attack, no one could see his fellow, they could not recognize each other in the torrent . . . even the gods were frightened by the Flood, and retreated to heaven . . . Ishtar shrieked like a woman in childbirth . . . Six days and seven nights came the wind and the flood, the storm flattening the land. When the seventh day arrived the storm was still pounding, the flood was a war, struggling with itself like a woman writhing in labour . . .'

When the storm subsided Utanapishtim opened a vent to feel the fresh air on his face.[1] He sent out a dove, but it returned because it could find no dry land. After a while he sent out a wren, and it too returned. But when he sent out a raven it did not return, so he knew that the floodwaters were subsiding.

For saving life on earth Utanapishtim was rewarded with immortality by the gods, who had repented their decision to exterminate everything. It was, however, a gift not available to Gilgamesh, who therefore had to set off back to Uruk disappointed. But there was a consolation; on the way he would find a herb that would make him young again.

Gilgamesh found the herb, but one day when he lay down to rest a snake stole it. The regenerative power of the herb was demonstrated by the snake sloughing off its skin to reveal a new skin beneath as it slithered away.[2]

Appendix III: *The Code of Hammurabi*

The following sketch of some of the *Code*'s provisions gives a flavour of the time in which it was written. It begins by deterring people from making false accusations, threatening them with the punishment that the accused would have received had the accusation been true. One test of innocence was a leap into the Euphrates to see if the accused sinks or swims; if the accused swims, the accuser will be put to death and the accused will take possession of his property. Judges are also deterred from error; if they are found to have decided a case wrongly they will be liable for twelve times the amount of the fine they had imposed, and will be dismissed from the bench. Those who steal from temples or the royal court, and those who 'fence' the stolen goods, will be put to death. Stealers of livestock will pay thirtyfold the value of what they stole, and if they cannot pay, they will be executed. High standards of evidence are required to prove that something in one's possession that is claimed by someone else is really the owner's own or that he acquired it properly. Kidnapping is punishable by death. Those who return runaway slaves to their masters will be paid a bounty of two shekels per slave. Heirs of those who, in serving the king, die or are captured in battle are assured of inheriting their property. If bad weather destroys the crops of a tenant farmer, it is he and not the landlord who must accept the burden of the loss. 'If anyone be too lazy to keep his dam in proper condition, and does not so keep it, then if the dam break and all the fields be flooded, then shall he in whose dam the break occurred be sold for money, and the money shall replace the corn which he has caused to be ruined.' If a man transforms waste land into arable land and returns it to its owner, the latter shall pay him for one year ten *gur* for ten *gan*. If a merchant entrusts money to an agent for some investment, and the broker suffers a loss in the place to which he goes, he shall make good the capital to the merchant. 'If a prisoner dies in prison from blows or maltreatment, the master of the prisoner shall convict the merchant before the judge. If he was a free-born man, the son of the merchant shall be put to death; if it was a slave, he shall pay one third of a *mina* of gold, and all that the master of the prisoner gave he shall forfeit.' If anyone fails to meet a claim for debt, and sells himself, his wife, his son and daughter for money or gives them away to forced labour, they shall work for three years in the house of the man who bought them, or the proprietor, and in the fourth year they shall be set free. If anyone stores corn in another man's house he shall pay him storage at the rate of one *gur* for every five *ka* of corn per year. If a man's wife is surprised in adultery with another man, both shall be tied and thrown into the river, but the husband may pardon his wife and the king his slaves. If a man is taken prisoner in war and there be no sustenance in his house and his wife goes to another house and bears children, if then he returns and comes to his home, then his wife shall return to her husband, but the children will follow their father. If a man takes a wife and she bears him no children, and he intends to take another wife, if he takes this second wife and brings her into his house, this second wife shall not be allowed equality with his wife. If a man takes a wife, and she be seized by disease, if he then desire to take a second wife, he shall not put away his wife who has been attacked by disease, but he shall keep her in the house and support her so long as she lives.

And so it goes on for 282 articles in all, covering everything from the hire of ferryboats to the purchase of slaves, from the penalty a physician must pay if he operates on a patient and makes a mess of it ('If a physician makes a large incision with the operating knife, and kill him . . . his hand shall be cut off'). Article 196 is the famous 'eye for an eye': 'If a man put out the eye of another man, his eye shall be put out.' Article 200 requires a tooth for a tooth. But 'If he put out the eye of a slave he shall pay one half of its value.'[1]

Notes

Introduction

1 'The Presocratic Philosophers: Thales', in A. C. Grayling, *The History of Philosophy* (London, 2019).

2 Ibid.

3 Histories of science, and A. C. Grayling, *The Age of Genius: The Seventeenth Century and the Birth of the Modern Mind* (London, 2016).

4 This is one of my own special interests as an academic philosopher; in two books and a number of essays I have explored the question of how we ('we' humans) construct justificatory schemes for knowledge claims about the world over which our perceptual experience and thought range.

5 Though even here a question arises, for 'knowledge' in the 'formal' systems of mathematics and logic might be merely a matter of definition – that is, a consequence of the way the terms and operations of the systems in question are defined. This is known as *a priori* knowledge – knowledge gained without recourse to empirical observation and experiment of how things actually are in the world, this latter being *a posteriori* knowledge.

6 The occasional rogue element in science, as in all things human, sometimes appears – falsifying experimental data, rushing publication before the required high degree of confidence can be achieved. But these disruptions are almost always quickly found out and put right. Any scientist who betrays the trust of the scientific community in this way is discredited and excluded, because science has no place for such.

7 I address questions about genetic medicine and AI in a forthcoming publication.

8 The name 'Big Bang' was coined by Fred Hoyle in a BBC radio talk in 1949; as a proponent of a 'Steady State Theory' of the universe (see Part I, Section 4), he was thought to have intended the name as an illustration

of the absurdity of the theory, but said it was merely a way of dramatizing what the theory entailed.

9 It is calculated (e.g., from lists of titles of lost works) that 90 per cent of the literature of classical antiquity has been lost, in significant part because of deliberate attempts by religious enthusiasts to efface the 'pagan' past following the Edict of Thessalonica in 380 CE making Christianity the official religion of the Roman Empire. See Catherine Nixey, *The Darkening Age* (London, 2017).

Part I: Science

1 Technology before Science

1 Human evolution is discussed in Part II, Section 2, below.

2 Though many have a relish for undercooked and raw flesh such as carpaccio; but these meats tend not to be positively rotten when consumed. Game hung until rotten in order to tenderize it is rarely eaten raw.

3 Interpretations of the past might involve a lot of 'reading-in', that is, construing what is found there in terms familiar to ourselves. This is controversial, as discussed in Part II, Section 3, below.

4 See Part II, Section 1, below for the Bronze Age Collapse.

5 *abjadic*: a writing system, such as Hebrew or Arabic, in which each character stands for a consonant, with the vowels inferred. *logophonetic*: a writing system, such as Chinese or Japanese, in which each character represents a word or morpheme (the smallest meaningful unit in a language).

6 Richard Bulliet, *The Wheel: Inventions and Reinventions* (New York, 2016).

7 Ibid., p. 1.

8 David W. Anthony, *The Horse, the Wheel, and Language: How Bronze-Age Riders from the Eurasian Steppes Shaped the Modern World* (Princeton, NJ, and Oxford, 2007).

9 Bulliet, Chapter 3, *passim*.

10 Note that generally speaking wagons are four-wheeled, carts two-wheeled and easier to steer than wagons unless the latter's front pair of wheels are on independent axles; for Lancelot and Guinevere, Chrétien de Troyes, *Le Chevalier de la charrete* (c. 1171).

11 See A. C. Grayling, *War: An Enquiry* (New Haven, Conn., and London, 2017), pp. 22–7.

12 Dava Sobel, *Longitude: The True Story of a Lone Genius Who Solved the Greatest Scientific Problem of His Time* (London, 1995).

13 See A. C. Grayling, *The Age of Genius: The Seventeenth Century and the Birth of the Modern Mind* (London, 2016).

14 See Grayling, *War*.

15 See A. C. Grayling, *Towards the Light: The Story of the Struggles for Liberty and Rights that Made the Modern West* (London, 2007), and *The Age of Genius*.

2 The Rise of Science

1 The 'reading-in' point is discussed more fully in Part II, Section 4, below.

2 See Grayling, *The History of Philosophy* (London, 2019), especially Part I, *passim*.

3 Cf. Greek *agein*, Sanskrit *ajati*, which derives from the Indo-European root *ag-*, 'drive, move, draw out or from'.

4 Grayling, *Towards the Light: The Story of the Struggles for Liberty and Rights that Made the Modern West* (London, 2007) and *The Age of Genius: The Seventeenth Century and the Birth of the Modern Mind* (London, 2016) discuss how, following the Reformation in the sixteenth century CE, Church organizations in the Protestant states were not powerful enough to control speculation and publication; this was not so in Catholic countries, in which (as the trial of Galileo shows) opposition to the advances of scientific enquiry continued for some time.

5 The pantheon of science's history *almost* exclusively consists of men, because women were denied the opportunity to participate. That has changed vastly for the better, especially since the mid twentieth century.

6 Benjamin Farrington, *Greek Science* (1944; London, 2nd edn 1949), p. 153.

7 Brahe made a detailed examination of the supernova in his *De nova et nullius aevi memoria prius visa stella* (*Concerning the Star, New and Never*

before Seen in Anyone's Life or Memory) of 1573, in which he recorded his own observations, and analyses of the observations of many others. It was reprinted twice by Kepler in the early seventeenth century.

8 Other giants more relevant to the calculus are René Descartes, Pierre de Fermat, and Newton's own patron at Cambridge and predecessor as Lucasian Professor, Isaac Barrow.

9 Strictly, *F* is the *vector sum* of the forces.

10 This account leaves out developments in other major sciences, especially chemistry and biology. The application of discoveries in chemistry to discoveries in biology had to wait their time; but the microscope was already a vital biological tool, and the classificatory system introduced by Linnaeus in the eighteenth century was another such in the organization of biological knowledge. The moment when science started to diverge into different specialisms was when Alessandro Volta invented the electric battery. This happened in 1800, and the importance of the event was that it made electrolysis possible, enabling the separation of compounds into their elements. Chemistry thus became a self-standing science apart from physics, and started to make great strides. The first version of a modern atomic theory owes itself to John Dalton's use of a concept of corpuscular atoms to explain chemical interactions.

3 *The Scientific World Picture*

1 See Grayling, *The History of Philosophy* (London, 2019), pp. 47–51.

2 See Steven Weinberg, 'The Making of the Standard Model', *European Physical Journal C*, vol. 34 (May 2004), pp. 5–13. https://cds.cern.ch/record/799984/files/0401010.pdf

3 Hadrons are composite entities consisting of *mesons* – force particles – as well as *baryons*.

4 Specifically but crucially, replacing 'Galilean transformations' with 'Lorentz transformations'.

4 *Through the Pinhole*

1 The Planck Mass is 22 mu/g; the energy of a fundamental particle is therefore huge: multiplied by c the result is 1.2×10^{28} eV, or 2 billion joules.

2 Richard Feynman, *QED: The Strange Theory of Light and Matter* (Princeton, NJ, and Oxford, 2014).

3 See Grayling, *The History of Philosophy* (London, 2019), pp. 256–67, for a sketch of the view.

4 These organizational principles, particularly the acquired ones, can instructively be called 'concepts'.

5 Eugene Wigner, 'The Unreasonable Effectiveness of Mathematics in the Natural Sciences', *Communications on Pure and Applied Mathematics*, vol. 13, no. 1 (February 1960).

6 R. W. Hamming, 'The Unreasonable Effectiveness of Mathematics', *American Mathematical Monthly*, vol. 87, no. 2 (February 1980), pp. 81–90.

7 Ibid., p. 88.

8 If the frequency of occurrence of any of the digits 1–9 as leading digit were uniform, they would each appear 11.1 per cent of the time. But 1 occurs 30 per cent of the time and 9 less than 5 per cent of the time.

9 Hamming, 'The Unreasonable Effectiveness of Mathematics', p. 89.

10 Ibid.

11 Ibid., p. 90.

12 To say this is not to subscribe to John Stuart Mill's theory that mathematics is empirical. But it is plausible to think that a prompt to understanding the basic idea of number can come from describing and comparing collections.

13 As noted, Mill thought that mathematics is rooted in experience; a *set* of things, a collection or group, is something we perceive, and we can tell by looking which is larger; he took it that mathematics is an elaboration of these basic intuitions.

14 Alexander L. Taylor, *The White Knight* (Edinburgh, 1952). Lewis Carroll's tricks with mathematics fill the *Wonderland* stories and are a delight. This one occurs in Chapter 2 of *Alice in Wonderland*.

15 See Douglas R. Hofstadter, *Gödel, Escher, Bach: An Eternal Golden Braid* (New York, 1979) for an instructive and entertaining survey of formal languages and structures.

16 Sundar Sarukkai, 'Revisiting the "Unreasonable Effectiveness" of Mathematics', *Current Science*, vol. 88, no. 3 (10 February 2005), pp. 415–23 (420).

17 John D. Barrow and Frank J. Tipler, *The Anthropic Cosmological Principle* (Oxford, 1986), pp. 16, 21–2.

18 David Deutsch, *The Fabric of Reality: Towards a Theory of Everything* (London, 1997).

19 See, e.g., Philippe Brax, 'What Makes the Universe Accelerate? A Review on What Dark Energy Could be and How to Test It', *Reports on Progress in Physics*, vol. 81, no. 1 (January 2018).

Part II: History

1 Min's cult is now dated to the fourth millennium BCE.

1 The Beginning of History

1 'Civilization' is a loaded term; here it means, at minimum, settled and populous urban life marked by the complex interactions of social and administrative organization, division of labour, decorative crafts, systems of exchange, and record-keeping evolving into writing.

2 J. Michael Rogers, 'To and Fro: Aspects of the Mediterranean Trade and Consumption in the Fifteenth and Sixteenth Centuries', *Revue des mondes musulmans et de la Méditerranée*, nos. 55–6 (1990), pp. 57–74.

3 Oliver Impey and Arthur MacGregor (eds.), *The Origins of Museums: The Cabinet of Curiosities in Sixteenth- and Seventeenth-century Europe* (London, 1985).

4 Alastair Hamilton, *Johann Michael Wansleben's Travels in the Levant, 1671–1674: An Annotated Edition of His Italian Report* (Leiden and Boston, 2018).

5 Rollin was a Jansenist and his academic career was blighted by sectarian hostilities; although elected Rector of the University of Paris he was barred from taking office because of his Jansenism.

6 Marc van de Mieroop, *A History of the Ancient Near East: c. 3000–323 BC* (2006; 3rd edn Oxford, 2016).

7   Genesis 10:10.

8   One motive for the draining of the marshes was the hostility of Saddam Hussein to the Shiite Marsh Arabs, who maintained an immemorial way of life there.

9   Gwendolyn Leick, *Mesopotamia: The Invention of the City* (London, 2001).

10 See Grayling, *War: An Enquiry* (New Haven, Conn., and London, 2017).

11 Birth narratives of figures such as Sargon, Moses, and Oedipus were studied by Otto Rank in *The Myth of the Birth of the Hero: A Psychological Interpretation of Mythology*, F. Robbins and Smith Ely Jelliffe (trs.) (New York, 1914).

12 On the theme of friendship in philosophy and literature, see A. C. Grayling, *Friendship* (New Haven, Conn., and London, 2013).

13 Eric H. Cline, *1177 BC: The Year Civilization Collapsed* (Princeton, NJ, and Oxford, 2014), p. 151.

14 Carol G. Thomas and Craig Conant, *Citadel to City-State: The Transformation of Greece, 1200–700 BCE* (Bloomington, Ind., 1999).

15 Joseph A. Tainter, *The Collapse of Complex Societies* (Cambridge, 1976).

16 Colin Renfrew, *Archaeology and Language: The Puzzle of Indo-European Origins* (1987; Cambridge, 1990).

17 Marija Gimbutas et al., *The Kurgan Culture and the Indo-Europeanization of Europe: Selected Articles from 1952 to 1993*, Washington D. C., 1997).

18 David W. Anthony, *The Horse, the Wheel, and Language: How Bronze-Age Riders from the Eurasian Steppes Shaped the Modern World* (Princeton, NJ, and Oxford, 2007).

19 Iñigo Olalde et al., 'The Beaker Phenomenon and the Genomic Transformation of North-west Europe', *Nature*, vol. 555, no. 7,695 (8 March 2018), pp. 190–96.

20 Wolfgang Haak et al., 'Massive Migration from the Steppe was a Source of Indo-European Languages in Europe', *Nature*, vol. 522, no. 7,555 (11 June 2015), pp. 207–11.

21 David Reich, *Who We are and How We Got Here: Ancient DNA and the New Science of the Human Past* (Oxford, 2018), pp. 99–121.

22 Marija Gimbutas, *The Goddesses and Gods of Old Europe* (London, 1974).

23 Reich, *Who We are and How We Got Here*, pp. 106–7.

24 Ibid., p. 102.

25 Anthony, *The Horse, the Wheel, and Language*.

26 Franco Nicolis (ed.), *Bell Beakers Today: Pottery People, Culture, Symbols in Prehistoric Europe: Proceedings of the International Colloquium, Riva Del Garda (Trento, Italy), 11–16 May 1998*, Vol. 2 (Trento, 1998).

27 William Jones, 'The Third Anniversary Discourse – on the Hindus', delivered 2 February 1786, *The Works of Sir William Jones*, Vol. 3 (Delhi, 1977), pp. 24–46.

28 *p* and *b* are very similar sounds made by the same formation of the lips, varying in the amount of breath expressed.

29 An irony is that most of the American settlers were British, and the army sent by Britain to contest the Declaration of Independence was over one third comprised of 'Hessians', hired German troops.

30 Debate about the origins of the Anatolian–Caucasus Steppe and PIE continues: see, e.g., Kristian Kristiansen, 'The Archaeology of Proto-Indo-European and Proto-Anatolian: Locating the Split', in M. Serangeli and Th. Olander (eds.), *Dispersals and Diversification: Linguistic and Archaeological Perspectives on the Early Stages of Indo-European* (Leiden and Boston, 2020).

2 The Coming of Humanity

1 There are good introductions to the story of human evolution: Louise Humphrey and Chris Stringer, *Our Human Story* (London, 2018); Francisco J. Ayala and Camilo J. Cela-Conde, *Processes in Human Evolution: The Journey from Early Hominins to Neanderthals and Modern Humans* (Oxford, 2017); *New Scientist, Human Origins: 7 Million Years and Counting* (London, 2018); Alice Roberts, *Evolution: The Human Story* (2nd edn London, 2018). The following pages draw on these among other sources (see further notes).

2 Although the diminutive *Homo floresiensis*, the 'Hobbit', was once thought to have died out on its island home of Flores as recently as twelve thousand years ago, subsequent investigations suggest that it went extinct about the same time that *Homo sapiens* appeared in the region of Indonesia, namely, about fifty thousand years ago.

3 Respectively *Geospiza magnirostris* and *Camarhynchus parvulus*.

4 Dmanisi Skulls. https://www.google.co.uk/search?source=hp&ei=Wmo DX_GoBqGXlwSAh7DIDw&q=dmanisi+skulls&oq=dmanisi+skulls &gs_lcp=CgZwc3ktYWIQAzICCAA6CAgAELEDEIMBOgUIABCx AzoECAAQAzoECAAQCjoGCAAQFhAeUP8iWIJDYLdGaABwA HgAgAFDiAGnBpIBAjEomAEAoAEBqgEHZ3dzLXdpeg&sclient= psy-ab&ved=oahUKEwjxvbXsnbnqAhWhy4UKHYADDPkQ4dUD CAw&uact=5

5 See Neus Martínez-Abadías et al., 'Heritability of Human Cranial Dimensions: Comparing the Evolvability of Different Cranial Regions', *Journal of Anatomy*, vol. 214, no. 1 (January 2009), pp. 19–35.

6 Lu Xun (Lu Hsün), *The True Story of Ah Q* (1921): 'Mr Chien's eldest son whom Ah Q despised . . . After studying in a foreign school in the city, it seemed he had gone to Japan. When he came home half a year later his legs were straight and his pigtail had disappeared . . . Ah Q . . . insisted on calling him "Imitation Foreign Devil".' Chapter 3. https:// www.marxists.org/archive/lu-xun/1921/12/ah-q/ch03.htm

7 The remarkable story of the *naledi* discovery, and the impressive science that accompanied it, is told at https://www.youtube.com/watch?v= 7mBIFFstNSo

8 'Usually' because in a small number of births per thousand some individuals can be born with only one sex chromosome (45X or 45Y, known as sex monosomies) and some with three or more sex chromosomes (47XXX or 47XXY or 47XYY or 49XXXXY, etc.; sex polysomies). See the WHO Genomic Resource Centre article 'Gender and Genetics'. https://www.who.int/genomics/gender/en/

3 The Problem of the Past

1 I shall use the term 'Near East' to denote the Levant, Mesopotamia, Egypt, and environs rather than 'Middle East'; this latter term is a Cold War coining reflecting the introduction of Soviet-dominated East Europe into the calculations of geopolitics and diplomatic history. 'The Orient' was 'the East' (that is what the word means), until China and Japan came into the more regular purview of European and American

travellers and traders, thus constituting the Far East, so 'the Orient' became the 'Near East'. With a new 'East' in Soviet-controlled Europe, the Near East had to become the 'Middle East'. It is now an otiose term.

2 Georg G. Iggers (ed.), *The Theory and Practice of History* (London, 2010).

3 However, slavery under other names and guises continues today; estimates suggest at least 12 million people live and work in slave conditions, the same number as had suffered in the North Atlantic slave trade between the fifteenth and nineteenth centuries.

4 Helen Fordham, 'Curating a Nation's Past: The Role of the Public Intellectual in Australia's History Wars', *M/C Journal*, vol. 18, no. 4 (2015).

5 Henry Reynolds, *Forgotten War* (Sydney, 2013).

6 Quoted in Reynolds, *Forgotten War*, p. 14.

7 Ryan Lyndall, 'List of Multiple Killings of Aborigines in Tasmania: 1804–1835', SciencesPo, *Violence de masse et Résistance – Réseau de recherche* (March 2008).

8 Ibid.

9 Ibid.

10 Ibid.

11 Henry Reynolds, YouTube talk. https://www.youtube.com/watch?v= ClS2gzn3QTg

12 Dee Brown, *Bury My Heart at Wounded Knee* (1970; New York, 2012).

13 Ibid., 'War Comes to the Cheyenne', pp. 86–7.

14 Ibid.

15 Brown here quotes Bent's words from the 39th US Congress, Second Session, Senate Report 156, pp. 73, 96.

16 Ibid.

17 https://www.google.com/search?q=United+States+Congress+Joint+Co mmittee+on+the+Conduct+of+the+War,+1865+(testimonies+and+report) %22&rls=com.microsoft:en-GB:IE-Address&ie=UTF-8&oe=UTF-8& sourceid=ie7&gws_rd=ssl#spf=1601690409362

18 Niall Ferguson, *Empire: How Britain Made the Modern World* (London, 2003).

19 Tom Engelhardt, 'Ambush at Kamikaze Pass', *Bulletin of Concerned Asian Scholars*, vol. 3, no. 1 (1971), pp. 64–84.

20 Christopher R. Browning, *The Origins of the Final Solution: The Evolution of Nazi Jewish Policy, September 1939–March 1942*. With contributions

by Jürgen Matthäus (Lincoln, Nebr., 2004; London, 2014 edn). A comprehensive history of the Holocaust.

21 Alex J. Kay, *The Making of an SS Killer: The Life of Colonel Alfred Filbert, 1905–1990* (Cambridge, 2016), pp. 57–62, 72.

22 Leni Yahil, *The Holocaust: The Fate of European Jewry, 1932–1945* (Oxford, 1991), p. 270.

23 Gas vans were tried at the Chełmno (in German, Kulmhof) extermination camp for Jews from the Łódź Ghetto.

24 Nestar Russell, 'The Nazi's Pursuit for a "Humane" Method of Killing', *Understanding Willing Participants: Milgram's Obedience Experiments and the Holocaust*, Vol. 2 (London, 2019), pp. 241–76.

25 Yisrael Gutman and Michael Berenbaum (eds.), *Anatomy of the Auschwitz Death Camp*, United States Holocaust Memorial Museum (Bloomington, Ind., 1998), p. 89.

26 Bone-crushing machines use at Holocaust death camps: https://collections.ushmm.org/search/catalog/pa10007

27 Paul Rassinier, *Holocaust Story and the Lies of Ulysses: Study of the German Concentration Camps and the Alleged Extermination of European Jewry* (republished 1978 by 'Legion for the Survival of Freedom, Inc.', based in California); Jean Norton Cru, *Witnesses: Tests, Analysis and Criticism of the Memories of Combatants (1915–1928)* (*Témoins: Essai d'analyse et de critique des souvenirs de combattants édités en français de 1915 à 1928*) (Paris, 1929; Nancy, 3rd edn 2006).

28 Elhanan Yakira, *Post-Zionism, Post-Holocaust* (Cambridge, 2010).

29 Ibid., p. 7.

30 Ibid., p. 8.

31 Deborah Lipstadt, *Denying the Holocaust: The Growing Assault on Truth and Memory* (New York, 1993), p. 75.

32 Ibid., in reference to Barnes's pamphlet *Revisionism and Brainwashing* (1961).

33 Ibid., p. 74.

34 Ibid., p. 214.

35. Arno Mayer, *Why Did the Heavens Not Darken?* (Verso, 2012), pp. 349, 452 and 453; Michael Shermer and Alex Grobman, *Denying History: Who Says the Holocaust Never Happened and Why Do They Say It?* (Berkeley, Calif., 2002), p. 126.

36. *David Irving v. Penguin Books and Deborah Lipstadt (2000)*, Section 13 (91).

37. See *Speculum*, vol. 65, no. 1 (January 1990), esp. Stephen G. Nichols's 'Philology in a Manuscript Culture'; and M. J. Driscoll, 'The Words on the Page', *Creating the Medieval Saga: Version, Variability and Editorial Interpretations of Old Norse Saga Literature*, Judy Quinn and Emily Lethbridge (eds.) (Odense, 2010).

38. 'Resolution of the Duke University History Department', printed in the *Duke Chronicle* (November 1991).

39. Shermer and Grobman, *Denying History*.

40. Christopher Hill, *The Intellectual Origins of the English Revolution Revisited* (Oxford, 1997).

41. My own book about the morality of area bombing in the Second World War, *Among the Dead Cities: Is the Targeting of Civilians in War Ever Justified?* (London, 2011), was attacked on the grounds of impugning the heroism of those who flew into danger over Germany; Canadian veterans were among the most vocal.

42. My co-author Xu You Yu of our *The Long March to the Fourth of June* (London, 1989), published under the joint pseudonym 'Li Xiao Jun', was one such.

43. Ge Jianxiong quoted in the *New York Times*, 6 December 2004.

44. Catherine Nixey, *The Darkening Age* (London, 2017).

45. See Grayling, *The History of Philosophy* (London, 2019), pp. 3–5.

46. This is a chief part of the argument of my *Towards the Light: The Story of the Struggles for Liberty and Rights that Made the Modern West* (London, 2007).

47. This is a chief part of the argument of my *The Age of Genius: The Seventeenth Century and the Birth of the Modern Mind* (London, 2016).

4 'Reading-in' to History

1 See, e.g., Frank Elwell, '*Verstehen*: The Sociology of Max Weber' (1996). https://www.faculty.rsu.edu/users/f/felwell/www/Theorists/Weber/Whome2.htm. The *loci classici* are Wilhelm Dilthey, *Das Wesen der Philosophie* (*The Essence of Philosophy*) (Berlin and Leipzig, 1907), and *Selected Works. Vol. 2: Understanding the Human World* (Princeton, NJ, 2010).

2 Homer, *Iliad*, Book 18, ll. 20–25, 33, A. T. Murray and W. F. Wyatt (trs.), (1924; 2003 Loeb edn).

3 https://www.smithsonianmag.com/history/gobekli-tepe-the-worlds-first-temple-83613665/

4 Ibid., and Klaus Schmidt, 'Göbekli Tepe – the Stone Age Sanctuaries: New Results of Ongoing Excavations with a Special Focus on Sculptures and High Reliefs', *Documenta Praehistorica*, vol. 37 (2010), pp. 239–56. https://web.archive.org/web/20120131114925/http://arheologija.ff.uni-lj.si/documenta/authors37/37_21.pdf

5 E. B. Banning, 'So Fair a House: Göbekli Tepe and the Identification of Temples in the Pre-Pottery Neolithic of the Near East', *Current Anthropology*, vol. 52, no. 5 (October 2011), pp. 619–60 (626).

6 See Maurice Bloch, *In and Out of Each Other's Bodies: Theories of Mind, Evolution, Truth, and the Nature of the Social* (Boulder, Col., 2013).

7 *Encyclopaedia of Ancient History.* https://www.ancient.eu/religion/

8 I have had occasion to draw attention to promoted reading-in before: see 'Children of God?', *Guardian*, 28 November 2008. https://www.theguardian.com/commentisfree/2008/nov/28/religion-children-innateness-barrett: 'The research is funded by the Templeton Foundation, an organisation keen to find, or to insert, religion into science and to promote belief in their compatibility . . . The Templeton Foundation is rich; it offers a very large money prize to any scientist or philosopher who will say things friendly to religion, and it supports "research" . . . into anything that will add credibility and respectability to religion.'

9 https://www.templeton.org/

10 Transcript of Suzan Mazur's interview with Ian Hodder, 'Çatalhöyük, Religion and Templeton's 25% Broadcast', *Huffington Post*, 28 April 2017.

11 Ian Hodder (ed.), *Religion in the Emergence of Civilization* (Cambridge, 2010); *Religion at Work in a Neolithic Society* (Cambridge, 2014); and *Religion, History and Place and the Origin of Settled Life* (Cambridge, 2018).

12 Some examples: Guillaume Lecointre, 'La Fondation Templeton', French National Center for Scientific Research; Libby A. Nelson, 'Some Philosophy Scholars Raise Concerns about Templeton Funding', *Inside Higher Ed*, 21 May 2013; Josh Rosenau, 'How Bad is the Templeton Foundation?', ScienceBlogs (5 March 2011); John Horgan, 'The Templeton Foundation: A Skeptic's Take', Edge.org., 2006. https://www.edge.org/conversation/john_horgan-the-templeton-foundation-a-skeptics-take;

Sean Carroll, 'The Templeton Foundation Distorts the Fundamental Nature of Reality: Why I Won't Take Money from the Templeton Foundation', Slate.com; Sunny Bains, 'Questioning the Integrity of the John Templeton Foundation', *Evolutionary Psychology*, vol. 9, no. 1 (2011), pp. 92–115. https://doi.org/10.1177%2F147470491100900111; Jerry Coyne, 'Martin Rees and the Templeton Travesty', *Guardian*, 6 April 2011, retrieved 8 April 2018; Donald Wiebe, 'Religious Biases in Funding Religious Studies Research?', *Religio: Revue Pro Religionistiku*, vol. 17, no. 2 (2009), pp. 125–140; Nathan Schneider, 'God, Science and Philanthropy', *Nation*, 3 June 2010; Sunny Bains, 'Keeping an Eye on the John Templeton Foundation', Association of British Science Writers, 6 April 2011.

13 Ian Hodder: Yes, well that's fine. That's Maurice's view. I just think he's wrong. He's one author. He's one author. I don't know how many authors came to Çatalhöyük to discuss this issue. It must be well over 30 by now. He's the only one who takes this extreme position. Suzan Mazur: Many of those authors are religious scholars.

14 https://templeton.org/

15 Iain Davidson, review of Hodder (ed.), *Religion at Work in a Neolithic Society*, *Australian Archaeology*, vol. 82, no. 2 (2016), pp. 192–5.

16 R. G. Klein, 'Out of Africa and the Evolution of Human Behavior', *Evolutionary Anthropology*, vol. 17, no. 6 (2008), pp. 267–81.

17 April Nowell, 'Defining Behavioral Modernity in the Context of Neandertal and Anatomically Modern Human Populations', *Annual Review of Anthropology*, vol. 39, no. 1 (2010), pp. 437–52.

18 Grayling, *War: An Enquiry* (New Haven, Conn., and London, 2017).

19 P. G. Chase, *The Emergence of Culture: The Evolution of a Uniquely Human Way of Life* (New York, 2006).

20 F. d'Errico and M. Vanhaeren, 'Evolution or Revolution? New Evidence for the Origins of Symbolic Behaviour In and Out of Africa', in P. Mellars et al., *Rethinking the Human Revolution: New Behavioural and Biological Perspectives on the Origin and Dispersal of Modern Humans* (Cambridge, 2007), pp. 275–86.

21 Nowell, 'Defining Behavioral Modernity'.

22 Lewis Binford et al. (eds.), *New Perspectives in Archeology* (Chicago, 1968). See Matthew Johnson, *Archaeological Theory: An Introduction* (1999; 2nd edn Oxford, 2010).

23　Bruce Trigger, *A History of Archaeological Thought* (1996; 2nd edn Cambridge, 2006).

24　Michael Shanks and Ian Hodder, 'Processual, Postprocessual, and Interpretive Archaeologies', in Ian Hodder et al. (eds.), *Interpreting Archaeology: Finding Meaning in the Past* (London, 1995).

25　Ibid.

Part III: The Brain and the Mind

1　The critics' 'like phrenology' charge is close to the bone in the case of transcranial magnetic stimulation (TMS) and electroencephalogram (EEG) investigations, which take readings from outside the skull. But they are based on a genuine understanding of intracranial electrochemistry and structure.

2　It is said that Descartes demonstrated the lack of animal consciousness by throwing a cat out of the upstairs window of his accommodation in Leiden in the Netherlands. It is not clear what the experiment proved on the point in question. See A. C. Grayling, *Descartes: The Life and Times of a Genius* (London, 2006).

3　See, for example, this argument for the brain being a digital device: James Tee and Desmond P. Taylor, 'Is Information in the Brain Represented in Continuous or Discrete Form?', *IEEE Transactions on Molecular, Biological, and Multi-Scale Communications* (21 September 2020), PDF at https://arxiv.org/ftp/arxiv/papers/1805/1805.01631.pdf

4　The success of Google's 'DeepMind' division and 'AlphaGo' is an early indicator of where this field of endeavour will go: see https://www.youtube.com/watch?v=WXuK6gekUiY

5　See Section 3 below for Roger Penrose's objections to thinking in computational terms.

1 Mind and Heart

1　Fyodor Dostoevsky, *The Brothers Karamazov*, Part 3, Book 7, Chapter 1, 'The Odour of Corruption', R. Pevear and L. Volokhonsky (trs.)

(London, 1992). See also A. C. Grayling, 'Neoplatonism', in *The History of Philosophy* (London, 2019), pp. 123–30.

2 The view that there is a sense in which the mind (of humanity as a whole, as it were) is located in its libraries (they would now say, on its hard disks) was held by Karl Popper and John Eccles. Jung had a different conception of a universal unconscious mind – the home of the archetypes: a sort of Platonic forms – of which individual minds partake. These are not contenders for discussion here.

3 Charles Gross, 'Aristotle on the Brain', *Neuroscientist*, vol. 1, no. 4 (July 1995), pp. 245ff.

4 Plato, *Timaeus* (Harmondsworth, 1965), Section 12.

5 Hippocrates, 'On the Sacred Disease'. See the sections 'On the Humours' and 'On the Heart' for further examples in selections from the Hippocratic corpus at https://oll.libertyfund.org/titles/hippocrates-the-writings-of-hippocrates-and-galen

6 The different reasons advanced by Aristotle in favour of the heart as the seat of mind occur principally in his biological works, *De partibus animalium* (*The Parts of Animals*), *Historia animalium* (*The History of Animals*), and the *Parva naturalia* (*Short Treatises on Nature*). See Gross, 'Aristotle on the Brain', pp. 247–8.

7 See Grayling, 'Aristotle', *The History of Philosophy*.

8 Herophilus was also one of the first to study the female reproductive system and to write a text on midwifery.

9 Quoted from Heinrich von Staden, *Herophilus: The Art of Medicine in Early Alexandria* (Cambridge, 1989), in Gross, 'Aristotle on the Brain', pp. 249–50.

10 Ibid.

11 Stavros J. Baloyannis, 'Galen as Neuroscientist and Neurophilosopher', *Encephalos*, vol. 53 (2016), pp. 1–10.

12 Ibid., p. 8.

13 David Ferrier, *The Functions of the Brain* (London, 1876).

14 See the respective entries in Grayling, *The History of Philosophy*.

15 Gilbert Ryle, *The Concept of Mind* (Chicago, 1949); see also, e.g., Steven Pinker, *How the Mind Works* (New York, 1997), 'minds are not animated by some godly vapor', Chapter 1: 'Standard Equipment'.

16 Flourens was a Creationist and opponent of Darwin, which might seem odd, given his scientific work. Like the naturalist Philip Henry Gosse,

he hoped that science might confirm the Creation account. Gosse, a member of the Plymouth Brethren, struggled with the geological and fossil evidence that contradicted his beliefs; the story both of this and his relationship with his son, the poet and critic Edmund Gosse, is told in the latter's poignant memoir *Father and Son* (1907).

17 The arcuate fasciculus also appears to connect with the motor centres of the parietal lobe and frontal lobes on either side of the central sulcus. In the right hemisphere the structure is associated with visuospatial processing. The original work by Broca and Wernicke appears in the following places: Paul Broca, 'Remarques sur le siège de la faculté du langage articulé, suivies d'une observation d'aphémie (perte de la parole)', *Bulletin de la Société Anatomique*, vol. 6, no. 36 (1861), pp. 330–37; Carl Wernicke, *Der aphasische Symptomencomplex: Eine psychologische Studie auf anatomischer Basis* (Breslau, 1874).

18 John Martyn Harlow, 'Passage of an Iron Rod through the Head' (1848). https://web.archive.org/web/20140523001027/https://www.countway. harvard.edu/menuNavigation/chom/warren/exhibits/HarlowBMSJ 1848.pdf

19 Ibid.

20 John Martyn Harlow, 'Recovery from the Passage of an Iron Bar through the Head' (1868), *Publications of the Massachusetts Medical Society*, vol. 2, no. 3, pp. 327–47. Reprinted in David Clapp & Son (1869). https:// en.wikisource.org/wiki/Recovery_from_the_passage_of_an_iron_bar_ through_the_head

21 Ibid.

22 On the hippocampus and memory, see W. B. Scoville and B. Milner, 'Loss of Recent Memory after Bilateral Hippocampal Lesions', *Journal of Neurology, Neurosurgery and Psychiatry*, vol. 20, no. 1 (1957), pp. 11–21. For observations of split-brain patients, see Roger W. Sperry, M. S. Gazzaniga, and J. E. Bogen, 'Interhemispheric Relationships: The Neocortical Commissures; Syndromes of Hemisphere Disconnection', in *Handbook of Clinical Neurology*, P. J. Vinken and G. W. Bruyn (eds.) (Amsterdam, 1969), pp. 273–90.

23 Roger W. Sperry, 'Cerebral Organization and Behavior', *Science*, vol. 133, no. 3,466 (2 June 1961), pp. 1,749–57. http://people.uncw.edu/puente/ sperry/sperrypapers/60s/85-1961.pdf

24 At 3T where T= *tesla*, the unit of measurement for magnetic flux density. Higher resolutions are possible: in 2019, 10.5T was safety-tested with human subjects, 21.5T in animal experiments.

25 Magnetoencephalography (MEG) and event-related potential (ERP) offer greater temporal, but poorer spatial, resolution.

2 *The Cognitive Brain*

1 A webpage for brain anatomy, e.g., https://www.webmd.com/brain/picture-of-the-brain#1

2 Suzana Herculano-Houzel and Roberto Lent, 'Isotropic Fractionator: A Simple, Rapid Method for the Quantification of Total Cell and Neuron Numbers in the Brain', *Journal of Neuroscience*, vol. 25, no. 10 (2010), pp. 2,518–21. Critics have pointed out that the work which produced this number used four male brains ranging in age from twenty to seventy, and that the standard deviation was 8 billion – meaning that the upper bound is not far off the '100 billion' so often cited.

3 This provides further support for the observation that less is more.

4 O. Sporns, 'The Human Connectome: A Complex Network', in M. B. Miller and A. Kingstone (eds.), *The Year in Cognitive Science*, Vol. 1,224 (Oxford, 2011), pp. 109–25.

5 See, e.g., Lisa Feldman Barrett and Ajay Satpute, 'Large Scale Brain Networks in Affective and Social Neuroscience', *Current Opinion in Neurobiology*, vol. 23, no. 3 (January 2013), pp. 361–71; Katherine Vytal and Stephen Hamann, 'Neuroimaging Support for Discrete Neural Correlates of Basic Emotions', *Journal of Cognitive Neuroscience*, vol. 22, no. 12 (December 2010), pp. 2,864–85.

6 Anton–Babinski Syndrome is regarded as a form of *anosognosia*, or lack of self-awareness about a disability, in which sufferers deny that they have it.

7 Gorilla basketball video. https://www.youtube.com/watch?v=vJG698 U2Mvo

8 Wrangler psychoneuroscience comes, so to say, direct from the horse's mouth: a week at a dude ranch near Tucson, Arizona, riding daily among the thorny cactuses of the Sonoran Desert, gave access to a fount of

opinion about horse psychology, with most of which equestrians on the other side of the Atlantic would disagree. An interesting contrast.

9 Neha Uppal and Patrick Hof, 'Discrete Cortical Neuropathology in Autism Spectrum Disorders', in *The Neuroscience of Autism Spectrum Disorders* (Amsterdam, 2013), pp. 313–25. https://doi.org/10.1016/B978-0-12-391924-3.00022-3

10 Thomas Grüter, Martina Grüter, and Claus-Christian Carbon, 'Neural and Genetic Foundations of Face Recognition and Prosopagnosia', *Journal of Neuropsychology*, vol. 2, no. 1 (2008), pp. 79–97.

11 Marlene Behrmann et al., 'Intact Visual Imagery and Impaired Visual Perception in a Patient with Visual Agnosia', *Journal of Experimental Psychology: Human Perception and Performance*, vol. 20, no. 5 (November 1994), pp. 1,068–87.

12 Philip J. Hilts, *Memory's Ghost* (New York, 1996); Suzanne Corkin, *Permanent Present Tense: The Unforgettable Life of the Amnesic Patient, H. M.* (New York, 2013).

13 Sarah K. Johnson and Michael C. Anderson, 'The Role of Inhibitory Control in Forgetting Semantic Knowledge', *Psychological Science*, vol. 15, no. 7 (July 2004), pp. 448–53.

14 Michael C. Anderson et al., 'Prefrontal-hippocampal Pathways Underlying Inhibitory Control Over Memory', *Neurobiology of Learning and Memory*, vol. 134, Part A (2016), pp. 145–61.

15 Henry L. Roediger III and Kathleen B. McDermott, 'Creating False Memories: Remembering Words Not Presented in Lists', *Journal of Experimental Psychology: Learning, Memory, and Cognition*, vol. 21, no. 4 (July 1995), pp. 803–14.

16 Sydney Brandon et al., 'Recovered Memories of Childhood Sexual Abuse: Implications for Clinical Practice', *British Journal of Psychiatry*, vol. 172, no. 4 (April 1998), pp. 296–307.

17 Monica Fabiani et al., 'True but Not False Memories Produce a Sensory Signature in Human Lateralized Brain Potentials', *Journal of Cognitive Neuroscience*, vol. 12, no. 6 (December 2000), pp. 941–9.

18 Patricia Churchland, *Neurophilosophy: Toward a Unified Science of the Mind/Brain* (1986; 2nd edn Cambridge, Mass., 1989); Paul Churchland, *Neurophilosophy at Work* (Cambridge, 2007).

19 The term 'theory of mind' was introduced in a famous paper by David Premack and Guy Woodruff, 'Does the Chimpanzee Have a Theory of Mind?', *Behavioral and Brain Sciences*, vol. 1, no. 4 (December 1978), pp. 515–26.

3 Neuroscience and Consciousness

1 John Locke, *An Essay Concerning Human Understanding*, Book 2, Chapter 27 (2nd edn London, 1691). The chapter 'Of Identity and Diversity' was added to the second edition at the suggestion of the Irish philosopher and science writer William Molyneux.

2 Daniel Dennett always told our students at the New College of the Humanities that the word 'surely' marks the weakest point in an argument.

3 See Grayling, 'Modern Philosophy II: The Empiricists', in A. C. Grayling (ed.), *Philosophy: A Guide through the Subject* (Oxford, 1995; 2nd edn 1998), and 'John Locke', *The History of Philosophy* (London, 2019), pp. 217–26.

4 Matthias Michel, 'Consciousness Science Underdetermined', *Ergo*, vol. 6, no. 28 (2019–20). http://dx.doi.org/10.3998/ergo.12405314.0006.028

5 Knowledge of mescaline and its effects was disseminated more widely by Aldous Huxley's *The Doors of Perception* (1954), which doubtless played a part in promoting interest in LSD in subsequent years.

6 H. Klüver and P. C. Bucy, ' "Psychic Blindness" and Other Symptoms following Bilateral Temporal Lobe Lobectomy in Rhesus Monkeys', *American Journal of Physiology*, vol. 119 (1937), pp. 352–3.

7 B. Milner, 'Intellectual Function of the Temporal Lobes', *Psychological Bulletin*, vol. 51, no. 1 (1954), pp. 42–62.

8 Francis Crick and Christof Koch, 'Towards a Neurobiological Theory of Consciousness', *Seminars in the Neurosciences*, vol. 2 (1990), pp. 263–75; Crick and Koch, 'Why Neuroscience May be Able to Explain Consciousness', *Scientific American*, vol. 273, no. 6 (1995), pp. 84–5.

9 David J. Chalmers, 'Facing Up to the Problem of Consciousness', *Journal of Consciousness Studies*, vol. 2, no. 3 (1995), pp. 200–219.

10 Quoted by Anil Seth in a Royal Institution lecture on consciousness in 2017. https://www.youtube.com/watch?v=xReliJKOEbI. Seth is

co-director with Hugo Critchley of the Sackler Centre for Consciousness Science at the University of Sussex; see Anil K. Seth, 'The Real Problem', *Aeon*, 10 November 2016. https://aeon.co/essays/the-hard-problem-of-consciousness-is-a-distraction-from-the-real-one

11 Bernard J. Baars, A *Cognitive Theory of Consciousness* (Cambridge, 1998).

12 Idem, 'The Global Brainweb: An Update on Global Workspace Theory', *Science and Consciousness Review* (October 2003). http://cogweb. ucla.edu/CogSci/Baars-update_03.html

13 Stanislas Dehaene and Lionel Naccache, 'Towards a Cognitive Neuroscience of Consciousness: Basic Evidence and a Workspace Framework', *Cognition*, vol. 79, nos. 1–2 (April 2001), pp. 1–37.

14 Victor A. F. Lamme, 'Separate Neural Definitions of Visual Consciousness and Visual Attention: A Case for Phenomenal Awareness', *Neural Networks*, vol. 17, nos. 5–6 (2004), pp. 861–72.

15 Giulio Tononi, 'An Information Integration Theory of Consciousness', *BMC Neuroscience*, vol. 5, no. 1 (November 2004), Article No. 42. https:// bmcneurosci.biomedcentral.com/articles/10.1186/1471-2202-5-42

16 Marcello Massimini et al., 'Breakdown of Cortical Effective Connectivity during Sleep', *Science*, vol. 309, no. 5,744 (2005), pp. 2,228–32; Adenauer G. Casali et al., 'A Theoretically Based Index of Consciousness Independent of Sensory Processing and Behavior', *Science Translational Medicine*, vol. 5, no. 198 (2013).

17 Adrian M. Owen et al., 'Detecting Awareness in the Vegetative State', *Science*, vol. 313, no. 5,792 (September 2006), p. 1,402; Anil K. Seth, Adam B. Barrett and Lionel Barnett, 'Causal Density and Integrated Information as Measures of Conscious Level', *Philosophical Transactions of the Royal Society A: Mathematical, Physical, and Engineering Sciences*, vol. 369 (2011), pp. 3,748–67.

18 John R. Ives et al., 'Method and Apparatus for Monitoring a Magnetic Resonance Image during Transcranial Magnetic Stimulation', US Patent No. 6,198,958 B1, 6 March 2001.

19 Dehaene and Naccache, 'Towards a Cognitive Neuroscience of Consciousness'; Lior Fisch et al., 'Neural "Ignition": Enhanced Activation Linked to Perceptual Awareness in Human Ventral Stream Visual Cortex', *Neuron*, vol. 64, no. 4 (2009), pp. 562–74; Raphaël Gaillard et al., 'Converging Intracranial Markers of Conscious Access', *PLoS Biology*, vol. 7, no. 3 (17 March 2009).

20 Karl Friston, 'The Free-energy Principle: A Rough Guide to the Brain?',
    Trends in Cognitive Sciences, vol. 13, no 7 (July 2009), pp. 293–301; Anil
    K. Seth, 'Interoceptive Inference, Emotion, and the Embodied Self',
    Trends in Cognitive Sciences, vol. 17, no. 11 (November 2013), pp. 565–73.
21 Stuart Hameroff and Roger Penrose, 'Consciousness in the Universe: A
    Review of the "Orch OR" Theory', *Physics of Life Reviews*, vol. 11, no. 1
    (2014), pp. 39–78 and 94–100 respectively.
22 Danko D. Georgiev, *Quantum Information and Consciousness: A Gentle
    Introduction* (Boca Raton, Flor., 2017), p. 177.
23 An accessible account of this by Chalmers himself occurs in his TED
    talk at https://www.youtube.com/watch?v=uhRhtFFhNzQ
24 Daniel Dennett, *Consciousness Explained* (Harmondsworth, 1992) and
    From Bacteria to Bach and Back (London, 2017).
25 Marvin Minsky, *The Society of Mind* (New York, 1986); Robert E. Orn-
    stein, *Evolution of Consciousness* (Upper Saddle River, NJ, 1991).

4 The Mind and the Self

 1 Paul Bloom, 'How Do Morals Change?', *Nature*, vol. 464, no. 7,288 (25
    March 2010), p. 490.
 2 'Wit' once mainly meant 'intelligence' and 'good sense' – it retains the
    latter meaning in 'he has his wits about him' and 'witless'; it is interest-
    ing to note that wit in a person is a good indication of intelligence.
 3 In the eighteenth century Tories were those who upheld the power of
    the Crown over Parliament; Whigs held the opposite view.
 4 The dilemmas relate to the poor-taste aspect: the twins are joined 'at the
    organ of generation', which means that Martinus's conjugal obligations
    simultaneously involve him in adultery and other crimes. The argu-
    ments about personhood and identity are deployed by the opposing
    barristers in the court case that follows.
 5 Antonio Damasio, *The Feeling of What Happens: Body, Emotion and the
    Making of Consciousness* (New York, 1999).
 6 Idem, *Descartes' Error: Emotion, Reason, and the Human Brain* (New York,
    1994).

7 Roger W. Sperry, M. S. Gazzaniga, and J. E. Bogen, 'Interhemispheric Relationships: The Neocortical Commissures; Syndromes of Hemisphere Disconnection', in *Handbook of Clinical Neurology*, P. J. Vinken and G. W. Bruyn (eds.) (Amsterdam, 1969); see also P. A. Reuter-Lorenz et al. (eds.), *The Cognitive Neuroscience of Mind: A Tribute to Michael S. Gazzaniga* (Cambridge, Mass., 2010).

8 'The Neuronal Platonist: Michael Gazzaniga in Conversation with Shaun Gallagher', *Journal of Consciousness Studies*, vol. 5, nos. 5–6 (1 May 1998), pp. 706–17 (12).

9 Michael Gazzaniga, *Who is in Charge? Free Will and the Science of the Brain* (New York, 2011).

10 For one example: Anil Seth says in his Royal Institution lecture that researchers at Glasgow University have made advances in this respect: https://www.youtube.com/watch?v=xRel1JKOEbI (28 minutes, 30 seconds, quoting L. Muckli et al., 'Contextual Feedback to Superficial Layers of V1', *Current Biology*, vol. 25, no. 20 (2015), pp. 2,690–95.

11 A. C. Grayling, *The Good of the World* (London, forthcoming 2022).

Conclusion: The View from Olympus

1 Cardinal John Henry Newman's *The Idea of a University* (1852) is a classic statement of the generalist view. http://www.newmanreader.org/works/idea/. Some of its themes are iterated in D. Daiches (ed.), *The Idea of a New University* (London,1964), setting out the aims of the newly founded University of Sussex, first of the 'Plate Glass Universities' of the 1960s.

2 See Grayling, *War: An Enquiry* (New Haven, Conn., and London, 2017).

Appendix II: The Epic of Gilgamesh

1 A literal translation of the text says that he felt the air 'fall upon the side of his nose', which must mean his cheek; either 'cheek' or 'face' conveys the idea better.

2 Quotations are from the translation by Maureen Gallery Kovacs, electronic edn published by Wolf Carnahan (1998). http://www.ancienttexts.org/library/mesopotamian/gilgamesh/

Appendix III: The Code of Hammurabi

1 The text, translated by L. W. King, is available at the Yale University Law website at https://avalon.law.yale.edu/ancient/hamframe.asp

Bibliography

Anderson, Michael C., et al., 'Prefrontal-hippocampal Pathways Underlying Inhibitory Control Over Memory', *Neurobiology of Learning and Memory*, vol. 134, Part A (2016), pp. 145–61

Anthony, David W., *The Horse, the Wheel, and Language: How Bronze-Age Riders from the Eurasian Steppes Shaped the Modern World* (Princeton, NJ, and Oxford, 2007)

Ayala, Francisco J., and Camilo J. Cela-Conde, *Processes in Human Evolution: The Journey from Early Hominins to Neanderthals and Modern Humans* (Oxford, 2017)

Bains, Sunny, 'Questioning the Integrity of the John Templeton Foundation', *Evolutionary Psychology*, vol. 9, no. 1 (2011), pp. 92–115. https://journals.sagepub.com/doi/10.1177/147470491100900111

—, 'Keeping an Eye on the John Templeton Foundation', Association of British Science Writers, 6 April 2011

Baloyannis, Stavros J., 'Galen as Neuroscientist and Neurophilosopher', *Encephalos*, vol. 53 (2016), pp. 1–10

Banning, E. B., 'So Fair a House: Göbekli Tepe and the Identification of Temples in the Pre-Pottery Neolithic of the Near East', *Current Anthropology*, vol. 52, no. 5 (October 2011), pp. 619–60

Barker, Roger A., et al., *Neuroanatomy and Neuroscience at a Glance* (1999; Chichester, 2018)

Barrett, Lisa Feldman, and Ajay Satpute, 'Large Scale Brain Networks in Affective and Social Neuroscience', *Current Opinion in Neurobiology*, vol. 23, no. 3 (January 2013), pp. 361–71

Barrow, John D., and Frank J. Tipler, *The Anthropic Cosmological Principle* (Oxford, 1986)

Behrmann, Marlene, et al., 'Intact Visual Imagery and Impaired Visual Perception in a Patient with Visual Agnosia', *Journal of Experimental Psychology: Human Perception and Performance*, vol. 20, no. 5 (November 1994), pp. 1,068–87

Bentley, Michael, *Companion to Historiography* (London, 2002)

Binford, Lewis, et al. (eds.), *New Perspectives in Archeology* (Chicago, 1968)

Blackmore, Susan, *Consciousness: An Introduction* (2003; London, 2018)

Bloch, Maurice, *In and Out of Each Other's Bodies: Theories of Mind, Evolution, Truth, and the Nature of the Social* (Boulder, Col., 2013)

Brandon, Sydney, et al., 'Recovered Memories of Childhood Sexual Abuse: Implications for Clinical Practice', *British Journal of Psychiatry*, vol. 172, no. 4 (April 1998), pp. 296–307

Brax, Philippe, 'What Makes the Universe Accelerate? A Review on What Dark Energy Could be and How to Test It', *Reports on Progress in Physics*, vol. 81, no. 1 (January 2018)

Broca, Paul, 'Remarques sur le siège de la faculté du langage articulé, suivies d'une observation d'aphémie (perte de la parole)', *Bulletin de la Société Anatomique*, vol. 6, no. 36 (1861), pp. 330–37

Broome, Richard, *Aboriginal Australians: A History since 1788* (St Leonards, NSW, 2020)

Brown, Dee, *Bury My Heart at Wounded Knee* (1970; New York, 2012)

Browning, Christopher R., *The Origins of the Final Solution: The Evolution of Nazi Jewish Policy, September 1939–March 1942*. With contributions by Jürgen Matthäus (Lincoln, Nebr., 2004; London, 2014)

Bulliet, Richard, *The Wheel: Inventions and Reinventions* (New York, 2016)

Carroll, Sean, 'The Templeton Foundation Distorts the Fundamental Nature of Reality: Why I Won't Take Money from the Templeton Foundation', Slate.com

—, *The Big Picture* (London, 2016)

Chase, P. G., *The Emergence of Culture: The Evolution of a Uniquely Human Way of Life* (New York, 2006)

Churchland, Patricia, *Neurophilosophy: Toward a Unified Science of the Mind/Brain* (1986; 2nd edn Cambridge, Mass., 1989)

Churchland, Paul, *Neurophilosophy at Work* (Cambridge, 2007)

Cline, Eric, *1177 BC: The Year Civilization Collapsed* (Princeton, NJ, and Oxford, 2014)

Corkin, Suzanne, *Permanent Present Tense: The Unforgettable Life of the Amnesic Patient, H. M.* (New York, 2013)

Coyne, Jerry, 'Martin Rees and the Templeton Travesty', *Guardian*, 6 April 2011

Crick, Francis, and Christof Koch, 'Towards a Neurobiological Theory of Consciousness', *Seminars in the Neurosciences*, vol. 2 (1990), pp. 263–75

—, 'Why Neuroscience May be Able to Explain Consciousness', *Scientific American*, vol. 273, no. 6 (1995), pp. 84–5

Cru, Jean Norton, *Witnesses: Tests, Analysis and Criticism of the Memories of Combatants (1915–1928)* (*Témoins: Essai d'analyse et de critique des souvenirs de combattants édités en français de 1915 à 1928*) (Paris, 1929; Nancy, 3rd edn 2006)

Curry, Andrew, 'Gobekli Tepe: The World's First Temple?', *Smithsonian Magazine* (November 2008). https://www.smithsonianmag.com/history/gobekli-tepe-the-worlds-first-temple-83613665/

Davidson, Iain, review of Ian Hodder (ed.), *Religion at Work in a Neolithic Society*, *Australian Archaeology*, vol. 82, no. 2 (2016), pp. 192–5

d'Errico, F., and M. Vanhaeren, 'Evolution or Revolution? New Evidence for the Origins of Symbolic Behaviour In and Out of Africa', in P. Mellars et al., *Rethinking the Human Revolution: New Behavioural and Biological Perspectives on the Origin and Dispersal of Modern Humans* (Cambridge, 2007), pp. 275–86

Deutsch, David, *The Fabric of Reality: The Science of Parallel Universes and Its Implications* (1997; Harmondsworth, 1998)

—, *The Beginning of Infinity: Explanations that Transform the World* (London, 2011)

Dilthey, Wilhelm, *Selected Works. Vol. 2: Understanding the Human World* (Princeton, NJ, 2010)

Driscoll, M. J., 'The Words on the Page', in *Creating the Medieval Saga: Version, Variability and Editorial Interpretations of Old Norse Saga Literature*, Judy Quinn and Emily Lethbridge (eds.) (Odense, 2010)

Eagleman, David, *The Brain: The Story of You* (Edinburgh, 2016)

Elton, G. R., *The Practice of History* (1967; new edn London, 1987)

Elwell, Frank, '*Verstehen*: The Sociology of Max Weber' (1996). https://www.faculty.rsu.edu/users/f/felwell/www/Theorists/Weber/Whome2.htm

Engelhardt, Tom, 'Ambush at Kamikaze Pass', *Bulletin of Concerned Asian Scholars*, vol. 3, no. 1 (1971), pp. 64–84

Evans, Richard J., *In Defence of History* (1997; London, 2018)

—, *The Third Reich and the Paranoid Imagination* (London, 2020)

Fabiani, Monica, et al., 'True but Not False Memories Produce a Sensory Signature in Human Lateralized Brain Potentials', *Journal of Cognitive Neuroscience*, vol. 12, no. 6 (December 2000), pp. 941–9

Farrington, Benjamin, *Greek Science* (1944; Harmondsworth, 2nd edn, 1949)

Ferguson, Niall, *Empire: How Britain Made the Modern World* (London, 2003)

Ferrier, David, *The Functions of the Brain* (London, 1876)

Feynman, Richard, *QED: The Strange Theory of Light and Matter* (Princeton, NJ, and Oxford, 2014)

Fordham, Helen, 'Curating a Nation's Past: The Role of the Public Intellectual in Australia's History Wars', *M/C Journal*, vol. 18, no. 4 (2015)

French, Howard W., 'China's Textbooks Twist and Omit History', *New York Times*, 6 December 2004

Gallery Kovacs, Maureen (trs.), *The Epic of Gilgamesh*, electronic edn Wolf Carnahan, 1998. https://uruk-warka.dk/Gilgamish/The%20Epic%20of%20Gilgamesh.pdf

Gazzaniga, Michael, *Who is in Charge? Free Will and the Science of the Brain* (2011; London, 2016)

Gimbutas, Marija, *The Goddesses and Gods of Old Europe* (London, 1974)

—, et al., *The Kurgan Culture and the Indo-Europeanization of Europe: Selected Articles from 1952 to 1993* (Washington D. C., 1997)

Gosse, Edmund, *Father and Son: A Study of Two Temperaments* (London, 1907)

Grayling, A. C., 'Modern Philosophy II: The Empiricists', in A. C. Grayling, *Philosophy: A Guide through the Subject* (1995; 2nd edn New York and Oxford, 1998)

—, *The Quarrel of the Age: The Life and Times of William Hazlitt* (London, 2000)

—, *Descartes: The Life of René Descartes and Its Place in His Times* (New York and London, 2005)

—, *Among the Dead Cities: Was the Allied Bombing of Civilians in WWII a Necessity or a Crime?* (London, 2006)

—, *Towards the Light: The Story of the Struggles for Liberty and Rights that Made the Modern West* (London, 2007)

—, 'Children of God?', *Guardian*, 28 November 2008. https://www.theguardian.com/commentisfree/2008/nov/28/religion-children-innateness-barrett

—, *Friendship* (New Haven, Conn., and London, 2013)

—, *The Age of Genius: The Seventeenth Century and the Birth of the Modern Mind* (London, 2016)

—, *War: An Enquiry* (New Haven, Conn., and London, 2017)

—, *The History of Philosophy* (London, 2019)

Greene, Brian, *The Fabric of the Cosmos: Space, Time and the Texture of Reality* (London, 2005)

—, *Until the End of Time: Mind, Matter, and Our Search for Meaning in an Evolving Universe* (London, 2020)

Gross, Charles, 'Aristotle on the Brain', *Neuroscientist*, vol. 1, no. 4 (July 1995)

Grüter, Thomas, Martina Grüter, and Claus-Christian Carbon, 'Neural and Genetic Foundations of Face Recognition and Prosopagnosia', *Journal of Neuropsychology*, vol. 2, no. 1 (2008), pp. 79–97

Gutman, Yisrael, and Michael Berenbaum (eds.), *Anatomy of the Auschwitz Death Camp*, United States Holocaust Memorial Museum (Bloomington, Ind., 1998)

Haak, Wolfgang, et al., 'Massive Migration from the Steppe was a Source of Indo-European Languages in Europe', *Nature*, vol. 522, no. 7,555 (11 June 2015), pp. 207–11

Hamilton, Alastair, *Johann Michael Wansleben's Travels in the Levant, 1671–1674: An Annotated Edition of His Italian Report* (Leiden and Boston, 2018)

Hamming, R. W., 'The Unreasonable Effectiveness of Mathematics', *American Mathematical Monthly*, vol. 87, no. 2 (February 1980), pp. 81–90

Harlow, John Martyn, 'Passage of an Iron Rod through the Head' (1848). https://web.archive.org/web/20140523001027/https:/www.countway.ha rvard.edu/menuNavigation/chom/warren/exhibits/HarlowBMSJ1848.pdf

—, 'Recovery from the Passage of an Iron Bar through the Head' (1868), *Publications of the Massachusetts Medical Society*, vol. 2, no. 3, pp. 327–47. Reprinted in David Clapp & Son (1869). https://en.wikisource.org/ wiki/Recovery_from_the_passage_of_an_iron_bar_through_the_head

Harris, Annaka, *Conscious: A Brief Guide to the Fundamental Mystery of the Mind* (London, illustrated edn 2019)

Herculano-Houzel, Suzana, and Roberto Lent, 'Isotropic Fractionator: A Simple, Rapid Method for the Quantification of Total Cell and Neuron Numbers in the Brain', *Journal of Neuroscience*, vol. 25, no. 10 (2010), pp. 2,518–21

Hill, Christopher, *The Intellectual Origins of the English Revolution Revisited* (Oxford, 1997)

Hilts, Philip J., *Memory's Ghost* (New York, 1996)

Hippocrates and Galen, *The Writings of Hippocrates and Galen*, John Redman Coxe (trs.) (Philadelphia, 1846). Available via the Online Library of Liberty. https://oll.libertyfund.org/titles/hippocrates-the-writings-of-hippocrates-and-galen

Hodder, Ian (ed.), *Religion in the Emergence of Civilization* (Cambridge, 2010)

—, *Religion at Work in a Neolithic Society* (Cambridge, 2014)

—, *Religion, History and Place and the Origin of Settled Life* (Cambridge, 2018)

Hoffman, Matthew, 'Picture of the Brain', WebMD, 2014. https://www.youtube.com/watch?v=WXuK6gekU1Y

Hofstadter, Douglas R., *Gödel, Escher, Bach: An Eternal Golden Braid* (New York, 1979)

Homer, *Iliad*, A. T. Murray and W. F. Wyatt (trs.), Vols. 1 and 2 (1924; 2003 Loeb edn)

Horgan, John, 'The Templeton Foundation: A Skeptic's Take', Edge.org., 2006. https://www.edge.org/conversation/john_horgan-the-templeton-foundation-a-skeptics-take

Humphrey, Louise, and Chris Stringer, *Our Human Story* (London, 2018)

Iggers, Georg G. (ed.), *The Theory and Practice of History* (London, 2010)

—, *Historiography in the Twentieth Century: From Scientific Objectivity to the Postmodern Challenge* (1997; Middleton, Conn., 2012)

Impey, Oliver, and Arthur MacGregor (eds.), *The Origins of Museums: The Cabinet of Curiosities in Sixteenth- and Seventeenth-century Europe* (London, 1985)

Johnson, Matthew, *Archaeological Theory: An Introduction* (1999; 2nd edn Oxford, 2010)

Johnson, Sarah K., and Michael C. Anderson, 'The Role of Inhibitory Control in Forgetting Semantic Knowledge', *Psychological Science*, vol. 15, no. 7 (July 2004), pp. 448–53

Jones, William, 'The Third Anniversary Discourse – on the Hindus', delivered 2 February 1786, *The Works of Sir William Jones*, Vol. 3 (Delhi, 1977), pp. 24–46

Jun, Li Xiao, *The Long March to the Fourth of June: The First Impartial Account by an Insider, Still Living in China, of the Background to the Events in Tian An Men Square* (London, 1989)

Kay, Alex J., *The Making of an SS Killer: The Life of Colonel Alfred Filbert, 1905–1990* (Cambridge, 2016)

King, L. W. (trs.), *The Code of Hammurabi* (The Avalon Project, Yale Law School, 2008). https://avalon.law.yale.edu/ancient/hamframe.asp

Klein, R. G., 'Out of Africa and the Evolution of Human Behavior', *Evolutionary Anthropology*, vol. 17, no. 6 (2008), pp. 267–81

Klüver, H., and P. C. Bucy, ' "Psychic Blindness" and Other Symptoms following Bilateral Temporal Lobe Lobectomy in Rhesus Monkeys', *American Journal of Physiology*, vol. 119 (1937), pp. 352–3

Krauss, Lawrence, *A Universe from Nothing* (London, 2012)

—, *The Greatest Story Ever Told* (London, 2017)

Kriwaczek, Paul, *Babylon: Mesopotamia and the Birth of Civilization* (London, 2012)

Leick, Gwendolyn, *Mesopotamia: The Invention of the City* (London, 2001)

Lipstadt, Deborah, *Denying the Holocaust: The Growing Assault on Truth and Memory* (New York, 1993)

Locke, John, *An Essay Concerning Human Understanding* (2nd edn London, 1691)

Lyndall, Ryan, 'List of Multiple Killings of Aborigines in Tasmania: 1804–1835', SciencesPo, *Violence de masse et Résistance – Réseau de recherche* (March 2008). https://www.sciencespo.fr/mass-violence-war-massacre-resistance/fr/document/list-multiple-killings-aborigines-tasmania-1804-1835.html

Manco, Jean, *Ancestral Journeys: The Peopling of Europe from the First Venturers to the Vikings* (London, 2015)

Mark, Joshua J., 'Religion in the Ancient World: Definition', *Ancient History Encyclopedia* (23 March 2018). https://www.ancient.eu/religion/

Martínez-Abadías, Neus, et al., 'Heritability of Human Cranial Dimensions: Comparing the Evolvability of Different Cranial Regions', *Journal of Anatomy*, vol. 214, no. 1 (January 2009), pp. 19–35

Mazur, Suzan, Ian Hodder, 'Çatalhöyük, Religion and Templeton's 25% Broadcast', *Huffington Post*, 28 April 2017. https://www.huffpost.com/entry/ian-hodder-%C3%A7atalh%C3%B6y%C3%BCk-religion-templetons-25_b_58fe2a64e4b0f02c3870ecf0?guccounter=1&guce_referrer=aHR0cH M6Ly93d3cuZ29vZ2xlLmNvbS8&guce_referrer_sig=AQAAAGnadsos 9ygn5gxHiXnw54czAGFTptG6z31jvVxGgU_OpiylkYnK6oKB8Z3gNe DHKqGZnkhWoiSSOb7bklaWZ_p3OFTZaru1wa5K_fFqv3Jx4fT3ViI4-IRRGn9U2BgctueOIpYorkAvBosjVkvV3Cr6FiIFo4DJogN1Y240-pi2

Michel, Matthias, 'Consciousness Science Underdetermined', *Ergo*, vol. 6, no. 28 (2019–20). http://dx.doi.org/10.3998/ergo.12405314.0006.028

Mieroop, Marc van de, *A History of the Ancient Near East: c. 3000–323 BC* (2006; 3rd edn Oxford, 2016)

Milner, B., 'Intellectual Function of the Temporal Lobes', *Psychological Bulletin*, vol. 51, no. 1 (1954), pp. 42–62

Nelson, Libby A., 'Some Philosophy Scholars Raise Concerns about Templeton Funding', *Inside Higher Ed*, 21 May 2013. https://www.insidehighered.com/news/2013/05/21/some-philosophy-scholars-raise-concerns-about-templeton-funding

New Scientist, Human Origins: 7 Million Years and Counting (London, 2018)

—, *How Numbers Work* (London, 2018)

Nichols, Stephen G., 'Introduction: Philology in a Manuscript Culture', *Speculum*, vol. 65, no. 1 (January 1990), pp. 1–10

Nicolis, Franco (ed.), *Bell Beakers Today: Pottery People, Culture, Symbols in Prehistoric Europe: Proceedings of the International Colloquium, Riva Del Garda (Trento, Italy), 11–16 May 1998*, Vol. 2 (Trento, 1998)

Nixey, Catherine, *The Darkening Age* (London, 2017)

Nowell, April, 'Defining Behavioral Modernity in the Context of Neandertal and Anatomically Modern Human Populations', *Annual Review of Anthropology*, vol. 39, no. 1 (2010), pp. 437–52

Olalde, Iñigo, et al., 'The Beaker Phenomenon and the Genomic Transformation of North-west Europe', *Nature*, vol. 555, no. 7,695 (8 March 2018), pp. 190–96

Penrose, Roger, *The Road to Reality* (London, 2004)

—, *The Emperor's New Mind* (Oxford, illustrated edn 2016)

Pinker, Steven, *How the Mind Works* (New York, 1997)

Premack, David, and Guy Woodruff, 'Does the Chimpanzee Have a Theory of Mind?', *Behavioral and Brain Sciences*, vol. 1, no. 4 (December 1978), pp. 515–26

Rank, Otto, *The Myth of the Birth of the Hero: A Psychological Interpretation of Mythology*, F. Robbins and Smith Ely Jelliffe (trs.) (New York, 1914)

Rassinier, Paul, *Holocaust Story and the Lies of Ulysses: Study of the German Concentration Camps and the Alleged Extermination of European Jewry* (republished 1978 by 'Legion for the Survival of Freedom, Inc.', based in California)

Rees, Laurence, *The Holocaust* (London, 2017)

Reich, David, *Who We are and How We Got Here: Ancient DNA and the New Science of the Human Past* (Oxford, 2018)

Renfrew, Colin, *Archaeology and Language: The Puzzle of Indo-European Origins* (1987; Cambridge, 1990)

Reynolds, Henry, *The Other Side of the Frontier: Aboriginal Resistance to the European Invasion of Australia* (Sydney, 2006)

—, *Forgotten War* (Sydney, 2013)

Roberts, Alice, *Evolution: The Human Story* (2nd edn London, 2018)

Roediger, Henry L., III, and Kathleen B. McDermott, 'Creating False Memories: Remembering Words Not Presented in Lists', *Journal of Experimental Psychology: Learning, Memory, and Cognition*, vol. 21, no. 4 (July 1995), pp. 803–14

Rogers, J. Michael, 'To and Fro: Aspects of the Mediterranean Trade and Consumption in the Fifteenth and Sixteenth Centuries', *Revue des mondes musulmans et de la Méditerranée*, nos. 55–6 (1990), pp. 57–74

Rosenau, Josh, 'How Bad is the Templeton Foundation?', ScienceBlogs (5 March 2011). https://scienceblogs.com/tfk/2011/03/05/how-bad-is-the-templeton-found

Rovelli, Carlo, *Reality is Not What It Seems* (London, 2017)

Russell, Bertrand, *Introduction to Mathematical Philosophy* (London, 1919)

Russell, Nestar, 'The Nazi's Pursuit for a "Humane" Method of Killing', *Understanding Willing Participants: Milgram's Obedience Experiments and the Holocaust*, Vol. 2 (London, 2019), pp. 241–76

Rutherford, Adam, *A Brief History of Everyone Who Ever Lived* (London, 2016)

Ryden, Barbara, *Introduction to Cosmology* (Cambridge, 2017)

Ryle, Gilbert, *The Concept of Mind* (Chicago, 1949)

Sarukkai, Sundar, 'Revisiting the "Unreasonable Effectiveness" of Mathematics', *Current Science*, vol. 88, no. 3 (10 February 2005), pp. 415–23

Schmidt, Klaus, 'Göbekli Tepe – the Stone Age Sanctuaries: New Results of Ongoing Excavations with a Special Focus on Sculptures and High Reliefs', *Documenta Praehistorica*, vol. 37 (2010), pp. 239–56. https://web.archive.org/web/20120131114925/http://arheologija.ff.uni-lj.si/documenta/authors37/37_21.pdf

Schneider, Nathan, 'God, Science and Philanthropy', *Nation*, 3 June 2010. https://www.thenation.com/article/archive/god-science-and-philanthropy/

Scoville, W. B., and B. Milner, 'Loss of Recent Memory after Bilateral Hippocampal Lesions', *Journal of Neurology, Neurosurgery and Psychiatry*, vol. 20, no. 1 (1957), pp. 11–21

Shanks, Michael, and Ian Hodder, 'Processual, Postprocessual, and Interpretive Archaeologies', in Ian Hodder et al. (eds.), *Interpreting Archaeology: Finding Meaning in the Past* (London, 1995)

Shermer, Michael, and Alex Grobman, *Denying History: Who Says the Holocaust Never Happened and Why Do They Say It?* (Berkeley, Calif., 2002)

Sobel, Dava, *Longitude: The True Story of a Lone Genius Who Solved the Greatest Scientific Problem of His Time* (London, 1995)

Sperry, Roger W., 'Cerebral Organization and Behavior', *Science*, vol. 133, no. 3,466 (2 June 1961), pp. 1,749–57. http://people.uncw.edu/puente/sperry/sperrypapers/60s/85-1961.pdf

Sporns, O., 'The Human Connectome: A Complex Network', in M. B. Miller and A. Kingstone (eds.), *The Year in Cognitive Science*, Vol. 1,224 (Oxford, 2011), pp. 109–25

Staden, Heinrich von, *Herophilus: The Art of Medicine in Early Alexandria* (Cambridge, 1989)

Susskind, Leonard, et al., *Quantum Mechanics: The Theoretical Minimum* (London, 2014)

Tainter, Joseph A., *The Collapse of Complex Societies* (Cambridge, 1976)

Taylor, Alexander L., *The White Knight* (Edinburgh, 1952)

Tee, James, and Desmond P. Taylor, 'Is Information in the Brain Represented in Continuous or Discrete Form?', *IEEE Transactions on Molecular, Biological, and Multi-Scale Communications* (21 September 2020). https://arxiv.org/ftp/arxiv/papers/1805/1805.01631.pdf

Thomas, Carol G., and Craig Conant, *Citadel to City-State: The Transformation of Greece, 1200–700 BCE* (Bloomington, Ind., 1999)

Trigger, Bruce, *A History of Archaeological Thought* (1996; 2nd edn Cambridge, 2006)

Uppal, Neha, and Patrick Hof, 'Discrete Cortical Neuropathology in Autism Spectrum Disorders', in *The Neuroscience of Autism Spectrum Disorders* (Amsterdam, 2013), pp. 313–25. https://doi.org/10.1016/B978-0-12-391924-3.00022-3

Vinken, P. J., and G. W. Bruyn (eds.), *Handbook of Clinical Neurology* (Amsterdam, 1969)

Vytal, Katherine, and Stephan Hamann, 'Neuroimaging Support for Discrete Neural Correlates of Basic Emotions', *Journal of Cognitive Neuroscience*, vol. 22, no. 12 (December 2010), pp. 2,864–85

Weinberg, Steven, 'The Making of the Standard Model', *European Physical Journal C*, vol. 34 (May 2004), pp. 5–13

Wernicke, Carl, *Der aphasische Symptomencomplex: Eine psychologische Studie auf anatomischer Basis* (Breslau, 1874)

Wiebe, Donald, 'Religious Biases in Funding Religious Studies Research?', *Religio: Revue Pro Religionistiku*, vol. 17, no. 2 (2009), pp. 125–140

Wigner, Eugene, 'The Unreasonable Effectiveness of Mathematics in the Natural Sciences', *Communications on Pure and Applied Mathematics*, vol. 13, no. 1 (February 1960)

Windschuttle, Keith, *The Fabrication of Aboriginal History* (3 vols.; Paddington, NSW, 2002)

Xun, Lu (Lu Hsün), *The True Story of Ah Q* (1921). https://www.marxists. org/archive/lu-xun/1921/12/ah-q/index.htm

Yahil, Leni, *The Holocaust: The Fate of European Jewry, 1932–1945* (Oxford, 1991)

Yakira, Elhanan, *Post-Zionism, Post-Holocaust* (Cambridge, 2010)

Zeman, Adam, *Consciousness: A User's Guide* (New Haven, Conn., and London, 1999)

—, *Portrait of the Brain* (New Haven, Conn., and London, 2017)

Miscellaneous

AlphaGo – The Movie I Full Documentary, YouTube, uploaded by DeepMind, 13 March 2020. https://www.youtube.com/watch?v=WXuK6gekU1Y

Dmanisi Skulls, Google Images. https://www.google.co.uk/search?source= hp&ei=WmoDX_GoBqGXlwSAh7DIDw&q=dmanisi+skulls&oq=dmanisi+skulls&gs_lcp=CgZwc3ktYWlQAzICCAA6CAgAELEDEIMBOg UIABCxAzoECAAQAzoECAAQCjoGCAAQFhAeUP8iWIJDYLdGa ABwAHgAgAFDiAGnBpIBAjEomAEAoAEBqgEHZ3dzLXdpeg&scli ent=psy-ab&ved=0ahUKEwjxvbXsnbnqAhWhy4UKHYADDPkQ4dUD CAw&uact=5

'Resolution of the Duke University History Department', printed in the *Duke Chronicle* (November 1991). https://dukelibraries.contentdm.oclc. org/digital/collection/p15957coll13/id/85692

French National Centre for Scientific Research, Wikipedia, 2020. https://en. wikipedia.org/wiki/French_National_Centre_for_Scientific_Research

'Gender and Genetics', World Health Organization. https://www.who.int/ genomics/gender/en/

David Irving v. *Penguin Books and Deborah Lipstadt* (2000), Section 13 (91), England and Wales High Court (Queen's Bench Division) Decision. http://www.bailii.org/ew/cases/EWHC/QB/2000/115.html

Report of the Joint Committee on the Conduct of the War at the Second Session, 39th US Congress, Senate Report 156, testimony of Robert Bent about the Sandy Creek Massacre

United States Holocaust Memorial Museum, 'Bone-crushing Machine Used by Sonderkommando to Grind the Bones of Victims after Their Bodies were Burned in the Janowska Camp, August 1944'. https:// encyclopedia.ushmm.org/content/en/photo/bone-crushing-machine-in-janowska

WN@TL – How New Discoveries of Homo naledi *are Changing Human Origins*, YouTube. https://www.youtube.com/watch?v=7mBIFFstNSo

Selective Attention Test, YouTube. https://www.youtube.com/watch?v= vJG698U2Mvo

Wikipedia, *John Templeton Foundation*, 2020. https://en.wikipedia.org/wiki/ John_Templeton_Foundation

Index